The Form of Becoming

The Form of Becoming

Embryology and the
Epistemology of Rhythm
1760–1830

Janina Wellmann

Translated by Kate Sturge

ZONE BOOKS · NEW YORK

2017

© 2017 Urzone, Inc.

ZONE BOOKS
633 Vanderbilt Street
Brooklyn, NY 11218

Originally published as *Die Form des Werdens* © Wallstein Verlag
Göttingen 2010. The translation of this work was funded by
Geisteswissenschaften International — Translation Funding for the
Humanities and Social Sciences from Germany, a joint initiative of
the Fritz Thyssen Foundation, the German Federal Foreign Office,
the collecting society VG WORT, and the Börsenverein des Deutschen
Buchhandels (German Publishers & Booksellers Association).

Printed in the United States of America.

Distributed by The MIT Press,
Cambridge, Massachusetts, and London, England.

Library of Congress Cataloging-in-Publication Data
Names: Wellmann, Janina. | Sturge, Kate, translator.
Title: The form of becoming : embryology and the epistemology of
    rhythm, 1760–1830 / Janina Wellmann ; translated by Kate Sturge.
Other titles: Form des Werdens. English
Description: New York : Zone Books, 2017. | Originally published
    as: Die Forme des Werdens (Göttingen : Wallstein Verlag,
    c2010). | Includes bibliographical references and index.
Identifiers: LCCN 2016043302 | ISBN 9781935408765 (hardcover)
Subjects: LCSH: Embryology—History—18th century. |
    Embryology—History—19th century.
Classification: LCC QL953 .W45 2017 | DDC 571.8 —dc23
LC record available at https://lccn.loc.gov/2016043302 9 9

Contents

Acknowledgments

I would like to thank all those who have accompanied this book over the years in its various forms — who read and engaged with it, liked or interrogated it, supported and helped fund it.

The beginnings of the book go back to the upper stories of the Czech Embassy on Wilhelmstrasse, Berlin, home of the Max Planck Institute for the History of Science at that time. For four years, I had the good fortune to be part of the institute's intellectual community and to participate in stimulating discussions with the neighboring Center for Literary and Cultural Research and the Center for French Studies at the Technical University Berlin. My heartfelt thanks go to Hans-Jörg Rheinberger, Etienne François, Michael Werner, Sigrid Weigel, and Jean-Paul Gaudillière. Iris Schröder and Hannah Landecker are friends from those days who shared moments of intellectual doubt and joyous recovery. They have remained a constant encouraging presence from afar.

This book would be inconceivable without the many journeys — geographical and intellectual — that have shaped it. The University of Berne's hospitality gave me access to the treasures of Swiss libraries and the serenity of the Alpine panorama facing me every day from my desk while I worked on the first part of *The Form of Becoming*. I am grateful to the German Historical Institute in London for funding several months of research that enabled me to explore the libraries of London, especially the Warburg Institute. It was on those shelves that I discovered the pictorial instructions for bodily movement.

In pursuit of the images for the third part of the book and their origins, I traveled the most diverse areas of knowledge. I wish to express warm thanks to the librarians who indefatigably searched for books untouched for decades and were curious to know what soldiers and embryos might have in common. Thanks to Urte Brauckmann and Esther Chen of the Max Planck Institute for the History of Science in Berlin; Christian Hogrefe and Felix Kommnick of the Herzog-August-Bibliothek Wolfenbüttel; Astrid Winde and the Rare Books Reading Room staff at the Staatsbibliothek zu Berlin; Birgit Frenz of the Staats- und Universitätsbibliothek Göttingen; Gerald Raab of the Staatsbibliothek Bamberg; Sophie Schrader of the Bayerische Staatsbibliothek; Thomas Schmid of the Burgerbibliothek Bern; Heidi Kupper of the Zentral- und Hochschulbibliothek Luzern; Siobhan Morris of the Institute of Historical Research, London; and Michele Losacco of the Biblioteca Braidense, Milan. Images were kindly made available for publication by these libraries and by Klassik Stiftung Weimar, Hamburg University's Martha-Muchow-Bibliothek, the Humboldt University Berlin, and the Universitäts- und Landesbibliothek Sachsen-Anhalt.

Fellowships at the Max Planck Institute for the History of Science, by this time based in leafy Dahlem, and the Cohn Institute for the History and Philosophy of Science and Ideas at Tel Aviv University gave me time to reflect, discuss, and rework the manuscript. Special thanks are due to Lorraine Daston, Leo Corry, and Rivka Feldhay and all my colleagues in Dahlem and Tel Aviv for sharing their ideas and different worlds.

My year as a fellow at the Wissenschaftskolleg zu Berlin in 2013–2014 still reverberates in my everyday life. Pierre-Laurent Aimard, Reinhart Meyer-Kalkus, and Klaus Ospald discussed biology and music there with elegance, artistry, and knowledge. I learned a lot. Disputation, sophistry, and laughter with Simon Teuscher, Yuri Slezkine, Jahnavi Phalkey, and Katharina Wiedemann cleared my thoughts and brightened my days.

More recently, I have been welcomed by colleagues and students at Leuphana University Lüneburg and the Technical University Berlin. The kindness, efficiency, and confidence of Claus Pias, Martin Warnke, and Jantje Sieling helped me to continue working on the book despite my other duties.

In 2012, the German Publishers' and Booksellers' Association awarded the book the Geisteswissenschaften International prize to cover the costs of translation. The professionalism and commitment of Luca Giuliani and Anke Simon of Geisteswissenschaften

International ensured that *The Form of Becoming* could set off on its voyage in English. It is an honor to have Zone Books as the publisher and a pleasure to have experienced Meighan Gale's and Ramona Naddaff's expert judgment and care in the preparation of the manuscript.

Kate Sturge was attentive to every single letter, never seemed to tire of my constantly discussing each and every word and idea, and was the most scrupulous and amiable translator imaginable.

This book has thrived and evolved through constant discussion with Ohad Parnes. It is dedicated to Ada and Noga, with whom I continue to learn and grow every day.

INTRODUCTION

The Form of Becoming

In his work on secretion, Ignaz Döllinger (1770–1841), professor of medicine at Würzburg and among the most influential German naturalists of his day, sought to set out the principles of a new science of life: biology. Döllinger argued that whereas physics builds causal links between its individual observations, biology cannot proceed that way because life not so much "is" as "becomes." Accordingly, it is not possible to observe the processes of life themselves, but only the structures that "mediate" them: "only the form of becoming remains, and this form is what we perceive as constant."[1]

Half a century later, the young Friedrich Nietzsche (1844–1900) worked on a "theory of quantitative rhythm," not published in his lifetime, which he also entitled "Rhythmic Investigations." In these notes — preliminary work for a study in classical philology — he tried to understand the "power of rhythm." Music and poetry affect us, writes Nietzsche, not only because they are themselves rhythmical, but because the rhythmic movement of the body is "restructured" by the movements of music or poetry operating upon it.[2] Rhythm as such is physiological because life is "a continuing rhythmical movement, of the pulse, of the gait, even of the cells."[3] Only through rhythm does physiological multiplicity attain structure and "individuation."[4] The power of rhythm thus lies in its "physiological grounding": "Rhythm is the form of becoming, and in general, the form of the world of appearances."[5]

This book works at the interface between Ignaz Döllinger's attempt to describe biological development as the "form of becoming" and Nietzsche's definition of that very form as rhythm. I argue that rhythm was a new category in science and culture at the turn of the eighteenth to the nineteenth century, one hitherto almost entirely

neglected by historiography. Around 1800, the living world, and especially development, was rethought through concepts of rhythmic
pattern, rhythmic movement, and rhythmic representation.

Döllinger and his work will be extensively discussed in the course
of this study, but Nietzsche and his deliberations are outside its time
frame, the period between 1760 and 1830. That is a significant point
to make, for my hypothesis is precisely that rhythm's "physiological grounding," as Nietzsche aptly called it, was not the product of
his own era. It was formulated as a central episteme far earlier, in
the final decades of the eighteenth century. Nietzsche's attention to
and fascination with rhythm dovetails with the conventional historiography, which identifies rhythm as a key cultural and scientific
category of the period around 1900.[6] It is this chronological focus,
however, that the present book sets out to query. From the perspective I develop here, Nietzsche appears less to anticipate modern theories of rhythm than to have recognized the epistemological roots of
the nineteenth century's new physiology and to have reflected on its
cultural and historical dimensions.

My claim that the rhythmic episteme was crucial to the founding of biology arises from many years of research and thinking on,
especially, the emergence of modern embryology. I was drawn to the
topic by an analysis of the "Haller-Wolff" debate. In the 1760s, the
influential Swiss naturalist Albrecht von Haller and a little-known
German physician called Caspar Friedrich Wolff embarked on a
fierce argument carried out by letter. Haller believed that organisms are generated out of preformed germs, the development of
which is merely a matter of growing and unpacking, whereas Wolff
espoused the "epigenetic" theory of generation in which development is a process of the gradual emergence of forms. Historians have
frequently highlighted this debate, regarding the dispute as the "last
stand" of preformationism and Wolff's theory as having instigated a
paradigm shift toward the modern, epigenetic notion of embryonic
development.

What originally interested me in the Haller-Wolff debate was
its iconographic dimension. Both Haller and Wolff used drawings
in their written texts; my question was how their different theories
of development were reflected in differing iconography.[7] Although
I believed this investigation had yielded some useful insights, my
concern with theories of generation seemed to lead me a step further

back, to the question: What actually was "development" around 1800? Was epigenesis simply an alternative theory that replaced an outdated theory of development, or did the notion of what development is take on a fundamentally new shape around 1800?

Scholars have generally regarded the emergence of the epigenetic theory of development around this time as part of a new view of the world, usually labeled "temporalization" or "dynamization."[8] Certainly, at the turn of the eighteenth to the nineteenth century, the image of nature underwent profound change. The years from 1760 to 1830 saw the advent of biology as the science of life.[9] The centuries-old practices of collecting, inventorizing, and classifying the world were now superseded by attempts to understand its dynamic relationships. Where the anatomist had seen a cadaver, the physiologist saw a living body in which blood circulates, glands labor, substances are continually transformed into other substances. In this dynamic vision of the world, the genesis of the embryo, too, became a process, one by which the body's structures gradually take form out of the formless mass of the egg. Existence was subjected to the inexorable onward movement of time, which left nothing as it found it and subjugated life to change and flux. Life had previously been timeless, the product of a single instant of Creation; it now had a beginning and an end.

It was in this context that, around 1800, embryology arose as a science. The beginnings of embryology have been studied primarily from two perspectives: the teleological dimension of life processes and the vital forces driving those processes. In the usual historiographical view, the epigenetic stance depended on the assumption that processes of life — in contrast to other natural processes — always move teleologically. This was necessary in order to explain why, once the idea of preformed germs had been jettisoned, developmental processes still would have a direction. The notion of vital forces resolved a second and crucial problem of emerging embryology: how living organisms were able to organize themselves independently. This capacity distinguished them radically from mere matter, and even if little could yet be said about the nature of vital forces, their existence seemed to offer an explanation for that distinction.[10]

So far, so good — but did these perspectives help to answer the question of what development was in the period around 1800? I did not feel that the conundrum of developmental thinking was resolved, but rather that it had disappeared from view. To talk about the directionality of processes and of the forces guiding them is already the second step. It takes for granted change, gradual progression, process itself. Yet what *is* a process?

Progression over time alone is not an adequate definition, given that theories of preformation had by no means denied progressive change — in preformationism, too, the organism was recognized as being profoundly different at the start of ontogenesis and at its end. Added to this, was it really a trivial matter to *see* the changing of the embryo? What did researchers around 1800 observe when they looked at chick embryos through their microscopes? And why was what they observed in this period different from what they had observed before?

Examining the role of images in the Haller-Wolff debate had shown me that pictures played a pivotal role in Wolff's blueprint for epigenetic developmental thinking, whereas they did not for Haller's theory of preformation. Wolff needed pictures in order to "see" development. So how exactly did he observe development? Could he see the changes in the embryo, or did he construct them on the basis of his observations? And what was the role of pictures in developmental thinking after Wolff, for example in the work of Christian Heinrich Pander and Karl Ernst von Baer, the founders of modern embryology?

A study of developmental thinking — and this is the premise of the present book — requires answers to these questions that go beyond the finding that the organism simply "becomes" a constantly changing, temporal entity. To understand the "form of becoming" that seemed to characterize living nature, a new conceptual framework had to be built, along with new experimental practices, new techniques of observation, and, crucially, new forms of visual representation. Reading the embryological treatises of the period with an eye to aspects other than their confrontation with the questions of teleology and forces, it becomes clear that a different problem also preoccupied naturalists around 1800. What they observed through the microscope was extremely difficult to grasp: living matter always changes, and the organism was a different one each time they looked through the lens. Nevertheless, it was a functioning whole, well ordered and coordinated. How could the organism continually change, yet still be ordered? How could the parts combine into a highly complex formation when they themselves were all changing as incessantly as did the whole?

In this book, I argue that the solution to the riddle was supplied by the episteme of rhythm. Around 1800, rhythm became a new way of imagining that played a critical role in the consolidation of the new science of biology. It may seem counterintuitive to describe a continuous change in time, a gradual transformation, as rhythmi-cal. Although rhythm is a temporal structure, after all, in contrast to the flow of time — to time's unbroken and amorphous flux — it signifies the restriction of fluidity in favor of a rule. In fact, however, this power of rhythm to *structure* temporal processes was precisely what made it such an important concept in the period around 1800. Rhythm did not abolish time; it subjugated time. It imposed an order, a rule, on the unceasing transformation undergone by everything organic. Establishing a science of organic life did not, then, mean regarding the organic as something governed by temporal change. Rather, it meant conceiving of the organic as an ordered structure under the condition of temporality. Rhythm described the emer-gence and formation of life not as a mere progression in time, but as an *ordering* of time.

By asserting that a new episteme of rhythm became established around 1800, this book differs from existing scholarship in several respects. First, it offers an analysis of rhythm that has so far been almost entirely lacking in the history of science. Second, it shifts the focus of historical research on rhythm back a hundred years, to the threshold of the nineteenth century. And third, it embeds the emergence of embryology in the context of aesthetics, poetics, and philosophy around 1800.

To flesh out the first two of these points, the following section sketches the existing historiography of rhythm and its dominant paradigms. I then briefly present the methodological and historio-graphical basis of my own argument, paying particular attention to the difficulties that arise from the search for a rhythmic episteme in such diverse areas of knowledge as music theory, poetology, and physiology.

The Paradigm of Rhythm around 1900

As mentioned, rhythm has previously been regarded first and fore-most as a feature of the threshold between the nineteenth and the twentieth century. Art history and cultural studies have frequently shown how rhythm became a quintessential cultural phenomenon

around 1900, responding to the challenges of modernity — to tech-
nologization and rationalization, to the redefinition of time, space,
and matter in physics, and to societal and political change. In the
arts, as well, rhythm became a novel creative principle, for example
in the work of Klee, Kirchner, and Mondrian, and it took up a central position in new understandings of corporeality. In this way, the
argument typically runs, rhythm came to emblematize the epoch
around 1900 and contributed importantly to the transition into the
twentieth century. As Janice Joan Schall has put it, rhythm was "the
key to a new world view," and in that capacity, it connected aesthetic
reorientations as divergent as art nouveau, Expressionism, Dada, and
Bauhaus.[11] In Geneva, the music professor Émile Jaques-Dalcroze
invented rhythmic gymnastics. The rise of eurhythmics entailed an
aesthetics of staging the voice and body that reached its artistic peak
with the operas performed in the garden city of Hellerau, near Dresden, between 1910 and 1914.[12] Historians of rhythm have also often
discussed the Leipzig economist Karl Bücher and his study *Arbeit und
Rhythmus* (Work and rhythm). Bücher argued that the rhythmical
structure of work is by no means a primitive form, but mirrors the
natural physical predispositions of the human being. He concluded
that the organization of work in civilized societies — which regard
human beings as indolent by nature and seek to counter that indo-
lence with ever more efficient motion sequences and mechanized
labor — amounts to humankind's estrangement from its innermost
rhythmic constitution.[13]

Most historical analyses of the era around 1900 concur in characterizing rhythm as the specific articulation of a particular historical
moment. In this view, rhythm was the historical response to the
need for "vitality, order, and unity."[14] Especially in Germany, writes
Schall (among others), the conjunction of this "irrational" model
with nationalist political positions ultimately, in the 1920s and 1930s,
fueled Nazi views of the human being and the aesthetic mise-en-
scène of National Socialism; in the person of Rudolf Bode, at least,
eurhythmics was united with Nazi physical education.[15]

Despite the wealth of rhythm research on the period around 1900,
historians of science have addressed the topic only sporadically.[16]
This is surprising, since there is little doubt that rhythm was an
object of interest in the natural sciences at this time. A paradigmatic
discipline was experimental psychology as it emerged around 1900,

especially in the work of Wilhelm Wundt. Although Wundt sought to understand rhythm through experimental methods and apparatuses, rhythm in his view was not only a technical and experimental category, but was also marked by its close connection to affect. The course of a rhythm, writes Wundt, is always the "expression of the course of a feeling," and accordingly itself generates feelings — it is "affect-arousing."[17] Similarly, physics around 1900 saw an increasing interest in rhythmic movements, for example with the work of Ernst Mach and Wilhelm Ostwald; the same is true of biology, for example in the "law of series" proposed by the Viennese biologist Paul Kammerer.[18] These approaches have attracted little or no attention from historians of science. But if specific historical research on rhythm in science around 1900 is sparse, broader contextualizations in the history and theory of science are far more so. There is no systematic or comprehensive study of the history of rhythm in the sciences or of rhythm as an epistemic concept for either the exact sciences or the life sciences.[19] This lacuna is all the more striking, given that rhythm is playing an increasingly prominent part in current experimental and theoretical biology — a new development that has not yet left its mark on the historiography of science.[20]

Rhythm circa 1800: A Neglected Topic

To all intents and purposes, a history of the concept of rhythm has never been written; accordingly, there has been practically no research on the cultural and scientific history of rhythm before 1900. Exceptions are the comparatively well-researched area of etymology (to which I return below) and the works of Wilhelm Seidel. In fact, Seidel's studies in music history remain the point of departure for any historical research on rhythm even today. His 1976 definition of the term, *Rhythmus: Eine Begriffsbestimmung*, is still the only book-length historical overview, which he subsequently modified in further publications.[21] Seidel is to be commended for addressing the changing fortunes of the concept of rhythm from antiquity into the twentieth century, but his contribution is limited by its almost exclusive focus on music. Chapter 2 of the present book, on music theory, returns to Seidel's findings, and here I will mention only one crucial result of his research. For Seidel, Johann Georg Sulzer's definition of rhythm in *Allgemeine Theorie der Schönen Künste* (General theory of the fine arts) of 1773–1775 marks a historical turning point in what

I call the rhythmic episteme. In the second half of the eighteenth century, Seidel argues, the origins, manifestations, and quality of rhythm were reformulated for the first time into an embracive concept of the rhythmic.[22]

Besides Seidel's studies, a brief essay on the nature and history of rhythm by Rudolf Steglich appeared in 1949 in a special issue of *Studium generale* dedicated to rhythm. Steglich differs from Seidel in including literary sources, although his comments remain little more than a sketch. Citing Schiller and Goethe, he describes the rhythm propounded by German classicism as "a forward movement that is experienced, in its rise and fall, with the 'inner and outer senses'," in which the concord of the "elemental and personal, of body and mind, founds the organic-harmonious, classical *Humanität* of rhythm."[23] For Steglich, too, the turn of the eighteenth to the nineteenth century brought a rupture in conceptions of rhythm. In Romanticism, rhythm lost its onward-striding character. Instead, it was caught in a circle that became the "encasement of the rhythm of the Romantic self as it hovers and oscillates around the core of its own soul."[24] The up-and-down motion that distinguished the old rhythm was now supplemented by the to-and-fro motion of oscillation. This doubled, alternating motion, Steglich finds, is the basis of the epoch's rhythmical richness.[25]

Only recently have new impulses arisen, this time from literary and cultural scholarship.[26] Although so far these are isolated studies, a renewed interest in rhythm is coming into view, accumulating evidence that rhythm increasingly became a key epistemic category from the second half of the eighteenth century onward. However, in most cases, these are marginal comments or occasional hints of a novel significance for rhythm at the epochal boundary around 1800. Franz Norbert Mennemeier, for example, mentions a "high, speculative level" and "new historical quality" of reflection on rhythm, which became a foundational poetological category in German idealism.[27] In his history of German verse, Wolfgang Kayser points to August Wilhelm Schlegel's new understanding of rhythm as the "originary phenomenon of the human spirit."[28] For Isabel Zollna, nascent Romanticism ushered in a liberation of poetics from the previous "classical models," resulting in a turn to the aesthetic and therefore especially the rhythmic dimension of language.[29]

While several of the more recent studies on the reciprocities of music, poetics, and aesthetics around 1800 largely ignore rhythm,[30] the work of Barbara Naumann — for example, on Novalis, August Wilhelm Schlegel, and F. W. J. Schelling — has revealed the foundational importance of rhythm for these writers' literary, philosophical, and aesthetic thinking. Naumann stresses August Wilhelm Schlegel's belief that rhythm is anchored in human nature, which she distinguishes from the stronger emphasis on art and reason in the explanations of rhythm offered by Schiller and by Schlegel's brother Friedrich. Analyzing the theoretical writings of Novalis, Naumann highlights the centrality of rhythm as a principle of both poetry and nature, but attends only peripherally to the physiological resonances that typify these poetic conceptions of rhythm.[31] An important study by Clémence Couturier-Heinrich on the discourse of rhythm in Germany between 1760 and 1820 identifies four different strands: rhythm as an anthropological given; conversely, rhythm as a specifically historical phenomenon of a past era, namely, classical antiquity; third, rhythm in the narrower philological sense; and fourth, rhythm in aesthetic theories. Couturier-Heinrich's investigation demonstrates not only the presence of ideas about rhythm around 1800, but also the expansion in meaning that the concept underwent in the Germanophone debate from the Enlightenment to Romanticism.[32]

However, all this historical research in literature and culture lacks a history of science perspective. The *physiological* dimension of rhythm is at most — for example in the work of Naumann or Couturier-Heinrich — a distant echo in historical studies of rhythm around 1800. Discussions of this theme and period have not addressed the discipline of biology that was emerging at the same time, the new concept of organic life, or the era's notions of the development and formation of organisms. My book's argument rests on the claim that around 1800, rhythm was precisely not "only" an aesthetic and philosophical category, but was reconceptualized as a biological figure of thought. This new character of rhythm as a law of the ordering and formation of organic nature is the object of my investigation.

What Is Rhythm?
This book argues that around 1800, the living world, especially organic development, was rethought in terms of rhythmic patterns, rhythmic motion, and rhythmic representation. The question thus arises: What is rhythm?

There is no simple answer to that question. One might say that it can be answered in as many different ways as it can be posed. Particular approaches may be roughly ascribed to particular disciplines, to anthropological, sociological, linguistic, or art-historical lines of questioning, or to shared historical contexts; however, this does little to clarify what the essential features of the phenomenon of rhythm actually are. If I now outline certain definitions of rhythm, this is not intended as an exhaustive account or as even the barest historical overview. Instead, I would like to highlight a particular current within rhythm research that is of special relevance for my own questions, namely, research on the ways that rhythm mediates between the spheres of biology and culture.

There is no consensus even on the etymology of the word "rhythm." The term comes from the Greek ρυθμος (*rhythmos*), which is generally derived from the root ρειν (*rhéein*, to flow), though alternative etymologies refer to the Greek ερυ or ρυ (*eru, ru*, to pull) and ερυσθαι (*erusthai*, to fend off, protect).[33] Émile Benveniste's authoritative study demonstrates the etymological proximity of the term "rhythm" to the meanings "form" or "figure." Benveniste calls rhythm "the form as improvised, momentary, changeable," as "a configuration of movements organized in time."[34] Indeed, human measure, understood both in an ethical sense and in that of the concrete perception of quantities, dominated thinking on rhythm in antiquity. In Plato's fourth-century BCE discussion of rhythm's ethical dimension, rhythm is not a matter of aesthetics alone, but above all a political issue. Plato defines rhythm as the "order of movement" and identifies this ordering function as the moral and normative force that is required for a community to settle on a shared measure of actions and values.[35]

The most wide-ranging theory of rhythm to survive from antiquity is Aristoxenus of Tarentum's *Elementa rhythmica*. For Aristoxenus, a pupil of Aristotle's, rhythm is the combination of motion and time, as he explains using the example of the human gait. In order to be rhythmical, the sequence of steps cannot be random; the individual steps must be related to each other — to the steps that preceded them and to the ones that follow. Aristoxenus asserts that this relationship is subject to a rule, that the length of strides or the interval at which the foot touches the ground obeys a law. The simplest such law describes the precise repetition of each step. Additionally, each step can be divided into an upward and a downward movement (*arsis* and *thesis*). As possible natural relations between these movements, Aristoxenus names even ratios of 1:1 and 1:2 or 2:1.

What he proposes for the physical motion of the step also applies to language. In the *pes*, or metrical foot, short and long (long being the sum of two short syllables) are ordered into different metrical units — iambs, dactyls, trochees, and so on. The ratio of upbeat to downbeat, lifting to sinking, *arsis* to *thesis*, determines the quality of the metrical foot. Even today, this remains the basic framework for describing language rhythms. Aristoxenus also made a theoretical move that would prove important for the subsequent history of ideas of rhythm: he posited that rhythm does not achieve expression by itself, but needs matter for its expression. Accordingly, in rhythm, formed material (the body of the dancer, for example, or the notes of music) encounters a forming principle. This distinction allows rhythm to be understood more abstractly, as the movement of a unit of time, which Aristoxenus describes as "primary time" (*chronos protos*). The movement of the formed material (that of a body or a syllable) merely manifests this.[36]

The reason that classical antiquity is so important for my interest in rhythm's mediation between the spheres of nature and art is, first, that classical thinkers made of rhythm an order of motion. As something spatiotemporal, rhythm remained bound to the measure of the human senses, even for St. Augustine (354–430 CE), whose ontology of rhythm as governed by numerical relationships was antiquity's last complex examination of rhythm. The second crucial aspect is the unity of the arts in the classical view of rhythm, which addressed pulse and pace, breath and music, the sway of dance and the stress of a line without distinction. In the Middle Ages, this unity dissolved as the arts drifted apart. Music, especially, adopted new perspectives. The new polyphony was interested in the consonance and layering of music, not in its temporal sequentiality.

The next radical transformation in thinking on rhythm came only in the second half of the eighteenth century. The aesthetics of Johann Georg Sulzer, as will be discussed in detail later in this book, reconnected the arts with the body, placing the physiological measure of the human being at the core of all aesthetic experience. This conception of rhythm endured well into the first half of the twentieth century, when twentieth-century music eroded rhythm's dependence on the aesthetics of measure. Rhythm became something difficult to define, the elastic description of an event's temporal structure.

In modern everyday language, rhythm has come to be regarded almost exclusively as a phenomenon of musical experience. However, in the last hundred years, there have been several systematic attempts to reveal rhythm's epistemological importance by stressing its mediating role between biology and culture. In his 1933 publication *Vom Wesen des Rhythmus* (On the essence of rhythm), a key text of the *Lebensphilosophie* movement, Ludwig Klages (1872–1956) gave a new turn to the topic by distinguishing rhythm, as a principle of life, from meter or cadence (*Takt*) as a principle of the intellect. Rhythm here is a "general phenomenon of life," meter a human act of rationality.[37] Structurally, meter and rhythm in Klages's account differ along the lines of their respective affiliations: meter keeps us alert, rhythm relaxes us; meter repeats, rhythm renews; meter brings forth the same, rhythm the similar;[38] meter is an "identically repeatable span," rhythm, with its alternation of rise and fall, is "polarized continuity."[39]

While Klages approached rhythm through his emphatic concept of life, Alfred North Whitehead (1861–1947) opened up analytical philosophy for the concept. In *An Enquiry concerning the Principles of Natural Knowledge* (1919), proposing a philosophical foundation for physics, Whitehead held that the organic world could not be described by means of the usual physical laws and therefore exceeded the horizons of his investigation. Because, however, he did not wish to exclude the organic from his deliberations, he found himself obliged to set out in a new form the special laws applicable to the living world. For this task, he chose the concept of rhythm, which the *Enquiry* introduces as follows. In place of the term "living objects" to describe organisms, Whitehead proposes to use the term "objects expressing life" or "life-bearing objects." He defines the state of being alive exhibited by these objects as the "relation of the object to the event which is its situation."[40] That relationship is a rhythmic one: "Life (as known to us) involves the completion of rhythmic parts within the life-bearing event which exhibits that object." Rhythm as a way of identifying living objects, Whitehead continues, is an exclusive condition: "wherever there is some rhythm, there is some life.... The rhythm is then the life."[41] In other words, life is coterminous with rhythm. Rhythm can be defined more precisely. It has a particular structure: "A rhythm involves a pattern and to that extent is always self-identical. But no rhythm can be a mere pattern; for the rhythmic quality depends equally upon the differences involved in each exhibition of the pattern." The order of rhythm is therefore a structure of deviating repetitions. Whitehead arrives at the following concise characterization of rhythm: "The essence of rhythm is the fusion of sameness and novelty; so that the whole never loses the essential unity of the pattern, while the parts exhibit the contrast arising from the novelty of their detail."[42]

The American philosopher John Dewey (1859-1952) took the aspect of the object–environment relationship in a different direction. For him, rhythm is the "interaction of the live creature with his surroundings"[43] and thus the origin of all human experiences. As the basis of every human activity, rhythm is also the crucial precondition for the products of the sciences and arts — it is "the tie which holds science and art in kinship."[44] In *Art as Experience* (1934), Dewey presents rhythm as a figure of recurrence, but not the recurrence of the same. Unlike mechanical recurrence, "esthetic recurrence" always involves variation, and it is an order of "recurring *relationships*."[45] As a result, rhythm in nature and art is always "novel as well as a reminder." The recurring relationships connect the parts to the whole in a mutually illuminating way to "constitute an object as a work of art."[46]

In her 1953 study *Feeling and Form*, Susanne Langer, following the lead of her teacher Ernst Cassirer, conceptualized music as a symbol of "living form."[47] Music as a symbolic form represents time not as the disparity between two instants or states,[48] but as passage — "an audible passage filled with motion that is just as illusory as the time it is measuring."[49] As passage, music can be experienced only "in terms of sensibilities, tensions, and emotions." In turn, since feeling is reserved for living beings, the logic of all symbolic forms must obey the logic of living organisms. That logic is rhythm.[50] Human experience, which appears to us as a unity, is actually the experience of a "rhythmic continuity." That is, rhythm continually produces new resolutions of the tensions we experience. It integrates those tensions into a unity only by resolving them into the new tensions of the future.[51]

The moment of discontinuity that Langer saw as bringing forth tension and thus rhythm was also the point of departure for Gaston Bachelard. Bachelard's *Dialectique de la durée*, published in 1950, is concerned with repose as "something to which thought has a right" and "an element of becoming."[52] Here, Bachelard opposes Henri Bergson's philosophy of *durée*, which demarcated lived time from the time of clocks and science. Bachelard, in contrast, imagines lived time as a rhythmic alternation between action and repose, since all the "phenomena of duration are constructed by rhythms."[53] In Bachelard's work, the point of rest, the pause, is thus constitutive of the experience of continuity. The figurations that integrate the flow of motion and its interruption are rhythms, which Bachelard also calls "systems of instants."[54] A fundamental component of human experience, rhythm brings the necessary element of order to the thinking and feeling of human beings: through rhythm, the disparate components of experience can be unified in the "reliability [*fidelité*] of rhythm."[55] In other words, human beings have to construct the unity of their experience using rhythms. Hence, for Bachelard — unlike for Bergson — duration is not simply given; it is produced.[56] No human experience of time can exist outside of rhythm.

Early French sociologists and anthropologists such as Émile Durkheim and Marcel Mauss also addressed rhythm as a fundamental element of human experience. As Durkheim put it in *The Elementary Forms of the Religious Life* (1912), "since a collective sentiment cannot express itself collectively except on the condition of observing a certain order permitting co-operation and movements in unison, these [collective] gestures and cries," as social practices, "naturally tend to become rhythmic and regular."[57] Writing in 1926, Mauss even regarded the human being as a "rhythmic animal,"[58] and rhythm as "the direct union" not only of the social and the psychological, but of the sociological and the physiological.[59] The physiological nature of rhythm assumed by Mauss was the starting point for the work of the French paleontologist André Leroi-Gourhan. In *Gesture and Speech* (originally published in 1964), Leroi-Gourhan proposed a theory of human behavior as a "physiological aesthetics," holding that all forms of interaction with the environment take place on the basis of "physiological cadences" as bodily rhythms "create a fabric upon which all activity is inscribed."[60] In the course of evolution, this formed the conditions for human beings' aesthetic interaction with the environment to arise. Artistic products, then, are also subject to the rhythmic constitution of the body. Regardless of human progress, all art remains bound to the physiological *dispositif*.

This tradition of French philosophy remains relevant in the postmodern era. A figure of rhythm can be identified, for example, in Jacques Derrida's notion of *différance*. The term *différance* brings together two images: that of "temporization" (in the sense of deferring) and that of spatialization or "spacing" (in the sense of differing). Derrida speaks of *différance* as an "'active,' moving discord of different forces, and of differences of forces."[61] Without wishing to become mired in the detail of Derridean terminology, it is fair to say that the concept of *différance*, which combines distinguishability or differentness with temporal deferral or detour, is thought in the categories of a rhythmic episteme. This interpretation seems useful, given Derrida's approach to *différance* as a "displaced and equivocal passage of one different thing to another," as "the sameness of difference and repetition in the eternal return,"[62] or as a "movement of signification," as long as "each so-called 'present' element, each element appearing on the scene of presence, is related to something other than itself, thereby keeping within itself the mark of the past element, and already letting itself be vitiated by the mark of its relation to the future element."[63]

In very recent times, there has been a revival of interest in rhythm.[64] New media and art forms are prompting the question "What is rhythm?" among the younger generation of media and art theorists, in particular. Because rhythm opens up a field of tension between "a quasi-rhythmic a priori in nature" and a "scientific or aesthetic construct,"[65] it is a privileged object of transdisciplinary study, fostering reflection on the interplay of art, science, and nature. In addition, the complexity, paradoxes, and very breadth of the concept of rhythm allow fundamental questions of aesthetics to be formulated and to be translated into new contexts: rhythm's oscillation between repetition and variation, uniformity and diversity, its tension between "continuity and disruption," can be discussed as a "creative potential" and a "condition of aesthetic experience" in light of new forms of performativity, of music, dance, and literature.[66] At the same time, the multiformity of rhythm permits the body to be located afresh — aesthetically, anthropologically, and sociopolitically — in the urban, artificial, and virtual world of the twenty-first century, between nature and culture, physiological disposition and sensual perception, individuality and collectivity, completion and provisionality, memory and blueprint.[67]

Nature and Culture: Methodological Remarks

This detour into the history of concepts indicates the potential inherent in the concept of rhythm as an epistemological and historical category. For the purposes of this book, however, "What is rhythm?" is the wrong question. Of interest here is not the concept's possible definition, but its historical substance and provenance. Put another way: this study asks not what rhythm is, but which actors conceptualized it around 1800, how, and in what contexts.

This highlights a peculiarity — and at the same time a difficulty — of my investigation. Rhythm as a *word* was conspicuously present in the field of aesthetic reflection, poetics, and philosophy around 1800, but without an agreed single meaning. In the music theory of the time, for example, the terms "rhythm" and "meter" were used differently by different authors, and not necessarily in the sense we use them today. In biological and physiological settings, the situation was even more complicated. Around 1800, the word "rhythm" was not common currency in physiology and in fact was used very rarely. Nonetheless, in this book, I will come back again and again to rhythm in biology, in the beginnings of developmental thinking, and in other contexts. In what sense, then, is "rhythm" meant here?

To speak of rhythm does not mean retracing the word's use within its various semantic fields, disciplinary contexts, or applications. I refer to a rhythmic episteme around 1800 on two other grounds. First, I use rhythm as an analytical category. My study shows that the modern concept of development, introduced to biology by epigenetic theory around 1800, is a fundamentally rhythmic one. The core elements of rhythm — repetition, variation, regularity, period, modification, alternation, relation — are also those of the new episteme of organic development circa 1800. They can be found in the work of historical actors such as Caspar Friedrich Wolff, Ignaz Döllinger, Carl Friedrich Kielmeyer, Christian Heinrich Pander, and Karl Ernst von Baer as ways of describing generation, formation or transformation, metamorphosis, and emergence or for physiological metabolic processes. Rhythmic structures also formed the basis of a new iconography of development, one still prevalent today. Second, I argue historically, showing that imagining living processes as fundamentally rhythmic structures was a defining feature of the historic configuration of the period around 1800. It was the rhythmic episteme that enabled a modern science of embryological development to emerge.

In search of the episteme of rhythm, this study travels between various and very different domains and forms of knowledge. They include music theory, literary theory, philosophy, aesthetics, and embryology, as well as physiology and botany. All these are arenas in which a new way of dealing with organic life was taking shape around 1800 — through description, illustration, and action. Methodologically, this book therefore applies a form of the history of concepts, or *Begriffsgeschichte*, to which Reinhard Koselleck contributed so importantly in the 1960s, but expands it by absorbing methods and epistemological questions from cultural history, visual studies, and the history of science.[68] Tracing the origins and changing meanings of concepts in the natural sciences leads us back to the experimental systems, research practices, and technologies of observation that both shape concepts and are shaped by them. Observation and experiment, text and image, concepts and material objects are all part of this understanding of how a concept is constituted as a category.

The category of rhythm indicates the unity between culture and nature before the nineteenth century split them into the separate spheres of science and the arts. Since my objective is not to identify the field of knowledge where the concept of rhythm originated, I do not address migrations, adaptations, or mutual influence — the episteme of rhythm was an event taking place in numerous theaters at the same time. My study attempts to go beyond the trope of transfer in discussing the relationships between literature and science, between aesthetic and scientific thinking, between biology and culture in Romanticism.[69] The emergence of the rhythmic episteme in the period around 1800 was not the transfer of an aesthetic perspective onto natural history, a "poetization of science" (Hegener) or "poetization of nature" (Mahoney), or a "procreative poetics" (Holland).[70] Neither was the poetological and philosophical discussion of rhythm merely a "reception" of scientific concepts or vice versa. Rather, the idea of rhythmically organized nature rested on an episteme of rhythm that *simultaneously* formed the foundation of new aesthetic concepts in literary and music theory and was articulated in scientific theories — whether in the epigenetic theory of generation,

Goethe's model of the metamorphosis of plants, or the growing currency of the physiological concept of the alternation or transformation of matter, in the contemporary German discourse *Wechsel der Materie*.

In other words, rhythm circa 1800 formed a deeper epistemic stratum. It responded to the quest for rules according to which both nature and human creativity — poetry, music, the visual arts — in equal measure bring forth their works, for the law according to which they are internally ordered and that governs their constantly changing configurations. With the help of the rhythmic episteme, the particularity of the living world seemed within reach for the very first time: its capacity for infinite plenitude while remaining bound to an existing framework. With rhythm, the temporal dimension of nature acquired a rule, development was ascribed to a law that gave rise to newness, to the multiplicity of nature, as a rhythmic repetition of what already existed. Development became both rule and variation — and that is what constituted its aesthetic and epistemological momentum.

Content

This book traces the emergence of the rhythmic episteme in three sections. The first of these shows how rhythm arose as a central epistemic category in theoretical writings on literature, art, and music in the period between 1760 and 1830 and how it was proposed as a physiological category. Chapter 1 examines the broad cultural discourse of rhythm around 1800.[71] Literary theory was key to this, starting with Friedrich Gottlieb Klopstock's theories of rhythm. The transitional figure of Klopstock is particularly interesting: he worked with existing classical rhythm theories, yet he was already beginning to advocate a physiological notion of rhythm. The turn becomes more obvious in the case of Friedrich Hölderlin, with his theory of the modulation or alternation of tones (*Wechsel der Töne*) and his thoughts on the "artistic and formative drive."

Karl Philipp Moritz's concept of the autonomy of artistic works was also modeled on the rhythmic ordering of nature. His aesthetics of autonomy is exemplified by language, which only as rhythmic language can become poetry and thus a work of art. In Moritz's view, just as rhythmic motion is the developmental law of poetic language, rhythmic motion in the organic world is the developmental law of emerging life. A few years later, August Wilhelm Schlegel formulated an anthropological theory of rhythm in which poetry was seen as an expression not of human artistic skill, but quite the contrary, of man's fundamental, physiological nature. For Schlegel, the arts — even in their most elaborate form — remained tied to the physiology of the body, and their basic ordering structure was rhythm. In the "universal poetry" proposed by Novalis, too, rhythm was seen as the order according to which nature constantly transforms itself, but also within which man integrates the fragmented knowledge of the disciplines into an ever-changing image of the world.

The second chapter addresses music theory. Music historians long ago identified the period around 1800 as a turning point in musical theories of rhythm. As I will show, this rupture was not the product of music theory alone. Theorists aimed to grasp the "vitality" of music correctly — and this was the source of the contemporary interest in rhythmic structures.

The role of rhythm in Schelling's system of absolute philosophy is the focus of the third chapter. Schelling is important for my argument because his system ascribes to rhythm the key function of mediating between the spheres of art and nature. For art, rhythm is a way of representing nature as it essentially is. Conversely, for nature, rhythm is a way of manifesting itself in the forms of art.

The second section of the book is dedicated to three aspects of the emerging biological sciences: theories of generation, botany, and physiology. I begin with Caspar Friedrich Wolff, who brought to the modern debate the concept of epigenesis as the gradual emergence of forms out of the egg's originally formless mass. Chapter 4's study of Wolff's groundbreaking works, from his 1759 dissertation *Theoria generationis* (Theory of generation) to *Über die Bildung des Darmkanals* (On the formation of the intestinal canal) of 1812, shows that he conceptualized the embryo's coming into being and subsequent formation as an interplay of repetition, regularity, and variation — thus, as a rhythmical process. These elements not only marked out the parameters of rhythm, but were also core components of the new episteme of organic development around 1800.

The focus of the fifth chapter, on the concept of metamorphosis, is Johann Wolfgang von Goethe's 1790 study *The Metamorphosis of Plants*, which describes the gradual formation of the plant out of a single leaf. Goethe's work is especially interesting in the present context because he approached the question of metamorphosis from a simultaneously scientific and aesthetic or poetological angle. Chapter 5 shows that rhythm played a pivotal role in his treatment of metamorphosis in terms of both scientific theory and poetic art.

Examining several scientific treatises that may be seen as exemplifying the physiological imagination of the period around 1800, Chapter 6 discusses the decisive role that the new way of understanding organic development played in physiology — for example, in Johann Christian Reil's notion of the life force, formulated in an influential text of 1795, or in the idea of physiological secretion proposed by Ignaz Döllinger at the beginning of the nineteenth century. If physiology in this period wished to become a science of life, it would have to make the "form of becoming" its object. Physiological knowledge was to be found not only in knowledge of the qualities of substances, but also in knowledge about what happens when one form changes into the next. Motion alone seemed to be what creates forms from formlessness, life from matter. And this organic motion was not a simple flow, but a rhythmical oscillation, named in the contemporary notion of *Wechsel der Materie*, in which rhythm orders the passage of forms, the states and processes of life, and gives them a temporal choreography.

The seventh and eighth chapters depart from this broadly chronological structure. They open the third section of the book, "Serial Iconography," which addresses the emergence of embryology in general and the use of pictorial series to depict development in particular. It is based on a conviction that the history of developmental thinking cannot be written without attending to the forms and conventions employed to visualize development. The establishment of embryology was thus inextricably entwined with a new iconography in the life sciences, which I will call "epigenetic iconography." By this, I mean the visual convention of depicting developmental processes in a series of images, each showing a different stage of embryogenesis, that taken together convey the complete process of development from homogeneous matter to a fully differentiated organism. It remains the standard iconography for visualizing organic processes even today.

Whereas these specific conventions arose in tandem with the new discipline of embryology, however, the serial form of representation more generally has its own long history. Chapter 7 locates the beginnings of serial representation in the visual format of pictorial instructions for bodily movements. Teaching the correct execution and training of physical moves, these can be found in treatises on military drill, fencing, riding, and dancing from the seventeenth century on. The culture of the drill, which defined movement behavior in seventeenth-century and eighteenth-century Europe, was an aesthetic culture of motion. More precisely, drills were rhythmic arts of movement. For embryology, my argument is that the serial form of representing human motion was adopted around 1800 and became constitutive for the explanation of epigenetic developmental processes. In Chapter 8, I consider the history of pictorial representations of biological development, beginning with the works of Fabricius and Malpighi in the seventeenth century and ending with the epigenetic theories of generation around 1800. I unpack the transition from a "chronological" pictorial tradition to the "epigenetic" iconography characterized by its deployment of rhythmic developmental sequences. This distinction is exemplified by the Haller-Wolff debate, here examined from the perspective of the two scientists' use of pictures.

The ninth and tenth chapters focus on Christian Heinrich Pander and Karl Ernst von Baer. Considered the founding fathers of modern embryology, both were pupils of Ignaz Döllinger, under whose aegis they began their first experimental investigations of chick embryos. The theories of Pander and von Baer, I argue, share a crucial epistemological dynamic that has hitherto attracted little scholarly attention: both regarded the entirety of embryonic formation as a rhythmic transformation of membranes through bending and folding. These movements and shifts of the membranes are complex formations in space — each movement of a membrane is repeated in the others; they are staggered in time and spatially differentiated so that the movements take place in various directions, on various levels, and at various points on the membranes. Rhythm is the rule that orchestrates a myriad of movements, coordinating them into an ordered course. In this case, again, the use of images is vital. I show how Pander (Chapter 9) and von Baer (Chapter 10) worked with pictures in their experimental studies of chick embryos and how they built their theory of embryological development on the foundation of pictures and pictorial series. In their work, the series constitutes a new observational regime where development is at once synthesis

THE FORM OF BECOMING

and analysis. It is both the individual form and the sequence of forms, both stasis and flux. This inherent order of time is the rhythm of the pictorial series.

In Conclusion

This book is dedicated to the idea of development around 1800 and the process by which development became framed within an episteme of rhythm. Each chapter describes one facet of the rhythmic episteme as it crystallized in various guises and fields of knowledge at this time. Following the sequence of the chapters and contemplating the emergence of embryology as embedded in the diversity of the era's knowledge, thinking and representation, seeing and describing, it becomes clear how rhythm — as an ordering of the human body's externally visible movements — moves inward into the body. There, it remains an order that can be grasped immediately by the human senses, but it is now also the measure of the body's physiology, of its interior, of its becoming, of a process of flow along the cadences of regularity and counterplay.

Recently, I have learned that for development to give rise to new forms, cells must not only divide, multiply, and move in a carefully orchestrated, rhythmical way — equally, they must die. Explaining his research on apoptosis or programmed cell death in *Tupaia* species, the anatomist and embryologist Wolfgang Knabe told me that modern three-dimensional visualization techniques reveal patterns of apoptosis within the developing embryo that are highly regulated in space and time. Apoptosis flows from dorsal to ventral parts of the embryo in waves, demarcating different placodes to lay down the foundations for the organism's future sense organs.[72] Clearly, rhythm is a theme for twenty-first-century developmental biology as much as it was for the eighteenth century. Not only does rhythm give order to flux and build the future upon the past, it also reconciles proliferation with apoptosis, life with death.

PART ONE

A New Epistemology of Rhythm

CHAPTER ONE

Literary Form

Thinking about literary language was at the core of the rhythmic episteme's emergence around 1800, as classical theories of versification gave way to the theoretical grounding of poetry in physiology. Major figures in this transition were Friedrich Klopstock, whose influential notion of poetry as a form of knowledge in its own right was underpinned by new terminologies and meters, and — most ambitiously — Friedrich Hölderlin, who raised poetic art to the status of a "better philosophy" and investigated the rhythmic organization of poetry and nature alike. Karl Philipp Moritz, too, made the bonding of nature and art through rhythm the foundation stone of a new aesthetic theory, the theory of aesthetic autonomy. Rhythm was also central to two other important late eighteenth-century poetic models that went beyond the sphere of aesthetics alone: those of Novalis and August Wilhelm Schlegel. Novalis attempted to unify the whole of human knowledge, the arts and sciences alike, by using rhythmic alternation to interweave the threads of his universal poetry. Schlegel went further. He proposed nothing less than that all artistic expression is corporeally bound to the rhythmic disposition of human nature.

Poetry as a Form of Thought: Klopstock
For Friedrich Gottlieb Klopstock (1724–1803), poetry was a distinct form of knowing,[1] because poetry (unlike philosophy, for example) appealed not merely to the truth of reason, but also to the truth of sensation and feeling. Only poetic form, he argued, could address all the psyche's faculties at once, and poetry's task was to move the human being:[2] "The essence of poetry is that, through the aid of language, it shows a *certain number* of objects that we *know*, or the existence of which we *suspect*, from an *aspect* that *employs* the *most refined* forces of our soul to such a high degree that one works upon

the other, and thus sets the *whole* soul in motion."[3] Encompassing the whole person in his or her thought, sensation, and feeling, poetry stands at the apex of a hierarchy of acts of knowing in which the position of the sciences, practicing only the "cold" thinking of reason, is far inferior. For the first time, Klopstock placed poetry at the center of the intellectual world.[4] By making rhythmic language an epistemological form of its own, he paved the way for rhythm to be disengaged from its bonds with language alone and rethought as a physical principle. A transitional figure in German poetics, he deployed established classical theories of verse while simultaneously pointing forward to a new, physiological grounding for rhythm.

Rhythmic Movement as Mitausdruck

Klopstock developed his theory of versification over the fifteen years between 1764 and 1779, from the fragments in "Abhandlung vom Sylbenmaße" (Treatise on poetic meter) to the essay "Vom deutschen Hexameter" (On German hexameter).[5] His goal was to define the relationship between poetry's content and its meter more precisely than anyone before him had managed to do. Rhythm was at the heart of Klopstock's deliberations, although he rejected the term itself, believing it could not accurately describe the complex links between meter and semantics: "The word rhythm (if I have used it, then in the sense of the relationship between sounds [*Tonverhalt*]) is one of those that show how, and how lastingly, words can sometimes seduce us into conceptual confusions."[6] Accordingly, Klopstock introduced new terms in his model of verse, which he created not least as a way of theorizing his own poetic meter.[7] In place of "rhythm," he proposed the concepts of *Zeitausdruck* ("temporal expression" — the fastness or slowness of syllables) and *Tonverhalt* ("sound relationship" or the acoustic relationship between syllables).[8] For Klopstock, *Zeitausdruck* and *Tonverhalt* are the two forces that make up the movement of words in a line. *Zeitausdruck* is based on the duration of the syllables: "The movement of words is either slow, or fast. Seen from this aspect, it has temporal expression."[9] In such expression, slowness or fastness is directly associated with the length or shortness of the syllables: "If a foot has more long than short syllables, then the temporal expression is slow, and if it has more short ones, fast."[10] Klopstock's concept of *Zeitausdruck* made slowness and speed — resulting from long and short syllables — the essential trait of movement in verse.

The term *Tonverhalt* is more difficult to explain: "The objects of the sound relationship are certain qualities of sensation and passion and that of the sensual which can be expressed by means of it."[11] The *Tonverhalt* of verse arises from the accentuation of syllables, the qualitative swelling or attenuation of the voice. The term thus supplements the vertical organization of verse through long and short syllables with a horizontal organization through the voice's rise and fall as short syllables alternate with long ones.[12] *Tonverhalt* directly ties rhythm to sensation. At the core of the rhythmic theory of *Zeitausdruck* and *Tonverhalt* is the postulate that meter's role is to amplify the content of poetry: "through the motion produced by their long and short syllables," words "mean even more, and more vividly, what they are meant to mean."[13] From 1779 onward, Klopstock described this phenomenon as "coexpression" (*Mitausdruck*), a term that he coined in "Vom deutschen Hexameter."[14] It was with the help of this word that, in 1794, Klopstock formulated his theory of versification most succinctly: "Poetic meter is coexpression through movement."[15] The concept of *Mitausdruck* is thus central to Klopstock's understanding of literary form. It assumes that the patterned movement of lines in poetry is not *additional* to what is expressed — to the content of the words — but forms part of that content, reinforcing it, elaborating it, intensifying it. This purpose is served not only by rhythm, but also by sound, the acoustic dimension of the spoken word. For the first time in poetic theory, poetry's task of "moving" the listener's feelings is relocated to the *formal* structure of the poem: its rhythm, tones, and musical sound. The work of moving the listener, making him feel, is delegated to the poem's metrical order.[16]

Not surprisingly, Klopstock himself believed that his chief contribution to poetry had been his creation of new forms of versification for the German language. In his poems, he sought to introduce classical meters (especially hexameter) into German or develop new ones in order to find a metrical order congruent with poetry's content.[17] But because Klopstock still regarded rhythm only as a coexpression, he had not fully escaped the importance of words; on the contrary, the "movement of words" remained "what matters most of all in the art of verse":[18] the words of poetry transport a kind of truth, a generally accessible insight into the world. For Klopstock, this truth was no longer restricted to the rational aspect of meaning alone, but had to be opened up to sensual experience and to feeling — and this was rhythm's role.[19]

   Although Klopstock still tied rhythm to semantics, his theory of the primacy of word meaning (*Wortbedeutung*) already contained the germ of its own destruction. It prepared the way for a transition from word to rhythm, from mind to body. It is for this reason that Menninghaus calls Klopstock's theory a theory of "the wordless movement of words"[20] — referring to the level of language that cannot be expressed in words, but *can* be expressed in the movement of language and that in Klopstock's view impacts on the listener more immediately, more directly, than does the semantics of the word.[21] Citing Klopstock's odes on skating, Menninghaus argues that they entrust expression, which is not communicable through words, entirely to the rhythm of the line. The urgency of sensation experienced while skating is enacted as a "metrical dance,"[22] for example in the ode "Skating": "Ich erfinde dem schlüpfenden Stahl seinen Tanz" ("The ice-dance, with the gliding steel / I trace, inventive"),[23] a line that maps the physical succession of moves on the ice as a rhythmical dance. Klopstock couples the emotional concept of being moved with the concrete movement of the body.[24]

   Klopstock's aim was above all to invoke ardent feelings of nature that would break through the limits of the word with intoxifying force, and not to replicate the inner organization of nature, the rhythmic quality of which is revealed in the body's dance. That step was taken only later, when rhythm became an epistemic category — neither an attribute of aesthetics nor an attribute of nature, but constituting the order of nature itself.

The Alternation of Tones: Hölderlin

Klopstock's thoughts on versification had aimed to raise the status of poetry to that of a form of knowledge on its own account. Knowing through poetry was to be more comprehensive than knowing through reason, because it would also embrace the human being's corporeality and sensuality. Shortly afterward, Friedrich Hölderlin (1770–1843) called poetry the "better philosophy" and proposed one of the eighteenth century's most ambitious theories of poetry, attempting to rethink the structural and developmental laws of the poetic process. Hölderlin made rhythm poetry's central organizing principle: it could bring together sensation, imagination, and reason — the three fundamental qualities of human knowledge — to form a higher meaning. Rhythmic language alone could grant man access to knowledge of the world. This power derived from the fact that the rhythmically organized "alternation of tones" (*Wechsel der Töne*), as Hölderlin also called his theory, was the formative law not only of poetry, but of nature itself. It is no coincidence that Hölderlin's formulation echoes the term *Wechsel der Materie*, the alternation or transformation of matter, that was current in contemporary physiology. In the rhythm of poetry, man is tied to nature's rhythm, and the sounding of language repeats the song of nature.[25] The parallel becomes particularly clear in Hölderlin's concept of the "artistic and formative drive" (*Kunst- und Bildungstrieb*). Proceeding from his poetics, Hölderlin drew up a developmental theory of nature that amounted to an original contribution to the physiology of his era. He took the debate on the formative drive (*Bildungstrieb*) of nature, key to late eighteenth-century biology, and turned it on its head. Because for Hölderlin the formative drive is also an artistic drive, nature does not simply move along a predetermined path, but makes use of man and his will to art in order to complete itself. Here, nature and art are more than merely interreferenced: they perfect each other.

Rhythmical Sounding: The Poet's Calculation

Hölderlin received his theological training at the Protestant seminary of the University of Tübingen.[26] During his studies, he maintained close friendships with Schelling and Hegel and was greatly influenced by the philosophy of Fichte, whose lectures he attended in Jena in 1794.[27] Hölderlin's poetry can be regarded as a response to the problems raised for him by transcendental philosophy, especially Fichte's philosophy of the "absolute I." The focal point of his reflections was the mediation between the subject and its surrounding world. Despite the philosophical nature of this question, Hölderlin did not seek answers in philosophy — for philosophy, as he wrote in a letter to his friend Immanuel Niethammer in 1796, is "a tyrant, and I endure its force more than I subject myself to it by free choice." Instead, "I want to discover the principle which explains to me the divisions in which we think and exist, yet which is also capable of dispelling the conflict between subject and object, between our self and the world, yes, also between reason and revelation." For this, the letter continues, "we need an aesthetic sense."[28] Hölderlin regarded poetry as a better form of philosophy — it should be the highest form of the mind's expression.[29] As a poet, therefore, he is also a philosopher. In contrast to Fichte, Hölderlin aimed to overcome the abstraction of thinking and guide it back into human experience through poetry: poetry is philosophy, and philosophy is poetry. Language, sound, and thought must be gathered together; each aspect alone is incomplete. In this sense, Hölderlin's principal focus was on thought under the conditions of the sounding of language. More than almost any other theorist of German literature, he was concerned with the form of language, its regularity, its precision and formal rigor, its construction, and its acoustic shape.[30]

In other words, Hölderlin's poetics straddles the dimensions of language as acoustic event, poetry, and philosophy, and in the place of philosophy's conceptual logic he sets the "calculable laws" of poetry.[31] Whereas philosophy is restricted to thinking by means of reason, poetry can tap the multifariousness of human experience — made up of imagination, sensation, and reason in equal measure — through the sounding of language. The capacity for poetry is one of man's most important faculties, if not the most important, for it is the only form of expression in which all human abilities are addressed simultaneously. Because of this pivotal status, poetry has a special need for "especially certain and characteristic principles and limits. Thereto, then, belongs that lawful calculation."[32] The lawful calculation (*gesetzlicher Kalkül*) of the poet is the body of rules that organize poetry, and the law of poetry is its rhythm.

Hölderlin worked out his concept of rhythm essentially through his reading of Sophocles. In "Remarks on 'Antigone,'" he demarcates philosophy from poetry as follows:

> The rule, the calculable law of "Antigone" . . . is one of the various successive modes through which representation [*Vorstellung*], sensation [*Empfindung*] and reason [*Räsonnement*] develop according to poetic logic. Just as philosophy always treats only one faculty of the soul, such that the presentation of this one faculty constitutes a whole and that the mere cohering of the parts of this one faculty is called logic, so poetry treats the various faculties of man, such that the depiction of these various faculties constitutes a whole and that the cohering of the more autonomous parts of the various faculties can be called rhythm, in the higher sense, or the calculable law.[33]

In poetry, the faculties of imagination, sensation, and reasoning come together without ceasing to exist independently. They are interwoven into the specific unity of poetic experience by means of the "calculable law," which Hölderlin also calls "poetic logic" or "rhythm in the higher sense." The rhythm of "various successive modes" thus means the alternation between the mind's various faculties, which rhythm does not dissolve, but orders in a law-like way. Hölderlin distinguishes this notion of rhythmic succession from the philosophical procedure of causal inference. Philosophy does not proceed rhythmically, because in reasoning, the individual steps function merely as links in a chain, not as independent elements or ones that are qualitatively distanced from each other.[34] In short, poetry is not a specific literary genre, a form of expression reserved for the subjective dimension of sensations, and neither is it inferior to philosophy—quite the contrary. Hölderlin described poetry as the "higher enlightenment,"[35] because in its calculable law it resolves the contradictions experienced in life, the conflicts between sensation, imagination, and reason, and unifies them into a higher meaning.

Hölderlin's theory of the "alternation of tones" built on this calculable law of poetry to distinguish the different "tones" of language, in a sense similar to musical keys. He identified a heroic, a naive, and an ideal tone, based on the Greek genres of epic, lyric, and tragic poetry.[36] Through the poetic law of rhythm, writes Hölderlin, poetry raises the constant alternation or modulation of these "tones" to a higher unity. In every poem, a base tone predominates—its naive, heroic, or idealistic mood.[37] But the basic tone does not exist simply as such, in an unalloyed form; it undergoes variation into an interplay of tones. The perfect alternation of tones is to be found in Greek tragedy, which is where the law of rhythm is manifested in its purest form:

> For indeed, the tragic transport is actually empty and the least restrained. Thereby, in the rhythmic sequence of the representations wherein transport presents itself, there becomes necessary what in poetic meter is called caesura, the pure word, the counter-rhythmic rupture [*gegenrhythmische Unterbrechung*]; namely, in order to meet the onrushing change [*Wechsel*] of representations at its highest point in such a manner that very soon there appears not the change of representation but the representation itself.[38]

The "counter-rhythmic rupture" or caesura is the crucial rhythmic element in the alternation of tones. It is only in the moment of standstill, the pause, that the changing of moods can reflect on itself and thus acquire its "lawful" order. For the genre of tragedy, this means that if the first presentations "are more pulled forward by the *following ones*, the caesura or counter-rhythmic rupture has to lie *from the beginning*, so that the first half is as it were protected against the second half"; if the later ones are more pressed by the earlier ones, "the equilibrium will incline from the end toward the beginning due to the counteracting caesura."[39] The caesura is thus no mere stylistic figure: it is a constitutive component of rhythmic movement. The caesura is what weaves the alternating tones into a web of relationships, an ordered sequence of reciprocal reference.[40] The "art-character" (*Kunstkarakter*) of poetry arises when a basic mood is dissected — like a musical chord — into the tensions of different tones, which nonetheless always sound together in their interplay.[41] It is the unity of contradictions or, as Hölderlin explained in an important theoretical text, "On the Operations of the Poetic Spirit," of the "harmoniously opposed" (*harmonischentgegengesetzt*):

> Thus, through this hyperbolic operation according to which the idealistic, harmoniously opposed and connected, is not merely considered as such as beautiful life, but also as life in general, hence also as capable of a different condition, and not of another harmoniously opposed [*harmonischentgegengesetzt*] one, to be sure, but of a directly opposed one, a most extreme, such that this new condition is comparable with the previous one only through the idea of life in general, — this is precisely how the poet provides the idealistic with a beginning, a direction, a significance.[42]

Hölderlin's theory of the alternation of tones relies on an epistemology that binds human cognition of the world to a rhythmic and musical view of language. Thinking and perceiving the world cannot be disentangled from language as a sound event: it is only through the sounding of language that, sensually and musically, the alternation of different and contradictory conditions of human existence can find expression. This belief in the autonomy of the sounding and rhythm of language vis-à-vis the logos of words anchors Hölderlin in the Greek tradition of language as μουσική τέχνη (*musiké téchne*). In classical views, language forges a unity out of rhythm, sound, sense, and body; such explanations reflected the experience that speaking is musical and that the onomatopoeia of speech unfolds its own meanings, independently of the semantics of the word.[43] The "poetic spirit" creates not a merely objective, but a "felt and tangible coherence and identity in the alternation of oppositions."[44]

The Alternation of Tones and the Artistic and Formative Drive
However, Hölderlin regarded rhythm as far more than an internal
structuring principle of poetry. His concept of rhythm placed him
at the heart of literary and philosophical debates around 1800, but
because he saw rhythm as part of a broader developmental principle
of nature, his work must also be considered in the context of his
period's biological and physiological thinking. For Hölderlin, it is
in rhythm that man's natural disposition reveals itself as also being
physical, corporeal existence. Hölderlin's understanding of poetry
and rhythm was schooled by classical antiquity, which associated
the rhythm of language directly with the biology of the body, so
that the corporeal aspect of rhythm had a specifically physiological
dimension. Hence the insistence in "On the Operations of the Poetic
Spirit" that not only "beautiful life," but "life in general" is expressed
in the rhythmic alternation of tones.[45] Just as human existence is
always subject to a conflict of the faculties, which never dissolve each
other but are brought by rhythm into a constantly changing order,
nature, too, is characterized by the ceaseless alternation of conflict-
ing forces in a continuing process of formation and transformation.
Being both poetic and physiological, rhythm is also what organizes
biological nature.

Hölderlin's approach here recalls contemporary physiological
debates on the alternation of matter, the *Wechsel der Materie*. This
important and recurrent concept in biological writings of the period
around 1800 referred to the continuing oscillation of organic matter
between solid and liquid states, something regarded as the defining quality of organic life.[46] An article entitled "Über den Wechsel
der thierischen Materie" (On the transformation of animal matter),
published in 1800 in Johann Christian Reil's *Archiv für die Physiologie*,
a journal founded just five years earlier, argued "that transformation
in general takes place in the solid and fluid parts, in all liquors and
organs without exception; that it is constant."[47] Around 1800, the
notion of the transformation of matter proposed a new image of the
body. As the Tübingen professor Johann Heinrich Ferdinand von
Autenrieth (1772–1835) put it, the organic body exists "solely in this
transitional state of neutral, shapeless fluidity, in separated gases and
rigid crystallization."[48] The human body is thus not a body formed
once and then remaining unchanged. Instead, at every moment of

its existence, it is a transitional phenomenon. The transformation of matter is what constitutes the physiology of the body. "The organs of animals work and vegetate, both at once, through a transformation of matter," wrote Reil in 1799, and "in this fashion, animal matter that would otherwise dissolve into its elements preserves itself as such."[49] Only by the transformation of matter being "ordered according to particular rules are the most admirable purposes in organic nature achieved."[50]

In 1781, Johann Friedrich Blumenbach (1752–1840) introduced the concept of the *Bildungstrieb*, or formative drive, into the biological debate. In his *Über den Bildungstrieb und das Zeugungsgeschäfte* (On the formative drive and the business of generation), an adapted English translation of which appeared in 1792 as *An Essay on Generation*, he explains how his studies have convinced him "that matter in all living creatures, from man to maggot and from cedar to mold ... takes on a particular action, or *nisus formativus* [*Bildungstrieb*], which *nisus* continues to act through the whole life of the animal, and that by it the first form of the animal, or plant is not only determined, but afterwards preserved, and when deranged, is again restored."[51] This formative drive, a plan inherent in organic life, steers the development, preservation, and regeneration of the organism. Blumenbach's drive is a teleological principle that organizes the living world, giving direction and goal to its constant re-formation, whereas Hölderlin turned the teleological drive into an "artistic and formative drive" (*Kunst- und Bildungstrieb*) that constitutes itself in rhythmic alternation. This was the move that expanded his poetological theory of the alternation of tones into a developmental theory of nature.[52]

As Hölderlin told his brother in a 1799 letter, it is a human drive "to nurture life, to accelerate the eternal passage of nature toward perfection — to perfect what he finds, to idealize it."[53] That same drive, however, is also the "something in us that is glad to keep our fetters," the "drive to be determined."[54] Humanity is torn between adherence to the boundaries that nature imposes on it and the urge to transcend those boundaries in its own works. For human beings, the formative drive is always a drive to art; these are not two separate things, but two sides of the same fundamental impulsion. Hölderlin calls it a "paradox" that the artistic and formative drive, "with all its modifications and deviations, is a real service that human beings render to nature."[55] While in Blumenbach's formative drive nature carries its own plan within itself and is thus capable of perfecting itself, Hölderlin reverses this view: for him, nature avails itself of man and his desire for art in order to achieve perfection. Nature needs mankind for its highest completion; at the same time, mankind needs the artistic and formative drive in order to attain an intensified existence — because in the works of the human mind, nature enlightens man about himself. It shows him "the path that men walk for the most part blindly, often with resentment and reluctance, and only too often meanly and ignobly . . . so that they can walk it with open eyes and with joyfulness and nobility."[56] The works of nature and the works of art, consequently, are more than closely interlinked — nature and art perfect each other reciprocally, and they do so by means of the artistic and formative drive. In man's artistic drive, the formative drive is released from its natural plan, exchanging, so to speak, the driven for the conscious, the controlled: "It makes a difference whether this drive to cultivate [*Bildungstrieb*] operates blindly or with consciousness, whether it knows from where it emerged and whereto it strives; for this is man's only mistake, that his formative drive goes astray, takes an unworthy, altogether mistaken direction or, at least, misses its proper place."[57] When the human "drive to idealize or encourage, rework, develop, perfect"[58] processes the "aorgic, the incomprehensible, the non-sensuous, the unlimited"[59] of formless nature into form, mistakes may occur. This is the danger of the artistic drive: by abrogating nature's plan, it can become flawed, arbitrary, misguided. Although the artistic and formative drive is ceaselessly at work, it is not an unbroken continuity, but a disparate succession of different right and wrong directions, a series of "detours or wrong ways that it can take."[60] It is here that the rhythmic nature of the drive becomes absolutely clear: 1. Art and nature are contingent upon each other (the artistic and formative drive being one and the same drive, though it manifests itself at both poles). 2. On the one hand, the drive is continuously at work, but on the other, it exists as a series of disparate "wrong" and "right" states. As a rhythmic principle, therefore, the artistic and formative drive is not fixed from the outset, but can change direction again and again, thanks to the disparity of its alternating elements.

Here we again find what Hölderlin's poetic theory referred to as the "counter-rhythmic rupture" or caesura. In nature, too, it is the interrupted movement of rhythm that prevents the drive — oscillating between the productions of nature and the productions of art — from running wild and taking a wrong path. Even if the *goal* of the drive is no longer known (as it was in Blumenbach's formative drive), rhythm still specifies the *direction*. For Hölderlin, the drive is preserved from taking the "wrong direction" or even acting completely randomly by the fact that we "give ourselves our own direction which is determined by the preceding pure and impure directions."[61] The law of rhythm continues the work of the drive onward into the future — not arbitrarily, but on the basis of its past effects. Even if man does not know the directions of the drive in detail, he does know its rhythmic law. He can observe it in "everything that emerged from that drive before and around us."[62]

For Hölderlin, nature and art relate to each other in rhythm. Nature cannot reach its highest form without the works of art, and art cannot become art at all without the help of nature. Karl Philipp Moritz took this inextricability of art and nature as the basis for an innovative aesthetics. His notion of the work of art's autonomy depended on the new, rhythmical conception of organic nature.

The Prosody of Development: Karl Philipp Moritz

Karl Philipp Moritz (1756–1793) and Johann Wolfgang von Goethe (1749–1832) met in 1786 in Rome. In the throes of an artistic crisis, Goethe had left Weimar for an extended trip through Italy, during which he worked out the beginnings of his theory of the metamorphosis of plants. Despite describing Moritz as "ravaged and injured by fate," Goethe found a like-minded brother in the suffering writer, who was recovering from a badly broken arm.[63] At this time, Moritz was working on a new aesthetics. He first published his thoughts on the autonomy of the work of art in a 1785 *Berlinische Monatsschrift* article entitled "Versuch einer Vereinigung aller schönen Künste und Wissenschaften unter dem Begriff des in sich selbst Vollendeten" (Attempt to unify all the fine arts and sciences under the concept of what is complete in itself), and continued the theme in 1788 with the essay "Über die bildende Nachahmung des Schönen" (On the creative imitation of the beautiful).[64] As Goethe wrote in a letter from Italy, he and the convalescent Moritz "talked through almost everything that I plan in art as a theory and description of nature, and I have drawn many benefits from his observations."[65] In *Travels in Italy*, he elaborated:

I have been spending very pleasant hours with Moritz explaining my botanical system to him. As we talk, I write down everything we say, so that I can see where I have got to. This is the only way in which I could get some of my ideas down on paper. But with my new pupil I have had the experience that even the most abstract ideas become intelligible to a receptive mind if they are presented properly. He enjoys our talks immensely and always anticipates my conclusions.[66]

Indeed, there are close links between Goethe's idea of development as set out in his theory of the metamorphosis of plants and Moritz's aesthetics of autonomy. For Moritz, the autonomy of the work of art consists in its "not having its purpose outside itself." The work of art is there "not because of the perfection of something else, but because of its own internal perfection."[67] Beauty exists "because the essence of the beautiful consists in its being complete in itself,"[68] and only by returning its attention to itself, to its "inner purposiveness,"[69] does the work of art achieve autonomy. At the same time, the work of art's inner constitution always reveals the way that it came into being: "The nature of the beautiful consists precisely in the fact that its inner being lies beyond the power of thought, in its emergence, in its own becoming."[70]

This brings the process by which the work emerges, its becoming, to the forefront of the idea of autonomy. In "Über die bildende Nachahmung des Schönen," Moritz insists that even just the very first moment of emergence can convey a "vivid idea of the creative imitation of the beautiful." At that moment, what is emerging reveals itself as simultaneously perfected and becoming; the work of art "steps before the soul in obscure intuition all at once, already completed, through all the degrees of its gradual becoming."[71] The completeness of the beautiful consists in the process of its formation, its own becoming, and in this sense, the emergence of art is like that of nature. For Moritz, art brings forth the inherently beautiful in the same way that nature brings forth the living world — it is only their formative laws that complete the works of nature and of art and produce them as autonomous objects. In other words, neither a work of art nor a work of nature can attain autonomy through the specific figuration of its finished shape; rather, it attains autonomy by virtue of its genesis, of its developmental law.

As I will now show, that law is a rhythmical one.[72] Moritz's aesthetics and Goethe's concept of development both rest upon this idea:

Goethe thought of the plant's development as a rhythmic interplay of expansion and contraction (see Chapter 5), while Moritz drew his notion of the autonomous artwork's formative and structural laws from an equally rhythmic model of the development of organic life in nature.

The Autonomy of Poetry through the Rhythm of Language

In his *Versuch einer deutschen Prosodie* (Attempt at a German prosody) of 1786, Moritz applied his aesthetics of autonomy to language, showing how language is perfected into a work of art by poetry. In poetry's transformation of language from not-art to art, Moritz attributes a constitutive role to rhythm.[73] *Versuch einer deutschen Prosodie*, his treatise on the rhythmic and metrical rules of the German language, is regarded as the first attempt to define German poetry's rhythm and versification following not the traditional rules of Greek and Latin, but the laws governing the German language.[74] Moritz starts from the assumption that rhythmically and poetically constituted language, in contradistinction to ordinary language, is a form of art. Whereas the purpose of unconstrained language lies in the particular tasks and functions that speakers try to fulfill by speaking, poetry is a linguistic body of rules. And the inner rule that changes language into poetry is rhythm, for in poetic language, the level of semantics is flanked by a new, entirely independent order — the order of rhythm.[75] Moritz's approach is original in detaching language's rhythmic order completely from its semantics. Key to this line of thinking is that rhythm *alone* accounts for the autonomy of the work of art. Poetry acquires its self-sufficiency as art solely through the specific regularity and relational ordering of rhythm. This process is twofold. First, individual syllables are isolated and separated off from the context of the word; this results in a leveling of the syllables, so that every syllable in principle has the same weight. Second, rhythm weaves the syllables into a new fabric of relationships, a whole that consists exclusively in the reciprocal relationships of its components.

Before the rhythmic structure of poetry can be established, then, the semantic order must first be broken open. At the level of meaning, the task of making sense falls to words; they are the units of understanding, of thought, of the meaning that the speaker wishes to express by speech. The intellect assigns value, hierarchizing words according to the speaker's intended message. For poetry, however, the units are not words, but their smaller elements, syllables:[76] "All the meaningful syllables that reason has subordinated to each other according to their stronger or weaker significance now become equal for sensation [*Empfindung*], and we seek only to bring symmetry to what is by nature the same."[77] The isolation and leveling of syllables performed by sensation is the starting point for the order of rhythm. Rhythm can now bring its own laws to speech — it can drive speech "back into itself."[78] Even in the individual words, sensation or sentiment "presses itself into a kind of melody, into a harmonious cadence of syllables that brings speech close to song."[79] Speech, in contrast, as a mere means to express ideas, strives "relentlessly toward those ideas without taking the trouble to cultivate its individual sounds properly. It neglects itself, so to speak, because its purpose lies more outside itself than within itself."[80] Only if all the notes of music, all the steps of dance, all the syllables of language are structurally equal in value, Moritz concludes, have they been generated purely "for their own sake."[81] Thus, what is special about rhythm is its dual structure. On the one hand, each separate element of rhythm forms a complete, coherent entity of its own. Where thought evaluates, hierarchizes, and prioritizes, sensation or feeling posits sameness. On the other hand, each individual component acquires meaning only in its relationship with all the others in the rhythmical series. To produce a new order, each of rhythm's components must be seen simultaneously as an independent entity and as part of the whole.[82] In this way, the poetic-rhythmic order adds a further factor to thoughts and sensations: the factor of development and flux, in which they are manifested and at the same time glide gently into new thoughts and sensations.

Isolating, Separating, Moving
The order of rhythm raises language to the rank of an autonomous work of art, but rhythm as a figure of knowledge also organizes Moritz's view of organic life. For him, the autonomy of artistic creations follows the autonomous order of nature — because in the organic world, rhythm is the developmental law of emerging life. Nature's greatest artwork is the embryo.[83] Moritz envisions a nature that is constantly destroying itself, constantly creating itself anew.[84] At the heart of the formation and transformation of organic life is the process of isolating and separating off, which is also the first stage in the process of artistic formation. "Isolating, separating from the

mass, is the perpetual occupation of man," writes Moritz, and "all the charm of poetry rests upon isolating, separating out from the whole, and on giving what is isolated its own center of gravity, by which means it makes itself into a whole once more."[85] Artistic activity is a developmental process that does not happen by chance, but begins with the singling out of a particular element. In poetry, breaking open the semantic unit of the word and isolating its syllables as independent entities are the preconditions for establishing a new rhythmical order. At the same time, it is on "isolation, secretion from the mass," that all formation depends, "and only in this does it differ from accident."[86] Life, too, continually arises anew as individual structures are first isolated from a whole, then reassembled into new sets of relationships.

The highest type of formation found in nature is sexual reproduction: "With the complete determinacy of formation, and the exclusion of everything accidental, through the necessary conjunction of two symmetrical halves, movement enters the embryo, which escapes the fetters of its immediate surroundings precisely because, by excluding all the randomness that presses it down to the earth, it now has its own center of gravity and the axis of its rotation is in itself."[87] For Moritz, life is movement, and embryonic development is a twofold movement. The embryo initially separates itself from an existing order (the mother's uterus). The beginning of the embryo's life as an autonomous organism arises from the independent movements of its body. But the independent physical movement of the embryo, separating it from its mother, is only the first step, for organic formation means development into ever higher forms of movement. The highest of these, the culmination of development, is no longer physical movement, but the movement of language (*Sprachbewegung*). The series thus stretches from physical motion up to the point "where the very gentlest movement, in the true tool of language, itself becomes language."[88] The embryo as an independent life becomes more meaningful by becoming more able to "speak" through itself — the body speaks through its motion.

As development proceeds, the language of movement, at first solely physical, turns into language itself, in which motion remains essential: "So in what merely grows, nothing but its formation is *determined*; in what lives and breathes, formation and movement are determined; but in what lives and thinks, formation, movement, and sound are determined — through which the whole resolves into harmony — the embracive embracing itself again."[89] Life begins with purely physical, physiological movements, which in the course of development lead seamlessly into linguistic movement. In human beings, the body's capacity to isolate itself from its environment — the precondition for all life — is consummated in the isolation of the sounds of language. Separation, movement, and sound: this is the triad of life's perfection, its inner order and determination, the rhythm of its development.

What Moritz describes for the embryo he proposes in just the same way for the poetic work of art. The separation of the syllables out of the original order of the word, the reordering of the linguistic movement in rhythm through which poetic language emerges as a work of art — these form the basis of art's autonomy, just as the separation of the embryo from its mother's womb, the movements of its body, then its heightening into poetic movement constitute the emerging being's growing independence from its environment. Rhythm is critically important to the idea of epigenetic development that this involves: embryonic life arises when, first, individual forms are isolated out of a homogeneous mass and established as entities that are independent, but structurally alike; movement then generates a new configuration of these structures, one that dissolves the previous connection with the mother as development progresses. An important aspect of this idea is that development cannot be deduced from the single form itself, but only from the rhythm driving it — from the physical movement, to the linguistic movement, up to the point of poetic completion, life is organized by rhythms. Following this logic, it also becomes clear why the embryo is the object of Moritz's 1788 treatise "Die Signatur des Schönen: In wie fern Kunstwerke beschrieben werden können" (The signature of beauty: The extent to which works of art can be described). The embryo is nature's quintessential work of art, and rhythm is the law of its development.

Being Is a Rhythmical Relation: Novalis

In the physiology that was taking shape around 1800, the concept of the *Wechsel der Materie* accounted for organic life's property of never being self-same, of always switching between different states: it was such alternation, such constant re-formation, that seemed to constitute the body and keep it alive. Novalis (the pseudonym of Friedrich von Hardenberg, 1771–1801) elaborated this physiological notion in numerous fragments and aphorisms.[90] His reflections on organic life circle around a recurrent idea: the organic world is a play of change that can be grasped only as a relationship of alternating states. Life exists not as something permanent, not as something flowing steadily or proceeding without interruption. Instead, what characterizes life is its tension, the ceaselessly changing accommodation of different states, each of which only presages the transition to the next: "Being expresses a permanence of positing, of change [*Wechsel*], of activity, of the productive action."[91] For Novalis, being could not be encapsulated either as stasis or as temporal duration alone: "I am — means I find myself in a universal relation, or I change."[92] But Novalis did not aspire to resolve life's tension into a higher synthesis in the sense of a dialectical movement. On the contrary, tension is what propels life on into the future. Life, presenting itself as a unity, is in fact a continued state of tension, the outcome of interruption and succession. It fluctuates between the poles of rest and motion.[93]

However, what interested Novalis was not the poles between which life oscillates, but the nature of the law connecting them. He went on to specify that connection: "In the temporal world, being is a rhythmical relation."[94] If the process of alternation comes to a standstill, death occurs; conversely, death also occurs if alternation strives to resolve itself, to become infinite: "Our life is imperfect because it has periods — If there was only one period, it would be infinite. The process of relation is the *substantial* relation. *Life* is there — where *enhancement* and *solidification* are combined."[95] Alternation must follow a rule, the law of rhythmic switching back and forth between the poles, in order to lead life back to itself again and again. Life does not only end when alternation is completely abolished, but also when rhythm is disrupted: then the organism becomes sick. At this point, Novalis proposes the interesting idea of a nosology based on disordered rhythm. Although he did not pursue this theme, his writings show that he arrived at the idea through his knowledge both of contemporary writings on medicine — such as the works of Andreas Röschlaub — and of recent developments in music, for example Ernst Florens Friedrich Chladni's 1787 study of acoustics, *Entdeckungen über die Theorie des Klanges*.[96] The tension of organic life finds temporary resolutions in rhythm, but always, in its onward aspiration, points forward into the future.

Poetry as the "Conduct of the Beautiful Rhythmic Soul"
Rhythm, for Novalis, is the law of alternation. Constantly recorrelat-
ing the accents of life, it is the basis upon which the "inexhaustible
amount of materials" provided by life can keep assembling afresh as
"new individual combinations."[97] Here, rhythm is the art of generating
relationships, creating new connections, dissolving old ones, and spell-
ing out organic life in every conceivable variation. This property of
the organic and rhythmic formed the foundations for Novalis's ambitious project of a universal poetry,[98] the aim of which was nothing less
than an "encyclopedistics" of knowledge. This would restore totality
to a knowledge that the individual sciences, poetry, and aesthetics had
produced in great abundance, but always in isolation from one another.
In the encyclopedic project — concerned with the relations and inter-
actions of the sciences, of knowledge and art, subject and object, the
sensual and the abstract — the notion of rhythm is the conceptual
linchpin: rhythm is the unifying principle, the connecting thread.
Following the example of organic nature, whose constant transformation and re-creation is regulated by the law of rhythm, universal
poetry makes rhythm the principle by which humankind can bring
together the fragmented knowledge of the disciplines into ever-new
images of the world.[99] Novalis's thinking on the rhythm of nature was
therefore critical to his project of understanding the world in a "universal" way. Nature showed him that such universal knowledge could
not be found in a single standpoint, that it could be derived only from a
unifying principle that subordinates the knowledge of all disciplines,
all genres, of science and art alike. Universal, encyclopedic knowledge
could consist only in the rule of connections between the multiply
differentiated, splintered knowledges of the individual domains. For
something new to emerge, existing knowledge had to be integrated
again and again in ever-different ways. And the task of universal poetry
was to define more precisely the laws governing those connections.

Like Klopstock and Hölderlin, Novalis regarded truly philosophical knowledge as being poetic in character. Only poetry could supply
the intended "universal" knowledge.[100] As Novalis put it in his *Fichte
Studies* fragments under the heading "Principal Rule," poetry is the
"conduct of the beautiful rhythmic soul — the voice accompanying
our developing [*bildende*] self."[101] By the developing or forming self,
Novalis does not mean the skilled training of the self, its aesthetic

exaltation in poetic art through the rules of rhythm. Instead, the forming self is organic, self-generating — it is life forming itself. Novalis condensed this idea in his well-known adage "Dichten ist zeugen": to write is to generate or beget.[102] It is no coincidence that his works equate the concept of formation with that of generation — the conflation directly reflects the epigenetic notions of development current at the time.[103] This is evident in Novalis's letter to August Wilhelm Schlegel on January 12, 1798: "Poetry is by nature fluid — omniformable — and unrestricted — Every stimulus moves it in all directions... — It becomes a kind of organic being — the whole construction of which betrays its emergence out of fluidity, its originally elastic nature, its unrestrictedness, its omnipotence."[104]

If poetry is an organic being, "Dichten ist zeugen" is not a unidirectional analogy in the sense of "Writing is like generating." Instead, it makes a strong assertion about the symmetry of the two spheres: writing *is* generating. The converse, "Generating is writing," is equally valid. The processes of both organic and poetic generation can be understood only through dynamic concepts — the dynamics by which the "omnipotent" emerges from the flux of the temporal world. For Novalis, both being and poetry are epigenetic phenomena. As we have seen, the organic manifests itself exclusively in the ceaseless changing of relationships, as a continual repositing or a rhythmical relation. Just as rhythm brings forth life, it perishes along with life, for "if we take away the rhythm of the world — then the world also disappears."[105] Outside the rhythmic relationship, and this is the kernel of Novalis's maxim "To write is to generate," there is no living world; outside knowledge of the rhythmical law of connection and relationships, there is no knowledge of that world.

The Physiological Origins of Language: August Wilhelm Schlegel

While Novalis hoped to encompass the whole of human knowledge and to draw the sciences and arts into a new, poetic order, August Wilhelm Schlegel (1767–1845) — elder brother of the more famous philologist and philosopher Friedrich Schlegel — aimed to restructure all the arts on the pattern of the developmental history of nature. His work is a paradigmatic case of the grounding of cultural theory in rhythm. Schlegel aspired to a "natural history of art" in which literature, dance, and music featured not as artistic genres, but as elemental poetry, as forms of expression for the physiological disposition of the human being.[106]

Urlanguage as Urpoetry

August Wilhelm Schlegel's work was innovative in associating the question of poetry's origins with the question of the origins of language in general. Schlegel began to discuss this theme in *Briefe über Poesie, Silbenmaß und Sprache* (Letters on poetry, poetic meter, and language) of 1795, and it remained the basis of his aesthetic theory in lectures he later gave at the universities of Jena and Berlin. Hitherto, Schlegel argued, the origins of language had been sought narrowly in history or philosophy alone.[107] Yet all that could be said about the beginning of language was that it arose "at the same time as the first stirring of a human existence," or rather, "it is the same thing."[108] The origins of language and the origins of humanness coincide, and language is a product equally of bodily life and of budding intellectual capacity. Language is not willfully invented, and neither is it solely the expression of sensations or of thoughts; it is "a formative [*bildende*] representation of both—that is, language is, in its origins, poetic."[109] Schlegel's initially poetological question is here absorbed into an anthropological one: language is poetry from the very beginnings of mankind; poetry and language are one. This is why it is more accurate to speak not of primal language, but of primal poetry, in which concept and sensation find shared expression.

At the same time, however, poetry does more than "impart . . . the total state of mind . . . by means of language".[110] "Everywhere that men have lived and breathed, received and spoken, they also made poetry and sang. This is attested by the oldest legend of the primeval world, which only speaks to us through the mouth of poetry; the observation of uncultivated, crude peoples demonstrates it every day. In its origins, poetry forms an indivisible whole with music and dance."[111] Living and breathing, singing and speaking, thus always go hand in hand, and poetry is always also music and dance. All are grounded in the corporeality of the human being, or more precisely in the rhythmic constitution of human physiology: "The rhythmic course of poetry is no less natural to man . . . than is poetry itself."[112] Rhythm is the natural foundation not only of language, but of humanness itself; thus, rhythmic order is a property not primarily of human reason, but of human corporeality. The expression of sensations obeys "the inner laws of the bodily structure,"[113] and the "expressive movements and sounds" of the body are always expressed in a tempo or measure of time (*Zeitmaß*):[114] "The soul, educated by nature alone and unaccustomed to fetters, demanded freedom in its external proclamation; the body, if it was not to succumb to the continued vehemence of that proclamation, needed a measure to which its inner arrangement palpably guided it. An ordered rhythm of the movements and sounds unified them both, and this is where its beneficent magic originally lay."[115]

According to Schlegel, the rule of rhythm also accounts for human life in society, since it was through rhythm that "expressive gestures and sounds, in which otherwise only unconstrained and obstinate caprice would have prevailed, were habituated to a peaceful coexistence, recreating them as sociability's bond and most beautiful symbol."[116] Their passions thus curbed, men were even "flattered" that "a measured rhythm" took on "a kind of mastery over them."[117] Rhythm, as "ordered expression in words, sounds, and gestures," is "the first step in taming and in human cultivation [Bildung]."[118] Later, lecturing in Berlin, Schlegel argued: "But man imprinted rhythm with the character of his freedom by binding his wild outbursts to a self-given rule. This rule was one and the same for gestures, sounds, and words: tempo, meter, rhythm."[119]

The Natural History of Art
In his Jena lectures,[120] Schlegel attempted a new taxonomy of the forms of art, from Homeric epic to landscape painting. He named his project a "natural history of art," by which he meant "a presentation and explanation of the necessary emergence of art out of the peculiar being and natural surroundings of man."[121] The natural history of art covers those arts in which the internal is revealed externally "in words, sounds, gestures" and thus "extends to poetry, vocal music, and mimic dance (where gestures are rhythmically ordered)."[122] In search of this natural history, Schlegel asked his students always to remember just one theoretical proposition as their axiom: "That every fine art, and especially poetry, is not an idle, chance invention of amusement, not a mere luxury of the spirit, but flows from an original, principal disposition of the human temper; that, consequently, it . . . arose with the human race and can never completely die out except with the human race."[123] Schlegel also referred to the unity of musical, poetic, and physical expression as *Urkunst*, primal art. This is not simply the sum of music, dance, and poetry — it is completely unique, because it alone provides the foundation of all human symbolic representation.

Rhythmical organization is initiated by "separating out the poetic successions inside us from other states by means of an external law of form, namely, *rhythm*."[124] This is an "arrangement of succession under the form of time."[125] A closer reading of Schlegel's argument shows that time is not what prescribes succession; in fact, time itself depends on a "physiological instigation" in the regularity of organic movements: "Thus, one could say that time is determined organically, for example in the galloping of horses."[126] This means that the basis of temporal order is the bodily constitution of the human being. Rather than the body working to the dictates of time, time works to the dictates of the body, since rhythm, "which may be defined as ordered alternation in succession, is a representation of what remains constant in the changing of ideas and therefore makes the arts of succession the expression of our whole intellectual and physical nature, of life, and of the personality."[127] In rhythm, accordingly, language defers not to time, but to the various movements of the body.[128]

Just as rhythm imposes its movements on time in the realm of organic life, in the realm of art rhythm is the means by which poetry finds its own temporal sequence. "Poetry must create its temporal sequence itself," because only in this way "will the listener be abstracted from reality and transported into an imaginative temporal series, so that he perceives in speech itself a lawlike distribution of successions, a tempo.... This law is tempo, meter, rhythm."[129] Schlegel's natural history defines the arts as a component of man's "principal disposition." That disposition is rhythm, in which art and nature, corporeality and consciousness, becoming and perishing, are directly tied to the physiology of the body: "A rhythmical series thus expresses first and foremost the outer, sensual life: tempo is its pulse, alternation its free movement. But then the tempo — which remains the same throughout — connects the successions into a unity; it is what remains constant in change; if we speak of the audible, this means an awareness of the sequence of sounds."[130]

To summarize: for August Wilhelm Schlegel, the arts came into being simultaneously with the human race and will truly perish only with it,[131] but art forms are not merely temporal forms, shaped and refined in the course of human cultural history. They are also an expression of the higher, rhythmic nature of humanity — as such, they are timeless. It is here that we find the "indescribable power of rhythm over the human soul," for rhythm "is compacted life."[132]

CHAPTER TWO

Epigenetic Music

Although music is only one of the fields in which the new conception of rhythm as a fundament of life developed in the period around
1800, the term "rhythm" tends to be associated first and foremost
with music.[1] Defining musical rhythm is no less problematic than
defining rhythm more generally. Almost every study concerned with
rhythm in music includes one or several definitions of the term, and
these differ considerably. In his classic *Rhythm and Tempo*, Curt Sachs
answered his own question "What is rhythm?" with some skepti-
cism: rhythm is "just — a word: a word without a generally accepted
meaning. Everybody believes himself entitled to usurp it for an arbitrary definition of his own."[2] As Sachs persuasively argues, even the
assertion that rhythm is what gives music its structure is too broad a
definition; at least in the eighteenth and nineteenth centuries, harmony had just as great a claim as rhythm to be the shaping principle
of music.[3] Nonetheless, Sachs does not quite avoid the temptation to
propose some kind of definition. To be rhythmic, he writes, something must be "kinetic, intermittent, and perceived through one of
the senses."[4] Grosvenor Cooper and Leonard Meyer's standard work
The Rhythmic Structure of Music argues that "to study rhythm is to
study all of music. Rhythm both organizes, and is itself organized by,
all the elements which create and shape musical processes."[5] However, Cooper and Meyer avoid a general definition, preferring to split
rhythm into smaller units such as pulse, meter, accent, and stress.
They thus approach the problem from the aesthetic aspect: "to experience rhythm is to group separate sounds into structured patterns."[6]

The studies by Sachs and by Cooper and Meyer are just two among
many, singled out here because of their influence on music historiography. Rhythm has always played an important role in perceptions of

music, and the concept has existed since antiquity in various forms and contexts. For the purposes I pursue here, it is not helpful to list or classify the plethora of definitions and theoretical approaches or to try to add a "thousand-and-first" to rhythm's "thousand definitions."[7] Instead, this chapter starts out from a different question: What did rhythm mean to music theorists around 1800, and what happened to rhythm in music at this time? At first sight, this seems a comparatively simple matter. Whatever rhythm may be, the form it takes in Beethoven's *Great Fugue* of 1825 differs enormously from that in Johann Sebastian Bach's 1751 *The Art of Fugue*. Music historians largely agree that a sea change in notions of rhythm occurred between these two points. As Wilhelm Seidel writes, during the sixteenth and seventeenth centuries, "rhythm in a comprehensive sense was the defining feature only of certain genres that specialized in it, especially dances and simple songs. That changed in the course of the eighteenth century. Rhythm — in the wide sense — then became a stylistic resource that shaped all genres, with a few exceptions such as the free fantasia. Not even the fugue was spared."[8]

The difficulties arise when it comes to tracing the exact nature of this transformation. Historians have regarded the advent of the *Akzenttheorie* or theory of accents as eighteenth-century music theory's most important innovation in the area of rhythm. How that innovation came about, however, either remains unexplained or the reasons are sought in music alone. Attempts to place the emergence of the new rhythmics within a broader historical and cultural context are rare. Seidel, for example, offers no explanation as to why music changed — he simply surmises that music theory reacted to a change by adopting a new theory of rhythm.[9] In *The Stratification of Musical Rhythm*, Maury Yeston notes that eighteenth-century discussions of rhythm "appear to have served a new curiosity about the aesthetic principles of rhythm as they apply to uniquely musical structures."[10] But Yeston neither names the sources of this curiosity nor considers other domains of knowledge where a similar interest in rhythm was taking shape. Sachs writes: "The nineteenth century opens with a true apotheosis of rhythm. In Beethoven's works, the hearer's attention is often forced to withdraw from melody, harmony, colour, and to concentrate upon the vigorous, all-dominating language of a rhythm that would persist as pattern throughout a whole movement or interfere with peaceful continuity in unexpected counterblows."[11]

But even Sachs concentrates on the ways in which individual composers worked with rhythmical structures, rather than addressing the background of the transformation.

This chapter will show that several important European music theorists of the period around 1800 regarded rhythm as the vital motor of music. They generally used "measure" and "beat" (*Takt*) only to describe the mechanical and predetermined aspects of a musical piece, whereas "rhythm" designated the living unity of music, its essential core, its life force. Rhythm as the inner organization of music — speaking to the human being with immediacy and addressing him as a unity of body and feeling — corresponded to rhythm as an internal structuring principle of organic nature. This new concept of rhythm evaded the grasp of exact description; instead, it tapped directly into the physiology of organic life. What was new in music theory around 1800, in other words, was not the musical concept of rhythm;[12] it was the fact that the vision of rhythm in music theory and biology had profoundly reordered knowledge in both domains. Rhythm was now the foundational structure of flowing movement, the guiding principle of what was understood as "development" — whether in the organism or in a musical composition.

Musical Rhythm as a Physiological Principle
In the final decades of the eighteenth century, music was increasingly understood as a representation of organic life. A musical piece was considered successful if it gave expression to the essential aspects of the living. Theoretical discussion gathered pace on the role of music in expressing bodily, even physiological, phenomena; accordingly, the categories that dominated the era's physiological debates can also be found in music theory. In 1795, for example, Christian Gottfried Körner wrote in his important essay "Ueber Charakterdarstellung in der Musik" (On the representation of character in music) that music is life, for "the sensory form in which the artist's thought is manifested is not dead, but animated."[13] The objective of every genuine work of art is therefore to create the impression that the objects represented are alive — an autonomous "vital force" must spread through all the components of the work: "In place of a puppet show, moved by an unknown power using invisible strings, come active persons. For each of these active persons, there is a particular sphere of influence

of which he is the focal point, and in this sphere of influence, there appear various states, which are named 'life.'"[14]

A successful work of art, then, possesses an independent life and its own sphere of influence. At the core of the literary and musical work is autonomous movement — what Körner calls rhythm: "In the *movement* of sound we observe partly the distinctions of *duration*, partly the distinctions of *kind*. The former are most important for the representation of character. The regularity in the alternation of note values — rhythm — is the autonomy of the movement. What we perceive in this rule is what remains constant in the living being, which asserts its independence despite all external changes."[15] Rhythm is the "autonomy of the movement"; equally, vital force and autonomous motion are the properties by which the emerging biology of the period defined the specificity of living nature.[16]

In his *Allgemeine Geschichte der Musik* (General history of music), a milestone in music history, Johann Nikolaus Forkel (1749–1818) also made rhythm the most important aspect of music. According to Forkel, without rhythm, music is unable to fulfill its prime task, the arousal of human sensation and feeling:

> No melody is capable of arousing a particular sensation [*Empfindung*], or a particular feeling, unless the individual notes it contains are ordered according to the tempo [*Zeitmaß*] or their duration, such that they have a particular relationship with and against one another. A single note or several notes without such a relationship essentially say nothing to our heart; they are sounds without meaning. The meaning and effect of several interconnected notes depends so strongly upon this, their rhythmic arrangement, that it is probably why the habit has arisen of calling the measure [*Takt*], which contains this arrangement, the soul of the whole of music.[17]

Forkel regarded rhythm as the inner power of music, its vital force. *Takt*, in contrast, is music's countable and analyzable aspect. Around this time, both musicians and music theorists were deeply concerned with understanding these two elements of musicality — on the one hand, its structured form, captured in notation; on the other, its character as flowing movement. Forkel believed that the differentiation of music's temporal constitution into a formal and a natural component was among the most important advances in the music theory of his day. Yet the terminological distinction between *Takt* and rhythm at this time was anything but unambiguous. The terms are used by different authors in different, even downright contradictory ways. On many occasions, the word *Takt* appears to mean what would today be called "rhythm"; other writers use it in the sense of a uniform metrical beat or internal metrical weight, in the sense of what today is called "meter."[18] In 1789, Daniel Gottlieb Türk addressed the diversity of usages. *Takt*, he wrote, was commonly understood as meaning "the correct disposition of a certain number of notes et cetera which are to be played within a particular time," but might also describe "the relationship according to which, in music, a number of notes are distributed in a certain time period."[19] Furthermore, Türk added, *Takt* could refer to "the measure [*Maß*] of the movement of a musical phrase" or indeed to "the soul of music."[20] If usage varied in its details, the concepts did have common ground in their reference to music's special characteristic of having a structure, yet not being completely determined by meter.

In this context, Forkel distinguished between inner and outer rhythm,[21] a dividing line that is key to understanding the role of rhythm in the epoch around 1800. External rhythm refers to formal counting, the precise specification of music's temporal structure. Internal rhythm, in contrast, cannot be measured. It constitutes music's inner life, its impetus, its vital force — just as rhythm directs the physiological processes of the body. In Forkel's view, this duality was produced by the introduction of *Takt*:

> Through the invention of several species of notes, or signs to determine several tempi, therefore, one not only came closer to the true nature of the relationships of syllables in languages, but also invented our measure [*Takt*], or at least the art of writing it down properly, through which the temporal division of our music acquired at once unity and multiplicity. The unity arose through the measure, which consists in the division of a whole sound piece into many components of the same length. But the multiplicity arose from the variety of the kinds of notes within each individual measure or in the larger and smaller compartments of time. It is clear that this recent arrangement has given our music a double rhythm, namely, an inner and an outer rhythm, or the rhythm of the melody and that of the meter.[22]

For Forkel, the invention of *Takt* was a revolution in music, placing it on a completely new footing: measure liberated music from the "slavery" of poetry and prosody. Only through this innovation did music become "the language of sensations and an art of its own."[23]

Not all musicians and music theorists shared Forkel's confidence that the formalization of *Takt* could capture at least certain aspects of rhythm. Instead, they feared that formal meter would destroy the authentic power of rhythm. This idea often went hand in hand with a belief in the purity of ancient Greek music, which was considered genuinely rhythmical and therefore particularly powerful. "Greek music was bare of all *Takt*, and knew only the rhythm of melody," asserted the philosopher Gottfried Hermann in 1799, "but all this multiplicity in our rhythm of melody is robbed of much of its effect by the rhythm of *Takt*."[24] Hermann therefore recommended caution in attempting to understand music's rhythm in terms of *Takt*, "for what the rhythm of *Takt* brings to our music is not only unity, but also uniformity."[25] Forkel's *Allgemeine Geschichte der Musik* brushes aside such doubts, denying that schematization dissociates music from its authenticity and vitality. On the contrary, the search for a particular structured quality in rhythm as elsewhere is the only way to understand music, because the rhythm of music replicates the rhythm of nature. Music, Forkel argues, is a representation of life, and life processes are rhythmical. This rhythmical order of nature is not synonymous with chaos, but is itself regular or even governed by laws: the underlying rhythm of nature is always also structured, full of *Takt*. Indeed, it is the ordered pulse that ensures the normal, healthy course of physiological processes:

> When our pulse begins to beat unevenly, then we are sick, or near to becoming so; when a man's gait keeps such an uneven and uncertain measure that, although we gaze after him, we never know if his next step will be faster or slower ... then we are inclined to think this man has become foolish. ... Why, then, should we regard as perfection, as beauty in music alone what everywhere in nature we regard as disorder, imperfection, and even as illness?[26]

Rhythm, asserts Forkel, is not just any order, but a healthy and vital order and the accurate representation of a normal physiological process.

Accordingly, he objects to the idea of "free" rhythm. For Forkel, real rhythm is never completely without structure. *Takt* is "a universal measure, and similar to the measure that nature observes in all its movements."[27] The variability of rhythmic structure should not lead us to believe that rhythm can change its structure abruptly and arbitrarily within a musical piece — that would be at odds with the course of physiological processes:

> Whoever wishes to obtain a clear idea of this need only watch what is called an obligatory recitative from the oldest operas, in which the tempo [*Zeitmaß*] is changed very frequently in the Greek style ... and he will soon notice how little this kind of music is suited to arousing and sustaining any feeling at all. The circulation of our blood, although it changes with every passion, or takes on its own particular rhythm, can nevertheless not change as rapidly and as often as would have to be the case if this kind of music were really suited to human nature.[28]

The new concept of rhythm was adopted by the majority of music theorists. As early as 1787, Heinrich Christoph Koch's *Versuch einer Anleitung zur Composition* (Attempt at a manual of composition) insisted on the centrality of rhythmic structure to musical composition. Koch's handbook was highly influential for the composers of his era, including Mozart and Beethoven, and was the first to charge the composer with breathing life into a musical piece through inner rhythm. It is the composer's task, writes Koch, to construct his works rhythmically in such a way that the rhythm touches the listeners according to the "movements" of sensation.[29] In his *Musikalisches Lexikon* (Musical encyclopedia) of 1802, Koch underlined this point: "Without doubt, rhythm was the first sensory expedient, with which in music's infancy men turned a series of sounds, meaningless in themselves, into something that entertained the feelings. When still at a very low level of culture, humanity must have discovered that a regular recurrence in the movement of otherwise perfectly indifferent things is somehow attractive to the feelings."[30]

Traditionally regarded as an attribute of spoken language, the concept of rhythm was associated most closely with rhetoric and poetics. But like Forkel, who aimed to liberate music from the tyranny of prosody, Koch held that the role of rhythm was far more significant in music than in language. In his view, verbal language is always rhythmic, and every spoken word is composed of syllables that determine its basic rhythm — but in music, notes cannot even exist without rhythm. Rhythm is not simply an aspect of music, but its driving and constitutive force: "For until they have been clothed in measure [*Takt*], there is nothing in the nature of notes themselves that would give an inner precedence to one note or another. Particular notes acquire this inner precedence above the others only once they are placed in a particular relationship of *Takt* [here in the sense of "rhythm"]."[31] The composer is "compelled to give a certain weighting, through which the notes attain a certain relationship to one another."[32] This compulsion to rhythm is grounded in the rhythmic constitution of nature, and especially of our sensory physiology:

> I say that the artist in sound is compelled to do this, because he is not capable of singing or playing such a number of notes of the same kind without placing them in a certain relationship to each other. The first reason why he cannot accomplish this even if he tries is, it seems to me, to be found not so much in the artist's feeling, which is already habituated to metrical movement, but chiefly in the nature of our senses and our imagination.[33]

The topos of the musicality of bodily processes was not entirely new in the late eighteenth century. It drew on a long tradition, going back to Greek antiquity, that focused on the human pulse.[34] In the third century BCE, the Greek physician Herophilos equated the movement of the pulse (diastole and systole) with the rise and fall of the poetic line (arsis and thesis) and of music. The medieval idea of the harmony of the spheres and the *musica humana*, in other words, the notion that the same numerical laws govern the cosmos, music, and mankind,[35] maintained the parallel between music and pulse — even if Galen, whose medical authority reached well into the modern era, did not accept the opinion that the pulse could be quantified using precisely determined numerical ratios. There had been calls in medical literature since the end of the Middle Ages for physicians to know musical theory in order to be able to measure the pulse, but in the mid-sixteenth century, the doctor Josephus Struthius (1510–1568) became the first to represent the pulse visually with the help of musical notes. He founded a tradition of musical pulse notation that ended only two centuries later, with François Nicolas Marquet (1687–1759) and his *Nouvelle méthode facile et curieuse pour apprendre par les Notes de Musique à connoître le Pous de l'Homme* of 1747. When the mathematical and cosmological concept of music faded in the eighteenth century along with the *musica humana*, the study of the pulse became increasingly empirical, and in the century's second half, attention also turned to the possible effects of different kinds of musical rhythms and tempi upon the pulse.[36]

Along with these numerous attempts since antiquity to diagnose and portray the human body's workings, the regularity of the pulse, physical illnesses, or health through the laws of music, the converse question also arose: To what extent do the laws of music follow those of the body? In the late fifteenth century, the Spanish music theorist Bartolomeo Ramos de Pareja (c. 1440–1491) proposed the human pulse as the measure of musical tempo; he was particularly interested in using the pulse's regularity as a pattern for the timekeeping movements of the hand. Only in the mid-eighteenth century did Johann Joachim Quantz set down a particular pulse rate, in the manner of a metronome, as the measure to describe musical tempi.[37] Far more common were attempts to use music to describe the pulse. Marquet, for example, chose the opposite route from Quantz at almost exactly the same time, defining the pulse in terms of the minuet.

Toward the end of the eighteenth century, the physiological anchoring of musical rhythm placed the relationship of music and medicine on a new foundation. The topos of the body's musicality now became the linchpin for both the theoretical explanation of music as a representation of life and the new science of physiology itself. In both domains, rhythm played a similar role, as the underlying structure of a flowing movement and the guiding principle of development. This is why physiological arguments and terminology so often feature in the musical debates of the period: rhythm as the inner organization of music corresponds with rhythm as the inner organization of organic nature.

The Theory of Accents
The idea of rhythm as a pattern, that is, as an independent ordering structure of musical time beyond the rigid limits of meter, also formed the basis of the *Akzenttheorie*. This "theory of accents" emerged toward the end of the century in the Berlin circle around Johann Georg Sulzer (1720–1789), to which the composer Johann Abraham Schulz (1747–1800) and Philipp Kirnberger (1721–1783), one of the period's leading music theorists, belonged. The *Akzenttheorie* imagined music as a series of like impulses that are diversified into the variety of musical expression through accentuation. It distinguished between meter, as a unit of counting and dividing, and rhythm, as a physiological ordering of organic life and of musical movement: meter in music represents the unbending, counted-out

hierarchy based on the division of musical duration, while rhythm is what transforms physical movements into the movements of music.[38]

In 1776-1779, Johann Philipp Kirnberger published his rulebook for student composers, *Die Kunst des reinen Satzes* (The art of pure composition). This work assumes that music is a "succession of notes that mean nothing by themselves";[39] the mere sequence of notes cannot yield musical melody "without precise regulation of speed, without accents, and without rest points."[40] These three elements have to interact in different ways for the series of sounds to be molded into musical expression. If melody, like language, is to be an "expression of various emotions and sentiments," then "individual notes must be turned into meaningful words and several words into comprehensible phrases."[41] The task of meter, or *Takt*, in this process is to set accents and determine the length or brevity of notes.[42] Once the notes are given accents and different lengths, the previously steady succession acquires structure. Meter comes about at the point when this organization appears in "the precise uniformity of accents" and the "completely regular distribution of long and short syllables."[43] Meter is thus characterized by the repetition of the same components and exact beats. The counting of the beats restricts the possible permutations of measures. More exactly, only those measures are conceivable that consist of groupings of "two, three, or four equal beats," because "besides these, there is no other natural type of measure."[44] The art of metrical design in a musical piece thus lies in carefully studying the possible combinations of beats. An initial distinction between different time signatures is that of even and uneven, simple and composite. On this arithmetical basis, the character of the individual time signatures can then be defined in more detail. The "large 4/4 time," for example, is "of extremely weighty tempo and execution and, because of its emphatic nature, is suited primarily to church pieces, choruses, and fugues." The expression of the uneven measures is quite different. Thus, the "3/8 meter" has the "lively tempo of a passepied; it is performed in a light but not an entirely playful manner and is widely used in chamber and theatrical music."[45] It is up to the composer to express the mood that he envisages for his piece by applying the characters of the various meters to appropriate effect.

Rhythm is a very different matter. It does not order musical duration by counting out individual segments; rather, "the flow of the melody is divided into larger or smaller phrases by the rhythm, without which the melody would progress monotonously; each of these phrases has a special meaning, like phrases in speech."[46] Whereas meter sets its accents according to fixed rules, rhythm is "not bound by such definite rules."[47] As Kirnberger contends, it is only rhythm's "external and somewhat mechanical nature" that can be described,[48] but this does not capture the essence of rhythm. Rhythm evades Kirnberger's attempts to assemble the principles of musical art as the transmissible rules of a craft. Instead, the special quality of rhythmical ordering — the quality making it a structuring order that goes beyond schematic timekeeping — is the fact that it is "immediately grasped" as a unit by the ear.[49] Kirnberger here emphasizes rhythm's embedment in the physiology of the senses and its immediately intelligible unity. Rhythm is physiologically intelligible, despite its lack of rules: the ear is capable of distinguishing with confidence between meter and rhythm.[50] This inextricable link between rhythm and the physiological disposition of the human being explains why "the greatest power of melody comes from rhythm."[51]

The "Natural Inclination" to Rhythm

In 1771–1774, the first edition of Johann Georg Sulzer's *Allgemeine Theorie der schönen Künste* (General theory of the fine arts) appeared.[52] As well as literature, rhetoric, and other arts, this much-reprinted work also addressed music. *Allgemeine Theorie der schönen Künste* was one of the key texts of German Enlightenment aesthetics and highly influential for the music and aesthetics of its era. But Sulzer was also an important psychologist.[53] In his *Allgemeine Theorie*, he situated rhythm in man's *physis*, by which he meant the physiological and psychological constitution in equal measure, physical and intellectual motion, man's existence as a feeling being whose perceptions spring from his sensory physiological disposition.[54]

Sulzer starts from the assumption that rhythm derives from a natural sentiment or sensation (*Empfindung*): "That rhythm is not something artificial that emerged from reflection, but is grounded in a natural sentiment, can be deduced from the fact that even semisavage peoples observe it in their dances and that all men bring something rhythmical to certain activities, without knowing why."[55] "Natural" here means that rhythm is not an intellectual skill requiring training, since it "is known or felt by the least reflective of men."[56] At the same time, "natural" means primal, elemental, physiological. The rhythms of movement ensue spontaneously, for example in dance, and they are perceived involuntarily. "As soon as the ear hears loud beats that succeed one another at equal intervals," audition begins to organize the uniform movement into rhythmic patterns.[57] Man has a "natural inclination to rhythm," which is present "wherever we experience several prolonged sensations of the same kind" — such as when marching, threshing, or hammering iron.[58] However, rhythm is a law not only of man's physical motions, but also of his psychological motions:

> Sensation [*Empfindung*] follows the laws of movement. The spinning top that a lad sets in motion turns for a short time and then falls: if its movement is to be sustained, the boy will have to give it new force from time to time by striking it repeatedly. If a passionate sensation is sustained because constantly fresh and different impressions renew it, it does not remain the same; although the mind remains constantly in motion, that movement becomes now stronger, now weaker, now directed at different objects, and may even change in kind. . . . From this we see that only the continued repetition of similar impressions has the power to sustain the same sensation for a period of time.[59]

Human actions that have a certain duration rely on their rhythmic repetition. That applies equally to physical movements and to movements of the mind. The rhythmicization of a uniform activity, argues Sulzer, is what makes it physically possible to perform that activity over a longer period of time. In just the same way, a feeling that is more than merely fleeting can constitute itself only through rhythmic repetition. Every sensation is dependent on its rhythmic recurrence, which is therefore a prerequisite for human experience in general. Sensory perceptions have to be maintained through constant physiological stimuli: "Thus, even before the preceding impression is fully exhausted, another one arrives, and in this way there occurs, so to speak, an accumulation, an amassing of sensation and of effect, as a result of which the mind is increasingly fired up and strengthened in its sensation. This can go so far that finally the whole system of the nerves begins to move."[60] Those practiced in the art of "observing psychological phenomena with some exactness," notes Sulzer, will join him in "completely understanding the effect of rhythm in facilitating prolonged repetitive work and in sustaining or gradually strengthening the sensations."[61]

Part and parcel of the physiological sensory apparatus of human movement and perception, rhythm also has its own specific features. The first and most important property of rhythm is that it does not repeat a structure in an identical form, but varies it in repetition. Rhythm, writes Sulzer, is nothing other than "the periodic division of a series of like things whereby the uniformity of these things is combined with multiplicity; such that a sustained sensation, which would have been identical (homogeneous), acquires variation and multiplicity through the rhythmic divisions."[62] In other words, the essence of rhythm is the conjunction of unity and diversity: a sensation perceived as unified can exist only over time through its rhythmical renewal, which in turn is not the repetition of the identical, but a continuing variation. It is precisely this tension between unity and diversity that rhythm perpetuates and by means of which it generates the diversity of experiences.

For music, and for aesthetics in general, a crucial point is that this structural tension of rhythm is also the source of our perception of beauty. Sulzer had proposed his definition of beauty as a state of tension as early as 1751–1752 in the essay "Untersuchungen über den Ursprung der angenehmen und unangenehmen Empfindungen" (Investigations on the origin of pleasant and unpleasant sensations). There, he asserted that because too much uniformity is tedious, while multiplicity without unity is confusing,[63] something we call beautiful consists of "unity in multiplicity . . . or multiplicity brought back to unity. A very simple object, within which no distinctions at all are possible, can never be beautiful: in order to be beautiful, an object must have many and multifarious parts . . . . The mere quantity of parts is not what constitutes beauty; there must also be multiplicity and connection." [64]

Sulzer was radical in his claim that rhythm *alone* is responsible for the beauty of music:

> It was not only the ancients who ascribed great aesthetic power to rhythm: even today, everyone admits that in melody and dance everything that is actually called beauty originates in rhythm; thus the investigation of the actual nature and effect of rhythm is directly pertinent here, and is all the more needful because, as far as I am aware, it has not yet been undertaken by any judge of art; as a result of which the musical composers themselves often have quite confused conceptions of rhythm, the necessity of which they sense without being able to name the slightest reason for it.[65]

This passage highlights Sulzer's demarcation of rhythm from other musical structures. Whereas the notes of music, for example, or the movements of dance may be "in themselves merry, joyful, tender, sad, and painful" and have the power to move us even "without any influence of art," "the beauty that arises from rhythm is something quite other; namely, it lies in things that are completely indifferent in themselves; that have no natural meaning of their own, no expression of joy, or of pain."[66]

Sulzer thus defines beauty as a quality of rhythm's specific *structure*. Put another way, the beauty of rhythm is exclusively structural because the individual elements of which it is composed possess no qualities of their own. He gives the example of dripping water, which means nothing as long as one hears only the "completely unordered noise of the drops." Yet as soon as one begins to distinguish the falling of individual drops, to perceive "that they always recur after the same period of time, or that after the same period of time two, three, or more drops always follow each other according to a certain order, and thus form something periodic, like the hammer blows of three or four blacksmiths; then the attention is drawn to observe this order. Now something of rhythm arises, namely, a regular recurrence of identical blows."[67] Rhythm arranges a monotonous sequence of individual components into a periodic succession. The components, previously uniform, thereby become structurally unequal — they are now characterized by their respective positions within the rhythmic order.

There are infinite possible variations on these rhythmic combinations of individual components, so that beauty is guaranteed the greatest diversity. Despite the breadth of variation, however, rhythm's ordering is not random. On the contrary, each component orders itself vis-à-vis the other in line with an inner regularity. That regularity, in turn, obeys our physiological disposition, our way of moving and feeling. As a result, rhythm is

> generally speaking, the division of the sequence into components of equal length, so that two, three, four, or more blows make up an element of this series, an element that is not merely arbitrary, but is distinguished from others by something one really feels. This is actually what in music is called beat [*Takt*] and in poetry meter [*Silbenmaaß*], and it is at once the first and the simplest kind of rhythm.[68]

Rhythm, then, is composed of individual elements, the varying order of which provides the conditions for the diversity of feelings of beauty. At the same time, rhythm preserves the unity of perception in full.

A third property is rhythm's forward drive, its extension into the future — for rhythm does not aim to resolve the tension of unity and diversity, of order and disorder, but to perpetuate it. In rhythm, "there emerges at every thesis, and at every entry of a new section, a new aspiration to initiate the next impression in the right way. Thus, even before the preceding impression is fully exhausted, another one arrives, and in this way there occurs, so to speak, an accumulation, an amassing of sensation and of effect."[69] Rhythm is a forward-looking, advancing, intensifying movement, each sequence of which always already evokes an anticipation of the next and continues the movement in repetitions that are, in principle, infinitely varied.

As the writings of Forkel, Koch, Kirnberger, and Sulzer have shown, music theory in the late eighteenth century coupled the notion of rhythm with the complex physiological disposition of the human being. It conceptualized a dimension of the musical organization of time that was not reflected in the strict patterns of meter, but was to become an essential element in a new view of music as the nineteenth century began.

CHAPTER THREE

Rhythmical Productivity in Schelling's
Philosophy of Nature and Art

Friedrich Wilhelm Joseph Schelling (1775–1854) completes my discussion of cultural facets of the rhythmic episteme in two respects. First, he elevated the unity of nature and art to the status of a systematic philosophy; second, his *The Philosophy of Art* elaborated on several key aspects of the thinking of Hölderlin, Moritz, and August Wilhelm Schlegel. *The Philosophy of Art*, based on lectures Schelling gave from 1802 to 1805, addresses the rhythmicity of the arts as a way of knowing nature in its absolute existence or productivity. Schelling's philosophy of identity posits that nature and art are one. Consequently, there is no epistemological opposition between artistic and conceptual, aesthetic and philosophical production: the absolute manifests itself in both, though in different ways. Art is not a means of depicting or representing the absolute, but on a par with philosophy as a nonconceptual instrument of knowledge. This is why Schelling could preface his study with the announcement that "for those already acquainted with my system of philosophy, the philosophy of art will be merely the repetition of that same philosophy in the highest potence."[1] To think philosophically about art means "to construe not first of all art as art, as this particular, but rather the universe in the form of art, and the philosophy of art is the science of the All in the form or potence of art."[2] All art, thus, is "the direct reflection of the absolute act of production or of the absolute self-affirmation."[3] It is in the products of art that we find the absolute in the highest form of its self-realization. Accordingly, the various forms of art, such as music, painting, and poetry, always render the "archetypes" behind the imperfect reproductions of the world,[4] which take objective form differently in the concepts of philosophy or the works of art: "the forms of art in general are the essential forms of things."[5]

In the following, I argue that rhythm was a crucial epistemic category for Schelling's philosophy of nature and art. For his philosophy of the absolute, rhythm fulfilled the vital epistemic function of enabling him to conceptualize the relationship between nature as productivity and the finite products in which it is realized. Rhythm offered Schelling a pattern that could define this reciprocal relationship as an ordered and continuing one. *The Philosophy of Art* explores the ways in which rhythm mediates between the real and the ideal, the finite and the infinite, nature as productivity and nature as product.

Schelling addresses the epistemic structure of rhythm more specifically for the art forms of music and poetry. Though the term "rhythm" was not yet current in the early days of *Naturphilosophie*, all the key components and lines of thought required for the rhythmic episteme were already in place. For Schelling, the notion of rhythm resolves a difficulty in his *Naturphilosophie* that he calls "the highest problem of all systematic science," namely, how the infinite is exhibited in finitude: "how an absolute activity (if there is such a thing in Nature) will present itself empirically, i.e., in the finite."[6]

Schelling's *Naturphilosophie* has been much studied. Philosophers and historians of philosophy have examined its grounding in his reception of Kant, Fichte, and Spinoza and the whole conceptual configuration of Schelling's philosophy of the absolute, and developments in modern biomedical sciences over the past twenty years have often put Schelling in a controversial spotlight.[7] His philosophy of art has received far less attention — although Schelling himself left no doubt that art played the central role in his system of philosophy, calling the philosophy of art "the universal organon of philosophy" and "the keystone of its entire arch."[8] Today, Schelling's philosophy of art remains overshadowed by Hegel's *Phenomenology of Spirit*, which appeared in 1807, shortly after the lectures that made up *The Philosophy of Art.*[9]

For the most part, scholarship has disregarded the great epistemic importance of rhythm to Schelling's conception of nature — the structure of rhythm as a law-governed framework for the mediation between nature's productivity and its products, in which the infinite remains tied to the finite and both attain their autonomy only in the ceaseless recalibration of their interdependence.[10] Schelling's notion of nature and art not only entails the much-discussed duality of finite and infinite, of product and productivity, but also formulates the concept of rhythm as the law by means of which that duality is negotiated, continues into the future in ever-new figurations, and thus becomes the wellspring of all productivity: the grounds of the creative opulence of both nature and art.

Absolute and Finite in the Play of Rhythm

Schelling's *Naturphilosophie* took shape during the short period between 1797 and 1800, in several works that appeared in quick succession. They include his first publication, *Ideas for a Philosophy of Nature: As Introduction to the Study of This Science* of 1797; the treatise "Von der Weltseele, eine Hypothese der höheren Physik zur Erklärung des allgemeinen Organismus" of 1798 (On the world-soul, a hypothesis of the higher physics for explaining the universal organism); *First Outline of the System of the Philosophy of Nature* (1799); and finally, also in 1799, the introductory essay to the *First Outline*.

Fundamental to Schelling's *Naturphilosophie* is the distinction between nature as subject and nature as object: "Nature as a mere *product* (*natura naturata*) we call Nature as object (with this alone all empiricism deals). Nature as *productivity* (*natura naturans*) we call Nature as subject (with this alone all theory deals)."[11] Whereas nature as productivity exhibits absolute existence, the object, as something conditioned, always already stands outside the absolute. This raises an important dilemma in the philosophy of identity: being absolute, the absolute must be identical with itself, but in order to recognize itself as absolute, it must also step outside itself and pass into a particular form. Yet the conditional finite and the unconditional absolute cannot simultaneously be identical and nonidentical with each other. Schelling sought the solution to this problem in the reciprocal *relationship* of nature as product and nature as productivity. Nature as subject is an infinite process of productivity that can never find adequate expression in its finite products, but only in the ceaseless becoming of nature as a whole.[12] Among the finite products, however, the organism is an exception. In the organism, nature tries to sublate, within a single product, the contradiction of the counteractive forces it contains. If it is to unite these opposites, the organism itself must, in turn, be productivity. On the level of the organism, then, the same structure of reciprocity between finiteness and infiniteness prevails as the one that prevails in nature as a whole.

Schelling explored various forms of this idea in the course of his work on *Naturphilosophie*. Common to all of them is that he thinks of the relationship between the absolute and the finite as rhythmical movement. This explains why Schelling rejected the preformation-ist theory of development — in his view, the organism is never present as a finished structure, so that in order to understand organic development, it is necessary to describe not its particular structures at particular moments, but the laws of its formation. By these Schelling meant the mechanisms that govern the *direction* of development. He used "epigenesis" and "metamorphosis" synonymously to describe this notion. Even in the earliest stages of development, writes Schelling, "the first seeds of all organic formation are themselves already products of the formative drive [*Bildungstrieb*]," and all the "multiplicity of organs and parts signifies nothing other than the multiplicity of *directions*."[13]

In a marginal note on his own copy of *First Outline*, Schelling recapitulates this point: "Various organs, parts, etc., signify nothing but different *directions* of the formative drive."[14] Development is therefore not the unfolding of already existing organs, but the potential unfolding of directions, and the laws of that unfolding are what must be sought. From this point on, he argues, development will be no longer solely a process of building structures, but also a process of dismantling them. Traditional preformationism explained "the *emergence of new parts* . . . . as an individual preformation." But if one observes the metamorphosis of the butterfly, for example, it becomes obvious that "nothing is lost from the pupa, and yet one does not find in the butterfly the organs that were in the pupa." To understand development thus also means asking: "How does one explain the *disappearance* of the parts that were there before?"[15] Development is incomparably more complex than was assumed by preformationism, for "that transition from one phase of the metamorphosis to the other is not at all just partial alteration, but a total one."[16]

The rule according to which such total alteration occurs in the organism is the rule that Schelling later named "rhythm." In fact, he had already set out the elements of rhythm in 1798, in the essay "Von der Weltseele," which contains his first detailed account of his concept of the organism and of organization in nature:

> To me, organization is nothing other than the arrested stream of causes and effects. Only where nature has not obstructed this stream does it flow forward

(in a straight line). Where she obstructs it, it returns (in a circular line) back into itself. It is not, thus, every succession of causes and effects that is excluded by the concept of the organism; the concept denotes only a succession that flows back into itself enclosed within certain limits.[17]

Schelling distinguished the mere succession of events in time from the processes that he defined as the signum of life, those that are "enclosed within certain limits." Among these limits are, first, the fact that the life process is constituted as a series of oppositions. Nature is "striving and resistance [*Streben und Widerstreben*]";[18] indeed, the possibility of life in general presupposes "a constant succession of processes of decomposition and restoration."[19] Nature is not solely the "continuous transformation [*Wechsel*] of animal material,"[20] but consists in the succession of individual processes — for which the vital point is that "each of them is the reverse or negative of the one that precedes it."[21] Second, the limits of these processes include the fact that they do not continue uninterruptedly into the future, for "nature cannot give the life process permanence except by always repeating it from the beginning."[22] Only if nature is constituted in the alternation of opposites as a necessary, nonrandom succession can its productivity be accounted for: "Nature must be free in her blind regularity, and conversely regular in her full freedom; in this confluence alone lies the concept of organization."[23]

"We have no other concept" for the unity of freedom and lawfulness, Schelling adds, "than that of the drive."[24] However continuous life may be, it actually constitutes itself as a series of disparate and contradictory states. These states relate to each other in a regular way, but they are variable and not teleologically determined. The notion of a law-governed, yet not merely "mechanical" succession here takes shape as the conjunction of freedom and regularity in the drive.

This idea of a drive whose goal and direction can be reset at any point in the chain of events, without being accidental or random, also featured in Hölderlin's work. Like Hölderlin, Schelling asserts that the interplay of freedom and regularity cannot be defined in detail or simply read off from the products of nature. It is pointless to try to penetrate an organization as "a whole complete in itself" at some random point with the aim of describing it in isolation, because everything is "at once." There is no before and after, and "the mechanical form of explanation" takes us nowhere.[25] Our knowledge is only knowledge of the relationship of each with the other and knowledge of the law that governs the oscillation between them: "Organization and life by no means express something that exists in itself, but only a particular form of being, something common to several causes working together."[26]

The law of reciprocal relationships that Schelling uses to describe the epigenetic developmental rule of the individual organism is also the general law determining how the productivity of the absolute unfolds into its finite products. One element of this rhythmic order is the inner completeness and autonomy of the process by which the absolute is realized in the finite. In one of Schelling's early writings in *Naturphilosophie*, "Treatise Explicatory of the Idealism in the *Science of Knowledge*" (1797–1798), Schelling writes of the human spirit as self-organizing nature:[27] "Hence, if the [spirit] is to have an intuition of itself as active in the succession of its representations, it will have to inspect itself as an *object* that contains an *inner principle of movement* within itself. Such is what we properly call a *living* being."[28] The movement of the spirit, as an example of self-organizing nature, has the following components. The spirit has to realize itself in disparate mental images; these must succeed each other ceaselessly; and to avoid becoming lost in mere succession, the spirit must fix on itself in particular states: "However, the soul is to have not only an intuition of this succession but also an intuition of *itself* within this succession, and (because it has only an intuition of its activity) it is to form an intuition *of itself* as *active* in this succession."[29] Only when the flow of mental representations is impeded can the spirit recognize itself as actively involved in its products. It is by passing through these "different *stages*" that "the spirit progressively attains an intuition of itself, *pure* self-consciousness."[30] On the one hand, it is "as though the soul were striving at each single moment to present something infinite; because it is not capable of this, it necessarily strives beyond each presence to present the infinite at least as a *succession* in *time*."[31] On the other hand, "the mere succession of the representations, considered externally, provides the concept of *mechanical movement*."[32]

The movement of realization thus describes an action in time, but not a merely mechanical temporal sequence. Every living being contains this movement, which is constituted from the ceaseless series of mental images and their inhibition and fixation into concrete states, as "an *inner principle of movement* within itself."[33] For the overall configuration of organic life, this means that the spirit "subsists only in the continuous transition from cause to effect, and the spirit no longer feels restricted by a discrete object but, rather, by a necessary sequence of successive appearances."[34] In other words, Schelling defines life not as an arbitrary succession, but as a necessary series of sequenced phenomena. Only a few years later, in *The Philosophy of Art*, he wrote of "the transformation of the accidental nature of a sequence into necessity" and called that process "rhythm" — thanks to which "the whole is no longer subjected to time but rather possesses time *within itself*."[35]

The Rhythm of the Absolute

Schelling worked out his philosophy of art in a series of lectures at the universities of Jena and Würzburg between 1802 and 1805. He began to construct his system of absolute philosophy from 1800 onward. Far from being peripheral to this process, the philosophy of art formed one of the foundation stones of Schelling's philosophical edifice. It was to emphasize the pivotal status of art in his system that Schelling chose the term "philosophy of art" in preference to "aesthetics." The aim of the new philosophy was to create a system of the arts within the framework of idealist philosophy. In it, Schelling deployed the concept of rhythm to explain how finite art forms and the absolute correspond in a reciprocal and autonomous movement — a movement in which absolute productivity can be glimpsed even in finitude.

In art, argues Schelling, the absolute presents itself in the phenomena of the various genres. His idealist system has a particular place for each form of art, determined by the particular relationship in which the infinite or absolute expresses itself in the finite form. Schelling draws a fundamental distinction between two kinds of arts: the real and the ideal. The first group comprises the formative (*bildende*) arts, in which the real, physical, objective features of the infinite become manifest. In the second group are the arts of the word, which in contrast to the formative arts take up the irreal, intellectual, subjective aspects of the infinite. Put another way, the absolute expresses itself objectively in nature and subjectively in art, but this is not a simple opposition — the various different forms of art contain the objective aspect of nature in varying proportions. As well as this two-way distinction, artistic genres are ordered hierarchically (in "potences") within each series.[36] Schelling classifies music as part of the series of the "real" arts. Within that order, it forms the first potence (followed by painting and sculpture). As the first potence of the "real" series, music is the art form of the "informing [*Einbildung*] of unity into multiplicity."[37] Through the plurality of tones and the diversity of their sequences, its task is to enable the absolute to be glimpsed as an antecedent, undifferentiated oneness. Specifically, this function is fulfilled by rhythm, which is, therefore, "one of the most wonderful mysteries of nature and art."[38] The "first prerequisite" of rhythm is

> unity within multiplicity. This multiplicity, however, does not inhere merely in the simple difference between the various units insofar as they take place arbitrarily or nonessentially, that is, simply within time, but rather insofar as they are simultaneously based on something real, essential, and qualitative. This quality resides only within the musical variability of the tones themselves. In this respect modulation is the art of maintaining the identity of the one tone that is the predominating one within the whole of a musical work, to maintain it in the *qualitative* difference just as through rhythm itself the same identity is observed in the quantitative difference.[39]

In rhythm, the absolute does not simply branch out into the innumerable conceivable combinations of sound sequences; rhythmic regularity simultaneously makes visible the original identity of the absolute. Rhythm *modulates* the absolute into the ordered, structurally interrelated sequence of notes. Schelling did not conceive of rhythm as a property specific to musical representation, however. Instead, music is "the perceived rhythm and the harmony of the visible universe,"[40] the "primal rhythm of nature and of the universe itself."[41] It presents "*pure* movement as such, separated from the object" and, in rhythm and harmony, "portrays the form of the movements of the cosmic bodies, the pure *form* . . . . liberated from the object or from matter."[42] For Schelling, then, music is the first potence of the "real" arts because of all the forms of art, it is the one in which nature communicates itself most elementally. This applies in a dual sense: music's sonority makes it closer than all the other arts to the materiality of nature, and its rhythm reveals most purely the abstract law of becoming in nature and art.

In the rhythm of music, the infinite "informs" itself into the finite while still shimmering through as infinite, but poetry goes one step further. Lyric poetry takes up the first place in the "ideal" series of the verbal arts (second and third being epic and drama). Unlike music, poetic language "allows that absolute act of knowledge to appear directly as cognitive act."[43] As a result, poetry is nature's highest form of knowledge, "to the extent that in the artistic image itself it yet maintains the nature and character of the ideal, of the essence, of the universal."[44] Poetry molds language, which "in itself is the chaos from which poesy is to construct the bodies of its ideas," into a "universe, a cosmic body."[45] But speech can be raised to a cosmic body "only if that speech possesses its own independent movement and for that reason its own time within itself, just as does any independent cosmic body."[46]

Thus, only if language is constituted like the absolute — like nature as productivity — can it be knowledge. And that is only the case when it is rhythmically ordered: in poetry. It is in poetry that the absolute reveals itself as rhythmical productivity:

> In art to the extent that it is music or verbal art, rhythm corresponds to that whereby a cosmic body is self-contained and possesses its own internal time. Since both music and speech are characterized by movement in time, their works would not be self-contained wholes if they were subject to time, and if they did not rather subject time to themselves and possess it internally. This control and subjugation of time = *rhythm*.[47]

Poetry rules over time by oscillating, as a rhythmic movement, between the individual and the universal. Poetic language is not a simple succession of sounds, syllables, and words. In its rhythmic autonomy, the assured ordering of its sequence, it overcomes the limitations that real time imposes on knowledge. Like music, poetry is ruled by the alternation of opposites. "The *essential nature* of all lyric poesy" lies in this alternation, and "*contrast* between the infinite and the finite thereby emerges as a kind of inner principle of life and movement."[48] In the rhythm of poetry, the oscillation between finite and infinite constitutes itself as "the opposite unity,"[49] and tension is sublated in "autonomously ordered change."[50] Rhythm thus encloses speech within itself, detaching it from the real sequence of time: "By means of rhythm, speech makes known that its own end inheres absolutely within it."[51]

Necessary Succession
Giving a general definition of its structure, Schelling describes rhythm as "nothing more than the periodic subdivision of homogeneity whereby the uniformity of the latter is combined with variety

and thus unity with multiplicity."[52] The simplest form of rhythm is that "in which the entire unity within a particular multiplicity depends only on the uniformity of the intervals within the sequence. An image of this might be equally large, equally separated points. That is the lowest level of rhythm."[53] The mere periodicity of divisions does not yet constitute rhythm, which is far more than mere sequence, the course of a movement over time. It is

> the transformation of an essentially meaningless succession into a meaningful one. Succession or sequence purely as such possesses the character of chance. The transformation of the accidental nature of a sequence into necessity = rhythm, whereby the whole is no longer subjected to time but rather possesses time *within itself*. Articulation within music is the forming of units into a series such that several tones together constitute yet another unit, one that is not accidentally or arbitrarily separated from others.[54]

Rhythm, then, is a structure that bestows inner regularity on a random succession of notes. It replaces the accidental series of events with an ordered one, the random series with a necessary one. Thanks to its rhythmic constitution, the temporal art of music becomes a formation of sound that unfolds in time and is nonetheless released from time. Rhythmically constituted music differs from a mere "forming of units into a series" in time because the series of individual units is determined by the inner laws of rhythm. This rhythmic order permeates music on all its various levels of sounds, measures, larger periods, and so on, up to the point where "this entire structure and composition still remains comprehensible to the inner poetic sense."[55] In other words, the bounds of rhythm are set as wide as the bounds of human perception.

To conclude, nature is not simply represented in art, but appears in the rhythm of music and poetry as it *is* in its inner regularity. The regularity of rhythm reveals nature as absolute nature in its ceaseless productivity and its perpetual becoming. And it prescribes the order of development in the organic world as contemplated by Schelling's *Naturphilosophie*.

PART TWO

Biological Rhythm

CHAPTER FOUR

Forms Out of Formlessness

In 1759, Caspar Friedrich Wolff (1734–1794) published his doctoral dissertation, *Theoria generationis*, investigating the emergence of the first structures in the egg out of the amorphous mass of the germ. It would be hard to overstate the importance of this study for the history of biology. *Theoria generationis* was nothing less than the blueprint for a new theory of organic development — the theory of epigenesis. Yet despite Wolff's immense significance for biology in general and embryology in particular, remarkably little has been written about him, even in more recent times.[1] His most innovative contribution has usually been regarded as the notion of the gradual emergence of forms out of initially homogeneous organic matter, coupled with the postulate of a force — Wolff called it the *vis essentialis* — guiding that development.[2] Indeed, for many scholars, the "essential force" is the real core of Wolff's concept of epigenesis[3] or even the key "topological feature" of epigenetic systems in general.[4] From this point of view, Wolff's recourse to a rather ill-defined "force" seems to present a gap in his theoretical edifice, making it "quite surprising how resolutely Wolff insists on this weak point of his theory."[5] But is Wolff's concept of epigenesis really adequately encapsulated in the two elements of the gradual emergence of forms and the *vis essentialis*? Are these the factors that bear the revolutionary potential of his thinking, which led the way to a profoundly new vision of development in the late eighteenth century? In my view, that summary gets to the heart not of epigenesis itself, but of its historiography.

What Is Epigenesis? A Historiographical Problem

In the fourth century BCE, Aristotle's *De generatione animalium* described the gradual emergence of the organism out of a homogeneous mass: each organ comes about as the precondition for the next. For Aristotle, this developmental series begins with the heart. As the most important organ — the seat of sensation and the source of blood and motion — the heart develops before all other organs, and in a chick embryo, it can already be seen on the third day, as a moving dot. The further differentiation of organic structures follows a teleological pattern. Once the impulse is given by the male semen, development moves onward to its goal, the complete formation of the living being.[6]

The English physician and anatomist William Harvey (1578–1657) is considered the founder of the modern version of epigenetic theory in the seventeenth century. His *Exercitationes de generatione animalium* of 1651 used the term "epigenesis" to describe the idea of development as successive new formations. The word is composed of the Greek prefix *epi* (upon, at, during) and *genesis* (origin, coming into being, formation), and literally means "new formation" or "growing upon."[7] Harvey writes:

> There are some [animals] in which one part is made before another, and then from the same material, afterwards receive at once nutrition, bulk, and form: that is to say, they have some parts made before, some after others, and these are at the same time increased in size and altered in form. The structure of these animals commences from some part as its nucleus and origin, by the instrumentality of which the rest of the limbs are joined on, and this we say takes place by the method of epigenesis, namely, by degrees, part after part; and this is ... generation properly called.[8]

Historians most often describe the origin of modern epigenetic theory as a kind of translation and appropriation of Aristotle's thinking by Harvey[9] and consequently the two pillars of epigenetic development as being that forms emerge successively out of formlessness and that they do so according to a teleological pattern.

The idea of successive development was by no means uncontested among theorists of generation. The late seventeenth century saw the beginnings of a theory of preformation, favored by, among other things, the Newtonian rejection of teleological thinking. Far from being clearly demarcated or monolithic, preformationist theory took many different shapes. What these shared was the underlying view that a complete, miniature embryo is already present in the germ. In ovist versions of preformationism, this complete (though infinitely small and therefore initially invisible) embryo was located in the female egg; animalculism, in contrast, located the tiny embryos in the male semen. Panspermists believed that they circulated freely in the air. In the eighteenth century, there was particular interest in the variant proposed by Nicolas Malebranche in 1674, *emboîtement*. This theory of "encasement" held that all the germs of future life, in an immeasurably small form, were already contained in the very first living being, like a Chinese box — for example, all the germs of later generations of women were contained in Eve. In all these variants, preformationism regarded development as the growth, unwrapping, or unfolding of structures that are already present in some form or another.[10]

In the second half of the eighteenth century, the interpretive hegemony of preformationism came under fire from alternative theories of development. The physicist and mathematician Pierre Louis Moreau de Maupertuis (1698–1759) was one of the first to attack preformationist views, in his *Vénus physique* of 1745. Georges Buffon (1707–1778) and Joseph Turberville Needham (1713–1781) also thought of development in new ways. Their corpuscular theories of generation agreed in attributing the embryo's formation to the union of rudimentary particles from the male and female body, brought together through the semen during fertilization. Buffon and Needham linked this view with the workings of a vital force or a vegetative principle that guides the process of combining the elements.[11] The different models of generation competed to explain the various aspects of development, one theory being capable of explaining some aspects, the next explaining others. Among the unresolved questions in the eighteenth-century debate about generation were those of regeneration, zoophytes, and spontaneous generation, along with the origins of abnormalities or monstrosity.[12]

While Caspar Friedrich Wolff's *Theoria generationis* of 1759 is widely regarded as the foundational text of the modern epigenetic concept of development, Wolff was not the only one to observe hen's eggs under the microscope at the time. The microscopic observations of the prominent physiologist and anatomist Albrecht von Haller (1708–1777) appeared almost simultaneously with Wolff's, in *Sur la formation du cœur dans le poulet* of 1758.[13] The resulting "Haller-Wolff debate" was probably one of the eighteenth century's most important scientific controversies. Shirley Roe's interpretation of the dispute and its aftermath has left deep marks on present-day views of epigenesis as a countermodel to preformation and most importantly as a model of generation driven by a particular ideological and philosophical agenda.

Roe argues that the quarrel between Haller and Wolff could not be resolved by empirical investigation, since the two men's stances on embryology were "in large measure dictated by their more hidden philosophical persuasions."[14] In fact, the observations made by Haller and Wolff coincided for the most part—but, according to Roe, they were interpreted on the basis of different answers to some fundamental philosophical questions: "the nature of matter and of forces, the applicability of mechanism to biology, the roles of the empirical and the logical in scientific explanation, the reducibility of life to nonlife."[15]

Those philosophical questions were answered in a radically new way in the decades following Wolff's dissertation. The turn of the nineteenth century was marked by profound ruptures, as Romanticism and *Naturphilosophie* brought dynamic, vitalist thinking both to early biology and to intellectual life as a whole. They paved the way for the emergence of biology as a scientific discipline and specifically of embryology. "Epigenesis," writes Roe, "was as compatible with the new progressivist view of human history and natural phenomena as preformation had been with the religious and mechanistic beliefs of the seventeenth and eighteenth centuries."[16] Roe is referring here to Elizabeth Gasking's thesis of a new developmental paradigm that, especially in Germany, was disseminated by *Naturphilosophie* and the Romantic movement. As Gasking puts it:

> The eighteenth century attempts at a causal account of generation failed, and preoccupation with the question gave rise to a welter of speculative theories. Scientists working with the new metaphysical framework were freed from this problem. Growth accompanied by change was now regarded as a fundamental feature of the Universe, and the growth of living things was the analogy in terms of which all other processes were to be understood. It, therefore, seemed a basic phenomenon requiring no further explanation. What had been an atypical, almost miraculous process from the seventeenth and eighteenth century point of view, became the paradigm of the natural for the nature philosophers.[17]

From this perspective, it was only in the wake of radical ideological changes at the threshold of the nineteenth century—shifting the whole world, not just biology, to a different intellectual horizon — that a new theory of development became possible. After 1800, in nothing less than a "complete break with the past,"[18] the new concept of a teleological epigenesis arose, which "*relied* on, rather than *explained*, embryonic organization."[19] According to Roe, the teleological epigenesis propounded by scholars such as Pander, Oken, and von Baer simply assumed what the eighteenth century had tried to explain: "No longer was the source of embryonic organization the central problem to be explained. Rather, organization became the one element of life that was taken for granted. Following Kant, these German embryologists uniformly embraced a teleological view of embryological development, based on a presupposed original state of organization in the generative material."[20] This liberation from the quest for causes ushered in the rise of descriptive embryology and its greatest discoveries.[21]

As the region most strongly influenced by Romanticism and *Naturphilosophie*, Germany was the center of the new embryology. Timothy Lenoir discusses a "Göttingen School" that clustered around Johann Friedrich Blumenbach and included Karl Friedrich Kielmeyer, Alexander von Humboldt, Georg Reinhold Treviranus, and Karl Ernst von Baer.[22] According to Lenoir, they formed a close-knit group of naturalists pursuing a shared and coherent research program over the course of several generations. He regards their agenda of "vital materialism" or "teleomechanism" as realizing the new framework of biological research in Germany between 1790 and 1840. Crucially influenced by the philosophy of Kant, teleomechanism assumed that living matter is organized purposively: "Biology cannot reduce life to physics or explain biological organization in terms of physical principles. Rather organization must be accepted as the primary given starting point of investigation within the organic realm. In order to conduct biological research it is necessary to assume the notion of *zweckmässig* or purposive agents as a regulative concept."[23] In Lenoir's view, the premise of an inherent purposiveness in organic life was what enabled an empirical and descriptive form of research to emerge.

In short, historians have so far, *grosso modo*, painted the following picture. For many centuries, the question of development was both fundamentally important and the object of vigorous debate. In the seventeenth and eighteenth centuries, preformationist, epigenetic, and corpuscular notions of generation in every conceivable permutation battled to answer this question, which was seen as crucial to an understanding of organic life. After 1800, the issue of development was resolved — but not by biology. Quite the contrary, the science of biology could come about only because a new, dynamic worldview took development for granted as a preassumed paradigm. Only within this framework could embryology take shape and begin its empirical research on the actual mechanisms and processes of transformation. It is here, historians have argued, that we find the epistemic rupture so widely noted for the period around 1800: while Wolff's epigenesis still struggled with the question of what development actually is, embryology after 1800 no longer needed to answer that question at all. Explaining development was no longer embryology's task; development was a philosophical and ideological matter requiring no empirical explanation. Put differently, Wolff's theory of epigenesis was a response to the question of what development is, while Pander's and von Baer's epigenesis was an answer to the question of how it proceeds. Or in yet other terms: in modern historiography, the theory of epigenesis is a developmental theory without development.

Certain doubts must be cast on this consensus. Did a rupture really occur during the fifty years between Wolff and Pander?[24] Were the questions they worked on and the answers they offered really so fundamentally different? Was development something debatable for Wolff, something given for Pander? And is it really possible to distinguish sharply between the question of what development is and the question of how it proceeds?

In the following, I try to answer these questions by looking differently at the issue of development. In order to trace the emergence of developmental thinking, it is not enough to consider embryological development alone, for embryogenesis is only one organic process among many — generation, regeneration, formation, transformation, transmutation, metamorphosis, motion, nutrition, metabolism — that came together in the thinking of the period around 1800. To expand the spectrum of phenomena thus is to shift the focus away from forces and teleology and toward the organization of the living. In all the processes I have named, the body proves to be a structure that is organized, yet changes constantly. How is that possible? How can the organic world's order be reconciled with its flux, its structure with its change, its constancy with its variability?

I start from the centrality of these questions to empirical research in the period between 1760 and 1830. When naturalists gazed into the microscope, made their observations of plants and animals, and dissected, prepared, and compared organisms, the duality of organic life became apparent again and again. On the one hand, what they saw seemed to suggest that life was well regulated, as it had to be in order to function; on the other, it was always changing, in mid-transition, every form simultaneously structure and the dissolution of structure.

How did naturalists deal with this ambivalence? The remaining chapters of this book will show that the concept of development was able to emerge only within the empirical encounter with several organic processes: physiological processes of metabolism and motion, embryological processes of development, botanical processes of metamorphosis. In these empirical studies, whether Wolff's observations of the hen's egg, Goethe's theory of metamorphosis, or Pander's embryology, we find a tireless struggle with the question of how to understand the mutability and variability of organic life in light of its simultaneous orderliness and lawfulness. The answer to that question, I argue, was also the answer to the question of what development is — and it was supplied by the episteme of rhythm.

In the mid-eighteenth century, Wolff's epigenetic theory was one of the first conceptualizations of rhythm in biology. It is important to note that by an "episteme of rhythm," I do not simply mean the use of the word "rhythm" in contemporary treatises; in fact, it was not until around 1900 that "rhythm" became a technical term in biology and medicine.[25] Around 1800, the word was limited mainly to discussions of music and language. Like most biologists of his day, Wolff used the term "rhythm" only occasionally, if at all. Instead, he described the formation of the embryo as the interplay of three moments: repetition, regularity, and variation. These factors were not specific to Wolff's thinking, nor were they randomly chosen. They formed the foundation of the new episteme of organic development arising at this time and the basis for articulating the parameters of rhythm as such.

Rule, period, spiral, temporality, alternation, repetition, pulse — treatises around 1800 used these concepts to characterize the organism as a specific ordered framework, or more precisely, as a structure that contains its own time within itself. Rhythm was the rule according to which the transformations of phenomena coalesced into a coherent structure and the organism's becoming could be thought in a new way.

Repetition, Pulse, Spiral: Wolff's Theory of Epigenesis

When *Theoria generationis* appeared in 1759, it met with disparagement and incomprehension. This prompted Wolff's publication, five years later, of *Theorie von der Generation* (1764) — not a German translation of the Latin dissertation, but a separate work intended primarily to expand and explain it. Wolff now addressed the criticisms of *Theoria generationis* in detail, refuting his opponents' key arguments and defending his own position. In 1768, he published a further groundbreaking work in Latin in the *Commentaries* of the Imperial Academy of Sciences, St. Petersburg: *De formatione intestinorum*, on the successive formation of organs out of particular structures. This study was translated by Friedrich Meckel in 1812 as *Über die Bildung des Darmkanals im bebrüteten Hühnchen* (On the formation of the chick's intestinal canal).[26] The first work of its kind in the history of biology, it traced a single organ's specific formation from an initial leaflike or layered structure right up to its final form. The "greatest masterpiece that we have seen in the field of the observational sciences," as von Baer called it,[27] Wolff's treatise provided the embryological groundwork that Pander and von Baer would build on after 1800.

Most studies of Wolff's theory of generation focus only on his doctoral dissertation, but if we widen the scope to cover his work on the intestinal tract, it becomes clear that he thought of the embryo's development as an interplay between three dynamics — repetition, pulsation, and spiral motion. Development is, first of all, repetition. For Wolff, generation means not ontogenesis alone, but all the fundamental organic processes, including nutrition and growth. In each of them, formation takes place in the ceaseless repetition of production and organization. First the organic material is produced, only then is it organized; and this process is repeated again and again during the individual's entire lifetime. Second, development is pulse: the pulsation between two states. Concretely, Wolff thought of the process of organic formation as a continual oscillation of the *Nahrungssäfte*, "nutritive fluids," between flow and solidification. For Wolff, the reciprocally conditioned interaction of movement and stasis in organic fluids sufficiently explains the formation of every structure in the organism. Importantly, that formation is not a dynamic process, but on the contrary occurs in the moments of interruption, when the nutritive fluids are at rest. The third element of development is spirality — the rhythm of repetition and variation. Wolff laid the foundations of modern embryology by recognizing that the organism does not produce each single organ separately, but always produces whole systems of organs. In the course of development, these systems then differentiate out as the same process is repeated again and again at different times and in different locations within the embryo. It is through this process of repetition, on the one hand, and variations, on the other, that the individual structures gradually take on definition.

"The Same Act Seems to Be Repeated Several Times"

In his dissertation, Wolff worked out his theory of epigenetic development in two parts, one devoted to plants and the other to animals. Fundamentally, he argued, formation occurs in the same way in both,[28] but the demonstration of his point for the case of plants contained "the whole type of my reasoning," serving as a kind of Ariadne's thread "to which one may hold when treating of the far more difficult circumstances" in the animal kingdom.[29]

Wolff describes his methodology as that of a "rational anatomy,"[30] meaning the "extremely precise" derivation of "the whole, formed body" from "principles and laws."[31] Unlike his predecessors — who, he says, gave no more than a description of the processes[32] — Wolff sees his theory's task as being to *explain*, in other words, to track what he has observed back to its fundamental principles: "Only he has truly explained the development of organic bodies who derives from the principles and laws (§ 2) that he propounds the components of the body and the manner of their composition (§ 1)."[33]

The first principle of generation is that organic material is first produced, then organized. Wolff also called this process "vegetation."[34] Looking back in 1789, he assessed this distinction between production and organization as the discovery most crucial to his theory of epigenesis: "I now saw what I would never before have dreamed of, that production and actual formation or organization were two distinct things, and that each part of a plant or an animal was first produced, and only then organized."[35] Only through organization, thus, does matter become organic: it is through the formation of vessels and vesicles (*Bläschen*) in the nutritive fluids that, "simultaneously, the part that was previously inorganic is organized, or changed into an organic body."[36]

All the physiological processes that occur within an organism over the course of its life do so as an uninterrupted series, a constant repetition of the two processes of producing and organizing the nutritive fluids. These fluids are the basic substance from which every plant and every animal is formed.[37] In this sense, Wolff's theory of generation was not limited to embryogenesis — to the one-off, initial forming of a living being — but was a comprehensive theory of all kinds of formation in all organic life:

> For the sake of brevity, in the following, the word "development" will generally designate the generation of new parts and any organization of those parts, both the organization effectuated by the formation of new separated or distinguished parts and the organization brought about by the expansion of substance through fluids; furthermore, simple growth and simple nutrition or preservation. The natural bodies in which one or other of these processes takes place I will describe as bodies that are developing.[38]

Procreation, nutrition, growth, development, in other words, are all ceaseless repetitions of the same fundamental motion of producing and organizing organic fluids. Nutrition serves to preserve the body; the organism is provided for in such a way "that the whole always remains similar to itself."[39] Growth, in contrast, is characterized by the body's accruing parts that did not exist before. This happens when the existing structures secrete new substance.[40] As a consequence of growth, "the whole composition is changed."[41] Alongside the preservation and augmentation of organic structures, finally, the propagation or formation of new organs occurs when whole structures "are replaced."[42] In propagation, nutritive fluids are reduced: it is a form of suppressed growth, which leads to the emergence of new structures such as blossoms, pollen, or seeds.[43] In his treatise on the nutritive force, Wolff summarized this point as follows: "All vegetable processes are of the same kind, all have been brought about in the same way, and also by one and the same force. It is only due to the different circumstances that this general development at one point becomes digestion, at another sanguification, at another secretion, and at yet another nutrition."[44] "The same act seems to be repeated several times," to use a sentence from his

study of the chick embryo.[45] It is because of the underlying principle of repetition, then, that all the physiological processes are "of
the same kind," as Wolff puts it. Yet repetition can produce very
different kinds of formation, depending on the particular local circumstances. This is made possible by two further factors: regularity
and variation.

"Now Driven Forward, Now Held Back"
Organic structures are formed by means of an alternating play
between flow and standstill in the nutritive fluids. This principle
of alternation — *Wechsel* — is the concrete mechanism of epigenetic
development, and it is the concept of pulsing alternation, at the very
core of Wolff's approach, that can reconcile novelty and repetition.
Alternation is regular because it repeats itself continually; at the
same time, it is variable in its repetition, enabling change and therefore the diversity of structures.

The process of building every organic structure begins in an
amorphous mass, the raw material of every plant and animal organism, which is "a clear, homogeneous, glassy substance, with no trace
of vesicles or vessels."[46] As the nutritive fluids stream into this substance, it expands more and more. The first structures to arise in
it are small, "roundish or cylindrical droplets of fluid"[47] that Wolff
describes as *Bläschen*, vesicles — in Wolff's Latin, *globuli, vesicula,
foraminula*.[48] A young leaf taken from the bud, for example, "shows
itself to be composed of many vesicles, completely devoid of fibers,
vessels, or streaks."[49] At the same time, the nutritive fluids continue
to move within the plant. They must "in part move through the substance of the vesicles, in part penetrate the vesicles themselves, but
for the greatest part creep through the spaces between the vesicles
and carve out their own paths."[50]

In this way, the flow of the liquids gives rise not only to vesicles,
but also to vessels. The vesicles are "mere little holes in the solid
vegetable substance"; the vessels are "channels furrowed into the
vegetable substance."[51] The nutritive fluid seeks out such channels
on its way through the plant, but it can also flow through the vesicles
themselves until the channels have become completely established.
Although the defining feature of the vesicles and vessels is the enclo-
sure of fluids, in the course of time, "delicate, solid substance"[52]
begins to be deposited in their interstices. This solid substance

within the amorphous mass is formed by the standstill of the fluids. In organic substance, therefore, there is always only one matter in two states: "All fluids that one may find distributed within the plant must either have been deposited and be in a state of rest or else they must be driven forward and advance."[53] If the fluids are sedimented out, vesicles are formed; if they continue flowing onward, vessels are formed.[54] The momentum of formation is thus to be found in the coagulation of the fluid into a solid form within the vesicles: "The nutritive fluid has the property that through its stasis and generally, over time, due to the evaporation of its aqueous parts, it is transformed first into a somewhat thick, then a viscous, and finally a solid substance. I will refer to this property as the capacity for solidification (*solidescibilitas*)."[55] At the moment of pause when nutritive fluid builds up "between the cellular tissue," or at least "moves through the vessels rather slowly,"[56] flow is reduced and the solid substance of the just-formed vessels and vesicles accumulates.[57]

Wolff describes the vesicles' morphological composition and structure in detail, since they are the setting of the interplay between solid and liquid matter. As for the fluids, he characterizes them as "thin, resinous or rubbery substances" that "can be pulled apart into filaments."[58] In his text on the nutritive force, Wolff also mentions "cellular tissue," a "continuing, uniform, almost semifluid" substance that is very supple and "can be dilated both easily and far."[59] Within this rubbery substance, the vesicle acts as a kind of container for the fluid: it is the form-giving envelope in which the nutritive juices are packed. Given that vesicles are like envelopes or mantles, it is possible to "1. change the vesicles' shape ad libitum, 2. move the vesicles from place to place, 3. combine two discrete, round vesicles into a single elongated one, 4. restore the combined vesicles to their previous state. And ... [the leaf] can be torn open and it will 5. release the fluid, the vesicles will collapse and disappear."[60]

Wolff's morphological description presents a porous substance that at some points forms enclosed spaces, at others elongated channels. His primary interest when describing the vesicles is their form and their spatial displacements. Both aspects are essential to the alternation of physical states, because the flux and consolidation of the nutritive fluids is reversible: "The processes (§ 28) must, however, be in reciprocity [*Wechselwirkung*] with the one discussed in § 21 and 22 such that ... the replenishing fluid distends the newly deposited vegetable substance again into channels and smaller cavities, they are perfected afresh, and in this way, vessels and vesicles are formed out of the newly created substance."[61] If fresh nutritive fluid flows through the now solidified fluids, it effects another distension into vesicles or channels.[62] This means that the structuration achieved within the organic material can be revoked at any time and replaced by a new distribution of solid and fluid matter.

The key processes in the organism are not continuous, a kind of unstructured temporal course, but are periodically repeated movements, in a double sense. First, the formation of organic structures occurs not through motion, but, quite the contrary, in the moment of pause. Without the interruption of flux, without standstill, there would be no solid forms at all, and the substance would always remain homogeneous. Second, the structure that has been formed can be dissolved again at any point. The alternation of two different states, solid and liquid, pulses through the organism again and again. The liquid and solid states condition each other — they are "two processes that are always found in combination, whether succeeding each other after no perceptible interval, or appearing simultaneously."[63] "Now driven forward, now held back,"[64] the rhythm of their alternation is what creates the specific forms of the organic structures: whereas a slow flow brings forth "regular" structures, the alternation of standstill and flux gives rise to "irregular cellular membranes."[65] In view of this, Wolff's vesicles are not cells in the modern sense of organic building blocks,[66] but temporary forms within the periodically alternating flow and stagnation of fluids. Those forms can transition into each other at any moment — that is to say, they are not finished and static formations. On the contrary, the alternation of their states is what drives the vital processes. Wolff himself regarded his distinction between solid and fluid, the interplay of which created forms, as a vital innovation in the physiological thought of his era, "because, as far as I know, no one has developed this concept or has, more generally, identified an essential difference between a fluid and a solid body."[67]

In *Theorie von der Generation*, the German sequel to his Latin dissertation, Wolff went one step further, proceeding from the effects of alternation (that is, the formation of structures) to the physical foundations of the different states. In a liquid, he writes, "every part attracts every other part with equal force; for this reason, let them all move among one another as much as they will, they always remain cohesive in the same way." In a solid, in contrast, "it is only certain, particular parts that attract one another with a particular force, whereas they do not attract other parts at all."[68] Wolff thus explains the solidity of materials in physical terms by the force of their internal cohesion. Something can be called an organic structure only if a point of formation has been reached at which the nutritive fluids can no longer gather randomly, but "are assembled in a particular manner and attract each other only in this manner."[69] In line with his theoretical claims, Wolff here attributes the formation of organic structures to their central principles and laws: in the interplay of solid and liquid and in the recurring sequence of nutrition, growth, and propagation, he finds a pulsing to and fro, a construction and destruction, a ceaseless forming and transforming of the organism. In the following, I discuss how he made this approach into the fundamental principle of ontogenesis and thus the cornerstone of embryology.

"Resembling Each Other, Even If They Differ in Their Essence"

In 1768, not long after working out the general patterns of the rhythmic episteme in his dissertation, Wolff published a study of the digestive tract in the chick embryo, *De formatione intestinorum*, which traced in detail the step-by-step development of a single organ. His objective was to "supply a complete portrayal of the manner in which [the digestive tract] takes shape from its very first beginnings and gradually develops to its full completion."[70] The focus of *De formatione*, unlike *Theoria generationis*, is no longer the beginnings of structures in the homogeneous substance of the egg, but how an organ develops once certain elementary structures are present. Before starting his research, Wolff conjectured that he would have to discover "precisely as many principles of generation and kinds of formation" in animal organs "as there are really essentially different components."[71] However, he very quickly realized that this was far from being the case. Wolff was one of the first to understand that individual organs originate in the generation of whole organ systems, which differentiate out only as development proceeds. The organs present in the adult organism, fully formed and diverse, therefore tell us nothing about the manner of their genesis. Instead, Wolff's study of the chick's embryogenesis shows the intestinal tract being repeatedly formed and destroyed in the repeated construction and dissolution of various membranes. The intestine is only one example—there are several such systems in the organism. What they all share is that they take shape in the same way, as a succession of generations: that is, by one and the same process being repeated at different points in time and different positions in the body. This is where we find the third element of the rhythmic episteme, the forward movement of development through the spiral of repetition and variation.

Wolff's observation of hens' eggs began at around the middle of the third day of incubation. He identified the origin of the intestine in two membranes that enclose the embryo during its first few days. At this point, the embryo is inside the area vasculosa, surrounded by membranes that Wolff calls an inner and an outer leaf or layer (*Blatt*).[72] Wolff now describes the genesis of the intestinal tract.[73] First, toward the end of the third day, the inner or lower layer folds over, creating a sac around the embryo. Wolff calls this the "false amnion" (figure 4.1). As the first shape, a pit forms in the middle of this sac, the fovea cardiaca (marked "k" in figure 4.2), which follows the shape and position of the embryo and the false amnion.

The fovea cardiaca is formed when the head sheath of the embryo fuses. It constitutes the beginnings of the stomach, or more precisely, the smaller, posterior part of what will become the stomach. On the third day, a groove appears, running lengthwise along this pit in the middle of the sac: the seam of the false amnion (marked "l" in figure 4.2), which divides the sac into two. It is formed when, on the third day, the previously separated edges of this channel fuse. The seam, at first running lengthwise, now follows the figure of the embryo and begins to wind along the embryo's twists and turns. As a connection between the two "lateral layers of the sac," the seam is the origin of the digestive tract, or the primal intestine (*Urdarm*). At first, the tract is not closed, but is formed by two sides that move closer and then grow together.

The third part of the intestinal system is the rectum. Like the stomach, it originates from a pit, which in this case forms in the lower part of the false amnion when the embryo's tail fold fuses.[74] The entire formation of the intestinal tract, then, is achieved by the two membranes or layers that initially surround the embryo bending, folding, and growing together at their edges. The three sections of the digestive system take shape as a result of this motion and fusion:

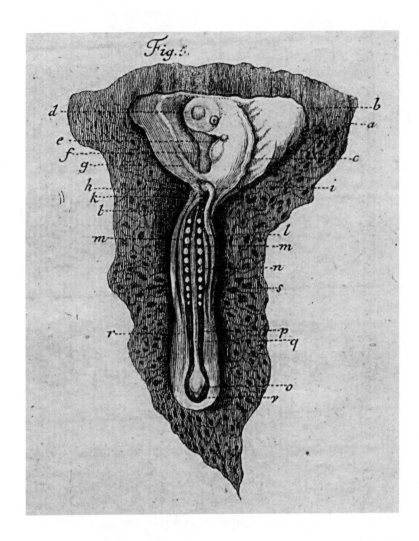

Figure 4.1 Embryo from a fertilized egg
at fifty-four hours. Detail of plate VII,
Wolff, *De formatione intestinorum*, 1768.

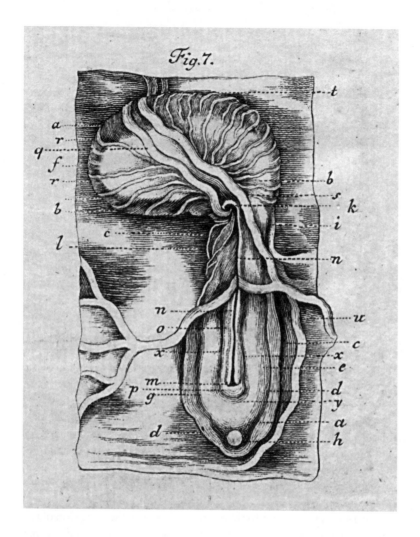

Figure 4.2 Embryo in the false amnion; "k" shows the fovea cardiaca, "l" the seam of the false amnion. Detail of plate VII, Wolff, *De formatione intestinorum*, 1768.

> At first, the intestine is a simple membrane, which gradually folds together in order to become double, and a new seam forms at some distance from that point and thereby forms a cavity between itself, the first seam, and the flat surfaces of the doubled membrane: thereupon the part of this doubled layer that runs lengthwise, which at first was flat for a while, finally swells into a tube and only then comes to resemble an intestine, and appears as the primal intestine [*Urdarm*].[75]

For Wolff, this process demonstrated epigenetic formation, since it showed "that this intestine has evidently been formed; it has not long lain completely hidden and only now come to light."[76]

But Wolff's next step was even more significant. He was surprised at the mismatch between the resemblance of the organs at the moment of their emergence and the diversity of their ultimate shape. Looking at the completed organs of an animal, such as the brain or the spinal cord, Wolff saw "no similarity; more generally, all these parts seem to me to have not the slightest thing in common." The case was very different at the moment of formation, when "the most striking similarity between them in respect to shape, position, and relationships" could be seen.[77] Wolff notes that the intestinal tract and the nervous system, for example, share the same origin: "In the first version of the embryonic nervous system, therefore, there appears very clearly the same shape, position, and arrangement of the parts that is shown by the seam of the false amnion, thus the first beginnings of the intestinal canal."[78] Wolff immediately recognized the great importance of these observations for his larger project of identifying the principles of formation in organs specific to animals, for which the observation of plants alone did not suffice: "This very marvelous analogy — not imagined, but based on the most secure observations — between parts that are so very different in their nature deserves the attention of the physiologist to the greatest degree, because one will easily admit that it has a deeper meaning and stands in the closest connection to the generation and nature of the animals."[79]

Wolff is here describing a type of epigenesis that we might call "spiral formation." This differs from the formation of the very first structures out of the homogeneous mass of the germ — which *Theoria generationis* had attributed to the pulsing of solid and liquid matter — by adding a further factor: the repetition and variation of particular membrane formations is now made responsible for the construction of whole organ systems. The conclusions that Wolff drew from his observations were crucial to the emergence of embryology: "In the generation of animals, thus, the same act appears to be repeated several times; or, if they are different ones, they nevertheless resemble one another in certain respects, and what happens here is not an uninterrupted series of causes and effects, nor several different series."[80] Rather than the number and diversity of the fully formed organs indicating an equivalent diversity in the manner of their genesis, embryonic development reveals a far more economical procedure. Nature works neither like a mechanical chain of "causes and effects" nor by a wasteful duplication of "different series" of developments. In fact, what we find is an efficient combination — a sequence that is both law-governed and variable. The "repeated generation" to which Wolff refers is not an exact repetition producing identical processes, for the nascent organ systems are only similar: "It seems as though at various moments and several times in succession, various systems were formed according to one and the same type, out of which a whole animal is then composed, and as though these systems resembled each other on that account, even if they differ in their essence."[81]

This kind of structuration typifies the rhythmic episteme. It is the repetition of a structure whose simultaneous variation ensures that the outcome is not simply the reproduction of the same forms again and again, but the genesis of new forms, which become more complex the more often the process is repeated. In the rhythmic episteme, nature neither acts as a simple causal chain nor does it create an entirely new process for each specific structure. Concretely, the various organ systems in the body take shape in the following sequence:

> The system that is generated first, that first takes on a particular shape of its own, is the nervous system. Once this is completed, the fleshy mass that really makes up the embryo is formed after the same type, almost like a second animal, similar to the first with respect to its outer shape, through the repetition of the same act of generation. There follows a third, the vascular system, which, although I have not yet discussed it, is certainly not so dissimilar to the first two that it would not be possible easily to see in it the form I have described as common to all the systems. After this follows the fourth, the intestinal canal, which once again is formed after the same type and, as a complete, self-contained whole, appears similar to the first three.[82]

Wolff here explains the formation of structures in embryogenesis as resulting from the periodic repetition of one and the same procedure. The nervous system, the embryo, the vascular system, and the intestinal system are repetitions *and variations* of a procedure that is structurally identical. Later, the germ layer theory of Christian Heinrich Pander and Karl Ernst von Baer would describe this process, more precisely than Wolff was yet able to do, as the folding of membranes. But Wolff had already recognized the principle of formation. The precondition for the structure of repetition that he identified is found in the formation of whole systems of organs, out of which the final organs take shape only in the further course of differentiation. This means that the animal body is "not, as is commonly believed, directly assembled out of various parts . . . but from whole systems of parts."[83] Each of these organ systems, finally, is a work in itself, a complete whole: "The similarity of these systems, which in the adult no longer resemble each other at all, proves that each is a complete, self-contained work independent of the other, and can certainly not be deemed a part of the other."[84]

This is to say that development means not only the triad of repetition, pulsation, and variation, but also the emergence of autonomous structures. In that sense, organic development fulfills all the criteria of rhythm I have already discussed for the genesis of works of art. In 1785, Karl Philipp Moritz described the work of art in almost identical terms, as "complete in itself, making up a whole in itself."[85] For Moritz, it was not the finished work of art, but exclusively its rhythmical becoming that explained its perfection and coherence. The same was true of the organs of the body for Wolff: he distinguished the various organs not on the basis of their different functions, positions, or composition as separate entities, but on the basis of their rhythmical genesis.

Returning now to Wolff's account of the development of organs, he has so far shown how, by the third day of incubation, the intestinal tract takes shape as a shared organ system of stomach, gut, and rectum in the process of initially separated membranes approaching each other and fusing and that this process is repeated in the embryo's upper (stomach), middle (gut), and lower (rectum) area. At the same time, parts of the membranes that form the intestinal system also contribute to the formation of other organ systems.[86] In the later days of embryogenesis, the principle of repetition turns out also to encompass an interplay of complete destruction and reconstruction.

The forms whose emergence Wolff has observed are by no means permanent; they are not direct routes to the digestive system in its final form. Rather, at the end of the fifth day, the entirety of previous development is replaced by completely new structures, and the process of formation begins from the beginning: "Finally, at the end of the fifth day, an extremely remarkable change in the egg occurs through which not only is the sac dissolved, but all things take on so new a shape that the observer believes he is regarding a quite new, completely unknown object, the ones formerly present and hitherto known having quite disappeared."[87] The new formation starts at the beginning of the sixth day, when the upper and lower membrane begin to separate, setting off a cascade of changes:

> But all these transformations — the contraction of the originally wide opening of the amnion into the fovea cardiaca, the seam, and the lower fovea, furthermore the continued amalgamation of these into a flat, roundish pit, ultimately their narrowing into a small microscopic point — are all consequences of one and the same act, a constriction [*Zusammenschnürung*], by means of which the first opening was closed and that continues to work uninterruptedly upon all of these parts.[88]

The sac that surrounded the embryo and that Wolff called the false amnion now dissolves, only to be succeeded by a new sac. This continues until the sequence of supersessions culminates in the true amnion:

> Now, the sac would become once again the only envelope of the embryo, were it not that at the same time as it disappears, within its cavity a third envelope, the true amnion, comes into being; thus, the sac that was previously the inner envelope begins to become the outer one, until the inner one, the true amnion, is completed. Then it, too, disappears, just as the transparent point did before, and gives way to the true amnion, which only now becomes apparent.[89]

The rhythmic repetition that Wolff observed as the key formative principle of organogenesis was thus to be found on several levels. First, the intestinal tract was formed by the repetition of the approach and fusion of initially separated membranes in three different regions of the embryo, bringing forth the different sections of the digestive system. Second, on the sixth day of incubation, the entire configuration, the construction of which had taken three days, was thrown over and replaced.

Describing Constant Change

In a methodological reflection that has attracted little scholarly attention, Wolff wrote:

> In general, one must remark with respect to the history of the fertilized egg that it differs from a mere anatomical description or a simple account of facts, which, if the two are not combined, may very easily be brought into a certain orderliness. The former requires merely a simple description of objects that are constantly present in a certain manner; the latter only a simple account of the facts according to the arrangement in which they succeeded each other; for the fertilized egg, however, both are united. New objects must be described and explained, and simultaneously their history must be given, even when they have not yet achieved their firm, lasting form, but are still continually changing.[90]

In other words, understanding epigenetic development calls for a new way of imagining. Neither historical nor anatomical description alone is sufficient to grasp the nature of organic changes; instead, Wolff seeks a vision that can unite the dual perspective of form and flux, of fixity and motion.

To theorize epigenetic generation, specific (and as yet unknown) components — states in the process of becoming — have to be conceptualized simultaneously with their interrelationships. Organic becoming and the products of nature are not intelligible in their individual stages without the whole context, and vice versa. Wolff's epigenesis is a theory about the laws of interaction, of order, according to which a structure is produced out of substance and structuration in concert. It is precisely this framework of the whole and its parts, of the ordering of time, that makes possible the episteme of rhythm I have traced in Wolff's work. In the rhythm of alternation between solid and liquid, formation and transformation, repetition and variation, the individual stages of development (such as the solidity of matter or the folding of a membrane) are always already thought within the entirety of formation (as components of an alternating play, of a repetition). A single link in the chain of development cannot be had without the complete series, and vice versa.

CHAPTER FIVE

Sense and Verse:

Goethe's *Metamorphosis of Plants*

Goethe's Italian journey of 1786–1788 was a turning point in his life. It ended his Weimar *Sturm und Drang* period, marked by the free exploration of subjective experience and the search for unconditional creativity. Goethe's interest in nature had been growing even before he left Weimar, but the encounter with Mediterranean flora and the art of antiquity completed his move toward a classicist concept of form that unified the "ideal" and "real" of intuition, to use Ernst Cassirer's terms. From now on, Goethe would see beauty as a manifestation of the laws of nature.[1]

It was also during his travels in Italy that Goethe became interested in metamorphosis, the subject of his 1790 essay *The Metamorphosis of Plants*. This chapter investigates Goethe's notion of metamorphosis as a manifestation of the rhythmic episteme. For Goethe, metamorphosis describes the rhythmical organization of nature, the law of its continual transformation, the pulse of expansion and contraction according to which organic life swells and declines. It is my contention that this rhythmic regularity is the truly innovative aspect of Goethe's thinking on metamorphosis. Whereas most existing scholarship interprets Goethe's metamorphosis as an expression of his dynamic view of nature,[2] I stress not the open dynamics of botanic development, but the regularity and constraints that rhythm imposes on change. Rhythm is the rule that governs transformation.[3]

This interpretation of metamorphosis as rhythm picks up on Cassirer's work on Goethe. Cassirer recognized metamorphosis as a rule, one that unites "strict tenacity with constant change."[4] The rule is perceived as "firm and eternal, yet simultaneously vital"; nothing can step outside it, but everything can be ceaselessly transformed within it.[5] Accordingly, the link between different forms is not some content that is "contained in the same way within every individual element"—what links them is the very rule of connection, and both intuition and thought must focus "on *how* this rule works, not on *what* a persistent, material substrate" actually "is."[6] At least in part, Cassirer also sees the rule of metamorphosis as a rhythmic one, at one point writing of a "pendulum swing" between idea and experience.[7] In *Das Erkenntnisproblem in der Philosophie und Wissenschaft der neueren Zeit* (The problem of knowledge in modern philosophy and science), he makes the point more trenchantly that Goethe "thinks in temporal formations and temporal rhythms,"[8] and again in *Freiheit und Form* (Freedom and form): "But for Goethe, since finding the idea of the archetypal plant and metamorphosis in Italy, the flowing, constant series of becoming was partitioned up by rhythmic motion, to vitalizing effect."[9] However, Cassirer did not regard metamorphosis as part of a broader rhythmic episteme in poetics, aesthetics, and biology, believing that Goethe's approach had little impact on his contemporaries.[10]

Goethe published his treatise *The Metamorphosis of Plants* in 1790, and in 1798, his elegy of the same name appeared. This poem has often been adduced to demonstrate the union of scientific and aesthetic thinking in Goethe's work. In this chapter, too, I read the treatise and the elegy as a unit, but not in the sense of simply correlating the essay's scientific content with the elegy's aesthetic form so as to transcend symbolically the separation of nature and art, knowledge and poetry. Such interpretations rest largely on semantic analysis. Instead, I highlight the meter of the elegy, proposing that the elegiac distich—the alternating rhythm of hexameter and pentameter—affords Goethe the sensually experienceable presence of the rhythmic episteme of metamorphosis. The rhythm of language, in other words, is the rhythm of becoming, the rule of metamorphosis. From this point of view, the two texts are interesting precisely because they do *not* form a duality—chronologically antecedent treatise and later elegy; prose and poetry; scholarly and aesthetic insight; scientific concept and symbolic form; abstract content and the rhythmic sounding of language. On the contrary, in metamorphosis, the formation of language and the formation of nature are one. Metamorphosis is at once a natural process and a rhythmical, poetic movement of language; nature's becoming takes linguistic form, and language is the nature of becoming.[11]

Goethe's deliberations on metamorphosis may be arranged into four rules, which constitute the core components of rhythm: first, formation is transformation; second, formation is alternating expansion and contraction; third, the time of formation is symmetrical; and fourth, nature is transformed into language and language into nature.

Continual Transformation

For Goethe, metamorphosis is the law according to which a basic botanical form differentiates.[12] All the structures that the plant produces, one after another, derive from the original figure or *Gestalt* of the leaf, a point encapsulated in Goethe's Italian journal by his famous dictum "all is leaf."[13]

Initially, the plant consists of nothing but raw matter. Its first structures are the seed leaves or cotyledons, which appear to the observer as "unformed, filled with a crude material."[14] Goethe also uses this image of filling, or stuffing (*stopfen*), whereby a mass with no shape of its own takes on the form of the envelope that holds it, to describe the germ as "bloated, as it were, with a crude sap," showing almost no structure and form, shapeless and only coarsely organized.[15] The treatise now begins to retrace the "successive development of the leaves,"[16] as Goethe finds the model and form of the leaf in the shoot, leaves, flower, and sexual organs of the plant right up to the fruit. He discusses only annual plants, on the grounds that because the annual "progresses continuously from seed to fruiting,"[17] it reveals the process of gradual development with particular clarity.

First, the sprout emerges from the cotyledons. The central rib pushes upward out of the leaf and "from node to node" forms new "side ribs."[18] Goethe observed this "successive and pronounced differentiation in the most simple leaf form" at its clearest in a date palm that he found in the Padua botanical gardens. With the leaf's differentiation, another new structure is formed: the leaf stalk, which equally "has a tendency to take on the form of a leaf."[19] As the plant now "refines its form step by step, it reaches the point ordained by nature. We finally see the leaves in their maximum size and form, and soon note a new phenomenon that tells us that the previous stage is over and the next is at hand, the stage of the flower."[20] Upon careful scrutiny, the leaves of the flower calyx prove to be "the same organs that appeared previously as the leaves of the stem," though they now "often have a very different form."[21] Nature forms the calyx, explains Goethe, by making the stem leaves "gently steal over, as it were, into the calyx."[22] Rather than emerging in succession and spatial separation, in the calyx, several leaves and nodes are formed at the same time and place. The petal, for example, is an expanded sepal, and the stamen is "a true petal, little changed,"[23] whereas the nectaries are "transitional forms in the change from petal to stamen."[24] This development out of the leaf is very obvious in the case of the iris flower, where we see the pistil and its stigma "in the full form of a flower leaf."[25] When the plant is regarded in this way, as the product of a step-by-step metamorphosis out of the leaf, growth and procreation are closely intertwined:

> If we consider the plant in terms of how it expresses its vitality, we will discover that this occurs in two ways: first through growth (production of stem and leaves); and secondly, through reproduction (culminating in the formation of flower and fruit). If we examine this growth more closely, we will find that as the plant continues from node to node, growing vegetatively from leaf to leaf, a kind of reproduction also takes place, but a reproduction unlike that of flower and fruit; whereas the latter occurs all at once, the former is successive and appears as a sequence of individual developments. The power [Kraft] shown in gradual vegetative growth is closely related to the power suddenly displayed in major reproduction.[26]

It is by "changing one form into another,"[27] then, that the plant's sprout, leaves, flower, and fruit take shape. Metamorphosis is transformation, with "regular and constant formations"[28] enabling the transformation of the parts so that "the process takes place unhindered."[29] For example, in the calyx, "nature does not create a new organ . . . it merely gathers and modifies the organs we are already familiar with, and thereby comes a step closer to its goal."[30] Nature does not have an absolute beginning from which it moves further and further away, but changes without end. It ceaselessly transforms existing forms, creating only — and only on the basis of — what already exists. This is how leaves successively change into ever finer and more differentiated forms of the leaf or into new forms such as those of the blossom or the stamens.

In 1817, Goethe had his essay reprinted, now with several additions, notes, and comments on the work.[31] These addenda return repeatedly to the idea of continual transformation. Thus, Goethe comments that his study of plants was motivated by a powerful emotional reaction to "the phenomena of formation and transformation."[32] In organic nature, "nowhere do we find permanence, repose, or termination. We find rather that everything is in ceaseless flux"; everything that is formed is immediately transformed.[33] Goethe's fascination with metamorphosis went beyond the "organic formation and transformation of the plant world" to include the life cycle of insects — an "unceasing transformation, visible to the eye and tangible to the hand."[34] In his later comparative anatomical and physiognomic studies, Goethe continued, "the method previously adopted in my study of plants and insects also guided me along this path: for when isolating and comparing their forms, the subject of formation and transformation necessarily came up for discussion."[35]

Through their shared reference to continuous transformation, all the parts of a plant are intimately connected and together form the precondition for each new structure. The plant "produces one part through another, creating a great variety of forms through the modification of a single organ."[36] The "laws of metamorphosis"[37] are thus to be found in the way that the parts relate to one another. Goethe makes this point using the example of the date palm, the striking shape of which makes it easy to observe the formation of the leaves. He observes how the central rib of the palm leaf lengthens, then further ribs extend laterally out of it; the different relationships between the ribs "are the principal cause of the manifold leaf forms."[38] The diversity of forms ensues from the diversity of relationships between the leaves and leaf parts, so that the basis of a form is the variation of relationships between the parts from which it is created.

For Goethe, then, to understand metamorphosis was to understand change in the relationship of the parts. This did not, however, explain the regularities or laws that govern the transition from one form into the next. How was the relationship of the parts to be described? Where was the precise point of transition from one leaf form to the next? And how could one identify the regular structure that underlay all the changes of shape manifested in the plant?

The Alternation of Expansion and Contraction
Goethe located the law of metamorphosis in the plant's pulsing alternation of expansion and contraction: the oscillation between these

antithetical processes is what gives rise to the plant's new forms — for example, the reproductive apparatus.

> The transition to flowering may occur quickly or slowly. In the latter case we usually find that the stem leaves begin to grow smaller again, and lose their various external divisions, although they expand somewhat at the base where they join the stem. At the same time we see that the area from node to node on the stem grows more delicate and slender in form; it may even become noticeably longer.[39]

Metamorphosis begins with the transformation of the stem leaves. Starting from the periphery, they contract inward. At the same time, *within* the leaves, the converse process runs its course: at their lower end, where they join the stem, they begin to expand. The petals of the corolla are now formed through an expansion of the leaves, while "secondary corollas are formed by contraction (that is, in the same way as the stamens)."[40] The leaf takes the contrary direction, expansion, to form the fruit.[41] The seed, the culmination of the plant's capacity for reproduction, reveals "the most extreme state of contraction and inner development."[42] This play of alternation is reiterated in the plant's overall structure:

> In moving up from the seed leaves, we have observed that a great expansion and development occurs in the leaves, especially in their periphery; from here to the calyx, a contraction takes place in their circumference. Now we note that the corolla is produced by another expansion; the petals are usually larger than the sepals. The organs were contracted in the calyx, but now we find that the purer juices, filtered further through the calyx, produce petals that expand in a quite refined form to present us with new, highly differentiated organs.[43]

The alternation of expansion and contraction undulates rhythmically through the plant, and in this wavelike movement it structures the individual plant components. The movement of expansion begins in the initial cotyledons, pushing forth the leaves. The calyx arises out of the contrary movement as the leaves contract, whereupon the petals of the flower are formed through expansion. Finally, the anthers emerge out of the expansion of the petals, "when the organs, which earlier expanded as petals, reappear in a highly contracted and refined state."[44] Observations such as this, writes Goethe, make us "even more aware of the alternating effects of contraction and expansion by which nature finally attains its goal."[45]

Nature "steadfastly does its eternal work of propagating vegetation by two genders"[46] through six steps that mark the switchbacks of
expansion and contraction, the peaks and troughs of rhythmic alternation. "We are convinced," says Goethe, "that with a little practice
the observer will find it easy to explain the various forms of flowers
and fruits in this way."[47] The tension between expansion and contraction permeates the plant on every level of its development. Not
only does it describe the annual life cycle, in which germination and
flowering are successive phases of expansion and contraction;[48] it is
also responsible for the leaf's refinement and differentiation and its
transformation into new forms, such as the petal, stamen, or fruit.
This is not a simple, linear succession of expansion and contraction,
but their rhythmic repetition, an alternation "by turn." Most importantly, it is always one and the same organ that contracts or expands:
"Whether the plant grows vegetatively, or flowers and bears fruit,
the same organs fulfill nature's laws throughout, although with different functions and often under different guises. The organ that
expanded on the stem as a leaf, assuming a variety of forms, is the
same organ that now contracts in the calyx, expands again in the
petal, contracts in the reproductive apparatus, only to expand finally
as the fruit."[49]

For Goethe, the rhythmicity of organic formation was central — he regarded it as his unique contribution, surpassing the work
of his "worthy forerunner," Caspar Friedrich Wolff.[50] Despite explicitly acknowledging "the identity of plant parts despite their variability," that is, the emergence of all structures out of the leaf, Wolff had
not taken "the final and decisive step." This consisted in recognizing
the leaf's metamorphosis as a rhythmic alternation of contrary processes: "In the transformation of plants he saw the same organ always
contracting, getting smaller. The fact that this contraction alternates with an expansion, he did not see." By this token, in Goethe's
view, Wolff had "himself closed off the path by which he might attain
directly to the metamorphosis of animals."[51]

"Going Backward or Forward in the Selfsame Way"
We have seen that germination is the phase of expansion, flowering
a phase of contraction. Between the two is the point at which one
movement tips into the opposite. If this moment does not material-
ize, the preceding movement is repeated,

As long as it remains necessary to draw off coarser juices, the potential organs of the plant must continue to develop as instruments for this need. With excessive nourishment this process must be repeated over and over; flowering is rendered impossible, as it were. When the plant is deprived of nourishment, nature can affect it more quickly and easily: the organs of the nodes are refined, the uncontaminated juices work with greater purity and strength, the transformation of the parts becomes possible, and the process takes place unhindered.[52]

If the nutrient liquid had continued to flow without hindrance, the plant would not have proceeded from germination to flowering. Its leaves would not have developed into new forms through contraction; they "would have appeared in separate locations and in their original form."[53] But as long as the conditions for the process driving the expansion of the leaves (an unhampered influx of nourishment) are given, the process is ceaselessly repeated. Repetition is the key impulse of the continuous transformation and modification of the leaf that Goethe calls metamorphosis, and it is the loop of repetition that keeps the cycle of the plant's formation contained. Formation thus takes on its own temporal structure. Its detachment from the conventional, advancing "line" of time is indicated by the fact that it makes as much sense backward as forward, "going backward or forward in the selfsame way," as Goethe put it in "On Morphology."[54] In fact, in order to understand the metamorphosis of plants, it is essential to trace the processes backward, to release them from their time:

> For the present . . . we must be satisfied with learning to relate these manifestations both forward and backward. Thus we can say that a stamen is a contracted petal or, with equal justification, that a petal is a stamen in a state of expansion; that a sepal is a contracted stem leaf with a certain degree of refinement, or that a stem leaf is a sepal expanded by an influx of cruder juices.[55]

In this passage, Goethe is not advocating that we regard metamorphosis as a process *in time*. Understanding the regularity of metamorphosis means something different, namely, observing the relationships between the parts and describing the laws that govern their changes. In other words, the plant must be regarded as a relationship *of time*. Nature can create its own temporal order — "sometimes nature skips completely over the organ of the calyx, as it were, and goes directly to the corolla."[56] Equally, "nature frequently shows us instances where it changes the styles and stigmas back into flower leaves. *Ranunculus asiaticus*, for example, becomes double by transforming the stigmas and pistils of the fruit vessel into true petals, while the anthers just behind the corolla are often unchanged."[57] The plant can also take a "backward step, reversing the order of growth."[58] In this case, the direction of metamorphosis doubles back in what Goethe calls an "irregular" or "*retrogressive* metamorphosis."[59] This implies that regular metamorphosis is a form of development in which the plant moves forward in time, but appropriates time for itself by making jumps and backward steps. Conversely, irregular or retrogressive metamorphosis allows us "to discover what is hidden in regular metamorphosis, to see clearly what we can only infer in regular metamorphosis."[60]

What exactly irregular metamorphosis might reveal about regular metamorphosis emerges in more detail in the collection of notes that Goethe presents in his *Morphologische Hefte* (Morphological notebooks).[61] There he discusses the conceptual couplets of formation and malformation, physiology and pathology, and deviation and norm, distinctions that are significant for the idea of metamorphosis as a rhythmic order. At stake is the question of what distinguishes a formation as normal development or its opposite, pathological development or abnormality. Goethe first insists that "the abnormal should not necessarily be regarded as diseased or pathological": "the words misdevelopment, malformation, crippling, and stunting should be used with care, for even though Nature operates with greatest freedom in this realm, she nevertheless may not depart from the fundamental laws of her being." Even in her deviations, thus, nature works within the framework of certain regularities.

> Nature fashions normally when she subjects innumerable particulars to a rule, defines and delimits them. Conversely, the phenomena are abnormal when the particulars carry the day, emerging in an arbitrary, indeed apparently accidental way. However, since both are closely related, and since the same spirit animates the regular and the irregular as well, an oscillation between the normal and the abnormal occurs, formation and transformation forever alternating, so that the abnormal seems to become normal, and the normal abnormal.[62]

This passage defines more precisely the rhythmical relationship of conformity and deviation. Formation means the individual structure in the plant being subjugated to the rule — that is, to the alternation between contraction and expansion. Rhythmic alteration prevents the "particulars" from prevailing. Formation as a rhythm of alternation determines the place of the individual elements in this process by aligning them with the succession of alternating states. Yet within this rhythmic order, nature always operates at the frontiers of randomness, of disorder, of chaos — for "every leaf, every bud, is in itself entitled to become a tree." But that is impossible, and to stop it happening, "the superior health of the stem" restrains them.[63] The leaf's expansion into a separate organism, thus, is prevented by the contrary mechanism of contraction to which the other parts of the plant subject it. The ordered and the disordered are so "closely related" in the process of constant transformation because each is defined only in relation to the other. Any formation can be restored to the norm through the rhythm of alternation or else continued into abnormality through disruptions to the rhythm. Regularity and randomness continually determine each other afresh, at every tipping point. Thus, "the abnormal seems to become normal, and the normal abnormal."

The tipping point is the point of ingress for the multiplicity of nature that Goethe so often evokes. It is only by operating at the edge of disrule — by always allowing for the loss of order — that nature can bring forth its multiplicity of form. This occurs within the constant renegotiation of norm and deviation, at the moment when the formative process finds its rhythmic apex: "In general, we find again and again that deformations return to the pattern, that Nature has no rule to which she might not make an exception, and makes no exception which she might not again turn back into a rule."[64] The truth is, argues Goethe, "that a complete view can never be attained unless the normal and the abnormal are regarded as constantly fluctuating and operating *toward* one another."[65] The creativity of transformation, the process of becoming and forming in nature and art, lies not in mere progression, but in the play of repetition, in the interplay between order and chaos, between rule and deviation, between the recurrent, yet different form that rhythm brings about.

A Brief Cultural History of Metamorphosis
Goethe worked on his concept of metamorphosis by observing annual plants. However, he regarded it as a universal principle that applied far beyond botany alone and promised to overcome the distinction

between thinking on art and thinking on nature. In Goethe's view, the rule of metamorphosis is not merely the law according to which nature transforms itself; it is the shared law of formation that applies in equal measure to the products of nature and of art. Just as metamorphosis relates nature's permanence to its mutability, form to formation, shape to transition, so it also relates nature to art. Goethe's notion of metamorphosis originated in his interest in nature, but he very soon began to expand it across the entire domain of natural and intellectual phenomena — to the products of art, the formation and education of man, and even to cosmic change. Christa Lichtenstern has called metamorphosis a new "knowledge ideal" for Goethe,[66] since he believed that it not only described the workings of nature, but constituted the "inner formative law and method" of art, as well.[67] To this extent, Cassirer's reading of a "parallelism" between the law of nature and the law of art falls short.[68] Goethe went further than that, making metamorphosis the law of human beings' inner formation and their cognition of themselves and the world, the principle that gives human existence an intellectual, moral, and creative dimension. Indeed, in the alternation of systole and diastole, he saw it permeating the whole universe. Metamorphosis as a "totality of idea and experience," Lichtenstern concludes, offered Goethe an "exemplum of creative activity per se."[69]

At the same time, Goethe's notion of metamorphosis also brought to late eighteenth-century botany one of the oldest and most powerful images in cultural and literary history.[70] Originating in classical myth, the concept of metamorphosis received its best-known and most influential (though not its first) formulation in Ovid.[71] Ovid's *Metamorphoses* starts with the beginning of the world. That beginning is not a creation *ex nihilo*, but the transformation of raw matter — already present eternally and eternally existent — into formed matter. Metamorphosis thus is the transformation of something that is already given. In the process, the given loses its borders, hesitates on the threshold of the other, of becoming, of an existence beyond its own frontiers. This liminality is at the heart of metamorphosis, which transforms gods, humans, and nature into different beings, animate or inanimate. It often occurs at a moment of jeopardy and marks the loss of identity, often enough the loss of humanity, by imposing a new animal or inorganic existence. Metamorphosis is the instant when the divine enters the everyday and vice versa: it shows the natural world to be magical and the magical world to be natural. As myth, metamorphosis is always also narrative, as much a linguistic and poetic transformation as a natural one. When literature thematizes the transformation of nature, it transforms itself. Metamorphosis therefore affects not only literature's object, but also its form; the transformation cannot be disentangled from its representation or the substance from its poetic shape.[72]

Toward the end of the eighteenth century, metamorphosis became an increasingly frequent theme in German literature. In theoretical debates around Johann Joachim Winckelmann's aesthetic dictum of "noble simplicity and quiet grandeur" and Gotthold Ephraim Lessing's distinctions between poetry and painting in his essay *Laocoon*, metamorphosis was edged out of the visual arts. Karl Philipp Moritz, August Wilhelm Schlegel, and Goethe all argued that there was no place in the visual arts for the physicality of metamorphosis, its dynamism, passion, and dissolution of boundaries. The poetic word was considered a more congenial vehicle for the idea of transformation, for example in Moritz's view of classical myth as a process of tireless poetic imagination and a battle between formless matter and its fashioning into art.

In terms of cultural history, metamorphosis is a liminal figure in both content and form: as a theme, because it evokes the tipping of known into unknown, of being into becoming; as a form, because it oscillates between described and description, nature and art, fact and fiction, aesthetic and scientific reflection.[73] In metamorphosis, nature and art lose their separate identities and become one. This reciprocal adaptation is located in language, for the metamorphosis of nature is always also the metamorphosis of the poetic form that produces it and in which alone it assumes existence.

Metamorphosis in Distichs

Goethe worked out his theory of metamorphosis in natural history during and immediately after his travels in Italy in 1786–1788, in dialogue with his literary work. He wrote to Charlotte von Stein from Italy: "I now look at art in the same way as I looked at nature; I am gaining what I have long striven for, a more complete conception of man's highest achievements, and my mind grows in this direction too and overlooks a wider expanse."[74] Later, he said more about this phase of his work:

I was simultaneously writing an essay on art, fashion, and style, one on the metamorphosis of plants; and "The Roman Carnival." Each of these reveals what was going on in my mind at the time and the position I had taken with respect to those three great fields of knowledge. I finished first the *Essay on the Metamorphosis of Plants*, which traces the manifold specific phenomena in the magnificent garden of the universe back to one simple general principle.[75]

Among the poems that Goethe wrote during his work on metamorphosis are the *Roman Elegies*, a cycle written between 1788 and spring 1790 and first published in 1795 in Schiller's journal *Die Horen*. The elegies combine love poetry with Goethe's experience of Rome and classical antiquity to articulate the "reunification of feeling, seeing, understanding, loving, and writing."[76] They draw on the love elegies of poets such as Catullus, Tibullus, or Propertius in the late Roman Republic, which used the distich form — couplets made up of one hexameter and one pentameter in a rise-and-fall rhythm that unites feeling and understanding, the sound and semantics of the word. Goethe's *Elegies*, too, are love poems about passion, parting, and death. They revolve around a key theme of Goethe's poetry of the 1790s: the alternation of eternity with temporality, permanence with change, law with transformation.[77]

In June 1798, eight years after *The Metamorphosis of Plants* and the *Roman Elegies*, Goethe wrote another elegy, "The Metamorphosis of Plants."[78] In it, he synthesized his experiments in natural history and poetry, promising to provide what he accused science of having forgotten: "Nowhere would anyone grant that science and poetry can be united. People forgot that science had developed from poetry and they failed to take into consideration that a swing of the pendulum might beneficently reunite the two, at a higher level and to mutual advantage."[79]

The Elegy "The Metamorphosis of Plants"
Goethe's elegy, written in classical elegiac couplets, formed part of a plan that Goethe devised in 1798–1799, but never carried out: to create a modern didactic poetry that would address his era's theory of nature.[80]

The Metamorphosis of Plants

THOU art confus'd, my beloved, at seeing the thousandfold union   1
Shown in this flowery troop, over the garden dispers'd;

Many a name dost thou hear assign'd; one after another   2
Falls on thy list'ning ear, with a barbarian sound.

None resembleth another, yet all their forms have a likeness;   3
Therefore, a mystical law is by the chorus proclaim'd;

Yes, a sacred enigma! Oh, dearest friend, could I only   4
Happily teach thee the word, which may the mystery solve!

Closely observe how the plant, by little and little progressing,   5
Step by step guided on, changeth to blossom and fruit!

First from the seed it unravels itself, as soon as the silent   6
Fruit-bearing womb of the earth kindly allows its escape,

And to the charms of the light, the holy, the ever-in-motion,   7
Trusteth the delicate leaves, feebly beginning to shoot.

Simply slumber'd the force in the seed; a germ of the future,   8
Peacefully lock'd in itself, 'neath the integument lay,

Leaf and root, and bud, still void of colour, and shapeless;   9
Thus doth the kernel, while dry, cover that motionless life.

Upward then strives it to swell, in gentle moisture confiding,   10
And, from the night where it dwelt, straightway ascendeth to light.

Yet still simple remaineth its figure, when first it appeareth;   11
And 'tis a token like this, points out the child 'mid the plants.

Soon a shoot, succeeding it, riseth on high, and reneweth,   12
Piling-up node upon node, ever the primitive form;

Yet not ever alike: for the following leaf, as thou seest,   13
Ever produceth itself, fashion'd in manifold ways.

Longer, more indented, in points and in parts more divided,   14
Which, all-deform'd until now, slept in the organ below,

So at length it attaineth the noble and destin'd perfection,   15
Which, in full many a tribe, fills thee with wondering awe.

Many ribb'd and tooth'd, on a surface juicy and swelling,   16
Free and unending the shoot seemeth in fullness to be;

Yet here Nature restraineth, with powerful hands, the formation,   17
And to a perfecter end, guideth with softness its growth,

Less abundantly yielding the sap, contracting the vessels,   18
So that the figure ere long gentler effects doth disclose.

Soon and in silence is check'd the growth of the vigorous branches,   19
And the rib of the stalk fuller becometh in form.

Leafless, however, and quick the tenderer stem then upspringeth,   20
And a miraculous sight doth the observer enchant.

Rang'd in a circle, in numbers that now are small, and now countless,   21
Gather the smaller-siz'd leaves, close by the side of their like.

Round the axis compress'd the shelt'ring calyx unfoldeth,   22
And, as the perfectest type, brilliant-hued coronals forms.

Thus doth Nature bloom, in glory still nobler and fuller,   23
Showing, in order arrang'd, member on member uprear'd.

Wonderment fresh dost thou feel, as soon as the stem rears the flower   24
Over the scaffolding frail of the alternating leaves.

But this glory is only the new creation's foreteller,   25
Yes, the leaf with its hues feeleth the hand all divine,

And on a sudden contracteth itself; the tenderest figures   26
Twofold as yet, hasten on, destin'd to blend into one.

Lovingly now the beauteous pairs are standing together,   27
Gather'd in countless array, there where the altar is rais'd.

Hymen hovereth o'er them, and scents delicious and mighty   28
Stream forth their fragrance so sweet, all things enliv'ning around.

Presently, parcell'd out, unnumber'd germs are seen swelling,   29
Sweetly conceal'd in the womb, where is made perfect the fruit.

Here doth Nature close the ring of her forces eternal;   30
Yet doth a new one, at once, cling to the one gone before,

So that the chain be prolong'd for ever through all generations,   31
And that the whole may have life, e'en as enjoy'd by each part.

Now, my belovèd one, turn thy gaze on the many-hued thousands   32
Which, confusing no more, gladden the mind as they wave.

Every plant unto thee proclaimeth the laws everlasting,   33
Every floweret speaks louder and louder to thee;

But if thou here canst decipher the mystic words of the goddess,   34
Everywhere will they be seen, e'en though the features are chang'd.

Creeping insects may linger, the eager butterfly hasten, —   35
Plastic and forming, may man change e'en the figure decreed!

Oh, then, bethink thee, as well, how out of the germ of acquaintance,   36
Kindly intercourse sprang, slowly unfolding its leaves;

Soon how friendship with might unveil'd itself in our bosoms,   37
And how Amor, at length, brought forth blossom and fruit!

Think of the manifold ways wherein Nature hath lent to our feelings,   38
Silently giving them birth, either the first or the last!

Yes, and rejoice in the present day! For love that is holy   39
Seeketh the noblest of fruits, — that where the thoughts are the same,

Where the opinions agree, —that the pair may, in rapt contemplation,   40
Lovingly blend into one, —find the more excellent world.[81]

The construction of the poem frames the discussion of plant metamorphosis within a concrete speech situation in which the narrator addresses a female beloved. The elegy begins with the beloved's problem of feeling "confused" by the plenitude of nature (distichs 1–4), to which the narrator responds in the central part by explaining metamorphosis (5–31). The plant's transformation is described from the seed to the germ and node, to the stem, to the flower and the fruit, then returning to the germ, with which nature closes the "ring of her forces eternal." This sequence corresponds to the structure of the *Metamorphosis* treatise,[82] but even in the central section, the beloved is addressed several times and invited to wonder and observe, so that the course of knowledge unfolds in direct interaction between the object of explanation and the knowing subject.[83] At the end of the elegy, metamorphosis is expanded to cover nature in general, human relationships, and not least, the human being as such, "plastic and forming," whom the poet calls upon to "change e'en the figure decreed" (35). Addressing the beloved again, the elegy now turns from reflection on the laws of nature to self-reflection on humanity (32–40), for knowledge of the laws of nature also offers human beings an understanding of themselves. Finally, the elegy is also a love poem about the human beloved and her interlocutor. This alignment makes metamorphosis the formative law not only of nature, but of humanity, in both human reflection and emotional life.

This is the last of the great elegies that Goethe wrote in a classical meter. It contains a striking tension between the formal and thematic design. In its subject matter, the text is a classic didactic poem in which the task of the poetic form is to provide a comprehensible explanation of a factual situation; in its distich form, it is a classic elegy. By coupling the elegy with the didactic poem, the poet multiplies our perspectives on the formative process of metamorphosis. Metamorphosis appears not merely as the course of the plant's growth, but also as the progression of knowledge in the woman to whom the explanation is addressed, then finally as the process of the couple's self-knowledge, which finds its completion only in love.[84]

Some critics have stressed the "sublation" of scientific discourse in the poem's form.[85] Reiner Wild, for example, argues that Goethe's is a utopian project, possible only in literary language, to overcome the separation between science and poetry and propound a unity of knowledge that can counter the emerging rift between the natural sciences and the humanities;[86] its outcome is to "anthropomorphize" nature[87] and sexualize metamorphosis.[88] The poem's proposed resolution to the polarization of nature and art is a symbolic one: in Günter Peters's view, Goethe's didactic poetry aims to "stage the cognition of nature as a symbolic drama."[89] Consequently, the convergence of aesthetics, literature, and science takes place in the elegy as a "natural aesthetics of scenic intuition,"[90] by which Peters means that cognition happens through memory triggered by a scenic visualization.[91] Accordingly, for Peters, Goethe's way of researching is also a "poetics and hermeneutics of natural spectacle."[92]

A contrasting, "morphological" interpretation of the elegy is offered by Günther Müller and Gertrud Overbeck.[93] Both focus on its metrical figuration, arguing that the elegiac couplet absorbs the formative principle of metamorphosis: the alternation of expansion and contraction. In this view, the recurrence of syllables corresponds to the continuity of metamorphosis, the elegiac rhythm to the alternation of contraction and expansion in the plant: "the inner motion of the poem flows through the constant regularity of the distichs in an animated alternation of contraction and expansion and, intensifying, achieves rest only at a final pinnacle."[94] The elegy is thus the mirror of metamorphosis and is suffused by "organic rhythm."[95] According to Overbeck, contraction and expansion also underlie the construction of the elegy as a whole. In the poem's various sections, a concentrated focus on the beloved, the single plant, or individual feelings alternates with an expansion of the view onto the regularity of motion, a universally guiding nature, and the general constitution of metamorphosis as a law that also governs the individual's development and the union of the couple.[96] Pascal Nicklas, finally, concludes that the poem's construction proceeds, so to speak, from the seed to the corolla in analogy with the transformations of the plant and itself becomes something "natural, organic."[97] Metamorphosis marks the point where natural development changes into poetic development,[98] and Goethe thus recapitulates "the link between poiesis and metamorphosis" that has been current since Ovid's day.[99]

Sense and the Senses in Metamorphosis
The previous section's short excursion into the cultural history of metamorphosis indicated the importance of linguistic form, the rhythmical organization of the movement of words through which alone the transformation of nature can be grasped. This is also the focus of certain perspectives in literary theory that regard rhythm not as a merely formal schema, but as something able to generate physical presence in poetic speech.[100] Hans-Ulrich Gumbrecht, for example, discussing Goethe's *Roman Elegies*, contests the prevailing interpretation that rhythm belongs to the semantic level of poetry. He shows that the semantic content of the word and the "bound," metrical form of the distich enter into an oscillating motion: for the duration of the poetic utterance, what is conceptually evoked is also tangible, experienceable. This means that the conceptual content of poetic speech is the position of the body by which it is spoken. The rhythm of the elegy, argues Gumbrecht, does not *reinforce* the content matter; rather, there is an oscillation between semantics and poetic form.[101]

In the following comments on Goethe's "The Metamorphosis of Plants," I start from the assumption that the treatise and the elegy form a unit. By this I mean that the elegy transforms natural change into a linguistic movement — in other words, that the rhythm of nature can be fully comprehended only through the rhythm of its linguistic description and becomes directly perceptible only in the sensual experience of poetic speech. Second, I focus on the tension or oscillation between meaning and rhythm that Gumbrecht identified in the *Roman Elegies* and show how Goethe deployed the rhythm of the elegiac distich to establish his theory of metamorphosis as a new botanical vision—both in opposition to and in dialogue with the dominant edifice of Linnaean thought.

As discussed, the distich form here is not art *supplementing* nature, nor does it merely amplify the semantic content of the elegy as a didactic poem. Rather, the rhythmical motion of the distich *is* the law of the emergence of forms, according to which nature and art in equal measure are produced, take shape, and, crossing and recrossing their own boundaries, turn into each other. As Goethe remarked of the carnival, metamorphosis means that out of the clash of "impetus and inertia" a "third thing emerges that is neither art nor nature, but both at the same time—inevitable and accidental, premeditated and fortuitous."[102]

Hexameter and Pentameter

We have seen that in the late eighteenth century, German literature and literary theory—as represented by writers such as Friedrich Klopstock, Karl Philipp Moritz, and August Wilhelm Schlegel—were intensely interested in the translation of classical meters into the German language.[103] In poetic theory, the elegiac distich was considered the metrical pattern that offers the greatest possibilities for variation by uniting constant reformulation with an invariable underlying pattern, the exigencies of the framework with freedom of content.[104] In the distich, a line of hexameter and a line of pentameter join to form a couplet. The gently decaying hexameter, ending on an unstressed syllable, "alternates in a rhythmic pulse" with the brusquely stressed ending of the pentameter line.[105] In the hexameter line, writes Gumbrecht, "rhythm recedes into the status of a sometimes only vaguely perceptible background resonance," whereas the pentameter "secures for the distich the poetic effects of an intensified present and of a spatial form."[106] The hexameter consists of five dactyls and one trochee, though the first four dactyls may be replaced by trochees; the couplet's second, pentameter line is like a hexameter in which the unstressed syllables of the third foot are deleted, and as a rule, the third foot is followed by a caesura. To illustrate this in the original German text (distich 1):

Dich verwirret Geliebte, die tausendfältige Mischung
Dieses Blumengewühls über den Garten umher
/ x | / x x | / x x | / x | / x x | / x
/ x | / x x | / || / x x | / x x | /

The elegiac distich is a compact metrical expression that favors rapid transition from one thought to another or a rapid sequence of variations on a single idea and thus the extended development of a line of thought.[107] Goethe took the rhythm that the ancient elegists had chosen to express mixed feelings and the shifting of their protagonists' passions and made of it the formative law of metamorphosis. Furthermore, he deployed the rhythmic structure of the distich to embed his theory of metamorphosis within botany while simultaneously stressing its distinctiveness. The botanical content of the elegy, making up the central section, is carefully organized in rhythmic terms. Goethe situates his description of the formative laws in the second line of each couplet, the pentameter, while the first, the hexameter, is reserved for a general introduction to botanical issues and to the Linnaean school.[108] Specifically, the first line of each distich addresses general botanical parameters such as seed, light, and force (distichs 6–8), the components of the plant (root, leaf, germ), and the sprouting, diversity, and form of the plant world (9, 12, 14–16). In the second half of the central part, the narrator adopts the botanical diction of Linnaeus, with references to stem and calyx, number, arrangement, and axis — the outer appearance of the plant, as the criterion of Linnaean description, is clearly intended here (20–23). In a third subsection, beginning at distich 25, the narrator moves on to the sexuality of plants, though not for the purposes of classification, as in the Linnaean system — instead, it is celebrated as the highest goal of nature. These lines describe at length the union of the plants to produce germs. The anthropomorphic language ("beauteous pairs," "lovingly," "altar") pronounces sexual procreation the highest perfection not only of vegetation, but of nature as a whole.

Goethe's elegy is a classical didactic poem to the extent that he uses the calmly flowing hexameter to present the botanical world in its essential components, then to introduce the conceptual building blocks of his era's most important botanical model, the Linnaean system. That changes in the second line of each distich, where Goethe proposes his theory of metamorphosis in contradistinction to the conventional botanical wisdom — we might say that Goethe clothes prevailing scientific opinion in the classical meter of the didactic poem, only to interrupt it on alternate lines with a description of the new conception of metamorphosis. Unlike the hexameter line, the pentameter line is doubly accented: once in the middle (by the absence of the unstressed syllables and the caesura) and once at the end. The open, more variable, and faster-moving rhythm of the pentameter that Goethe reserves for his own proposal contrasts with the even, predictable rhythm of the hexameter. The two different meters in the couplet form an interlocking structure that emphasizes not only the rhythmicity of metamorphosis, but also the tension between established botanic teachings and the approach propounded by Goethe, who was disappointed at the lack of recognition for his original treatise eight years earlier.

The pentameter lines portray the laws of metamorphosis as follows. Distichs 5–11 thematize what nature conceals, the germ (the "child"), which is triply hidden — in the womb of the earth, in the darkness of the night, and in the hull of the seed — as it waits to leave its "motionless life," step into the light, and initiate its autonomous existence with motion. In the subsequent lines, this movement of the seedling is described in more detail as Goethe sets out its laws, in other words, the laws of metamorphic transformation. The explanation of metamorphosis begins with a description of the plant's shape as a piled-up tower of nodes (12). These differentiate into the various parts of the plant through a series of different types of movement: succession ("the following leaf," 13), the assumption of different forms (which prompts "wondering awe," 15), perfection (17), and refinement (18). The movements that create this "miraculous sight" (20) are then specified: they are similar ("of their like," 21), "arranged" (23), tiered ("member on member," in the original, *gestuft*; 23), and form a "scaffolding" that is nevertheless mobile ("rears," 24).

In Goethe's metamorphosis, then, the form of the plant develops as a law of successive movements. These are further emphasized by the pentameter's rhythm, with its central caesura and its stressed final syllable. The "type" is not definitive, but an ephemeral form that is released ("unfoldeth," in the original, *entläßt*, 22); the movements are not random, but "arranged" in order (23), and the regularity of this "scaffolding" remains variable (24). Finally, in the topic of sexual union that opens the third part of the elegy, ontogenetic movement is heightened into the universal movement of nature.

Based on my contention that Goethe presents his new conception of metamorphosis in the pentameter line of each couplet and conventional botany in the hexameter line, what is new about this movement of the whole of nature is not that it closes "the ring of her forces eternal" (30) or that it is continued like a chain "for ever through all generations" (31) — since these statements are made in hexameter lines. Goethe's novel contribution is his claim that nature returns back into itself through its creations only to repeat itself, though not in an identical form (30). The significance of this chain of becoming is not in the way it connects and prolongs the individual parts, but in the way it defines the whole and the parts solely in relation to each other, so that "the whole may have life, e'en as enjoy'd by each part" (31).

Whereas established botanical doctrine is formulated in the slower flow of hexameter, the motion of metamorphosis — clothed in pentameter — is faster and more flexible. But the motion is also ordered; it is tiered and arranged in its flux. Its order is defined by its motion, its repetition by its variation. This multifaceted movement determines the genesis both of natural forms and of the forms of art. As such, it also determines Goethe's theory of metamorphosis itself and its place in the biology of the period. For if Goethe introduced metamorphosis to botany as an innovation, he used the interlocking hexameter and pentameter of his elegy to bind that innovation into existing knowledge. The two models relate to each other in a form of reciprocal *Wechselwirkung*. In other words, metamorphosis as a theory was itself a transformation, a variation of familiar knowledge, and arose both in interaction with existing theories and in distinction from them. Far from articulating distance and the "art character" of the work, as Wild claims,[109] the elegiac couplet constitutes the direct, sensual, experienced presence of the rhythmic episteme.

Much more could be said about Goethe's notion of metamorphosis. The points I have made here are mainly restricted to the treatise and the elegy, whereas the theme pervades Goethe's entire oeuvre and would repay extensive study in other perspectives and contexts.[110] For the purposes of the present book, I have shown that Goethe is relevant to the rhythmic episteme in two ways. First, he uses the plant realm to describe how nature perpetuates itself in rhythmic alternation, always operating at the edges of its own dissolution. The transformation of nature is a fragile defining of boundaries between what exists and what does not yet exist, between the past and a projected future, between reliable reproduction and random variation. As a process, metamorphosis is continuous, but it is not permanence as such — indeed, permanence is change, and it is constituted only at the moment of tipping into the opposite, of contraction into expansion and vice versa. In this sense, Goethe addresses questions and contexts that were occupying both embryology and physiology at the time. At stake was the organization and temporality of organic life and, above all, the ordering of time within the organism. At its core, Goethe's concept of metamorphosis aimed not to account for the temporal dynamism of nature, but to wrest a certain order from a nature experienced as unstable and mutable — to give mutability structure through rhythm.

Second, Goethe's material is language. His two works offer an instructive opportunity to address the relationship of nature and aesthetics in ways that go beyond mere parallelism. The idea of metamorphosis erases the very distinction, for nature and language are fused in rhythm. This implies neither biologism nor aestheticism, nor an attempt to examine one using the categories and tools of the other. The notion's productivity is to be found in the "in between," the overlap of the dichotomies, which makes them possible in the first place, but also annihilates them, the space in which all is transition. For the duration of the poetic utterance, the metrical form of the distich is just such an in-between space: between sense and sensation, nature and language, word and voice, concept and body.

CHAPTER SIX

The Rhythm of the Living World:

Physiology circa 1800

At the end of the eighteenth century, there was much attention to the question of whether the living world is driven by particular vital principles and forces. It seemed incontrovertible that certain forces in the organic world must exist; without them, life, its emergence, preservation, and reproduction could not be explained. When it came to the nature and composition of these forces, however, views differed considerably. As divergent as the labels chosen — Casimir Medicus wrote of a *Lebenskraft*, Paul-Joseph Barthez of a *principe vital*, Blumenbach of a *Bildungstrieb* or *nisus formativus* — were the logics that underlay them. Almost every question concerning the vital forces was still unanswered. Do they actually exist, or are they purely heuristic? Are they material (for example, chemical or physical) forces, or immaterial ones? How many forces are there — five, as Carl Friedrich Kielmeyer postulated in his speech *Über die Verhältnisse der organischen Kräfte untereinander* (On the relationships of the organic forces)?[1] Are there separate forces for generation, vegetation, muscles, and nerves, or is there (as in Blumenbach's *Bildungstrieb*) a kind of architectural principle, "a force that effects action,"[2] to which the other forces are subordinated? And how do the force or forces work in the body?[3]

The Temper of the Life Force
The first issue of the physiological journal *Archiv für die Physiologie*, founded and edited by the physician and anatomist Johann Christian Reil (1759–1813) of Halle, appeared in 1795. Its opening essay, Reil's "Von der Lebenskraft" (On the life force), held a special place in the contemporary debate about vital forces and has gone down in the history of biology as one of the foundational texts in that discourse.[4]

In it, Reil analyzed not only the nature and role of the vital force as the specific characteristic of the organic world, but also, and most importantly, the concrete action of the life force — or rather, its different actions, depending on the different rhythms that in his view organize nature: the phases of life, the seasons, or the lunar cycle.

In the debate on how to demarcate living from nonliving matter, Reil propounded a universal vital force characterizing all organic substances.[5] The *Lebenskraft* that he describes in "On the Life Force" is not an immaterial principle. Instead, "the forces of the human body" are "properties of its matter, and its particular forces the results of its peculiar matter."[6] Every force, including the life force, is something "inseparable from matter, a property of matter by means of which matter brings forth phenomena."[7] Not only is the life force a property of matter, but, conversely, matter itself is "nothing other than a force, its accidents are its effects, its being is action, and its specific being is its specific way of acting."[8] As a property of matter, the life force is more precisely a property of the organic material's specific form and composition (*Form und Mischung*).[9]

For Reil, what is special about the organic world, equipped with vital force, is that it organizes *itself*. This autonomy of organic life, the focus of his scientific research, explains why "the laws of the effects of animal bodies are so difficult to discover."[10] All the changes in the organism are expressions of the organic structure's independent action: "The forces of the animal (the form and composition of its matter) are forever modifying themselves of their own accord. As time progresses, they never remain the same, but are something different at every moment; and just as changeable are their effects."[11]

Organisms' ability to modify their own forces increases with the increasing perfection of the animal organization. As an autonomous organization, every organic being "generates itself as an individual, develops and preserves itself."[12] Accordingly, its individual parts constantly reorganize themselves as well. Although "the preservation of one part depends, turn and turn about, on the preservation of the next,"[13] so that the interaction of the body's different organs and parts is a precondition for its life, each individual organ is nevertheless a coherent and independently functioning whole. Every part or organ of the body lives "on its own account, possesses the force for its own perpetuation in itself, and preserves itself by its own means";[14] the "immediate cause of all phenomena that it brings forth" is "directly contained within it."[15]

The body is thus organized like a "republic," within which each component stands in a defined relationship to the next, yet still operates "through its own forces," possessing its "own perfections, deficiencies, and infirmities."[16] At no moment is this republican physical body ever self-identical: it changes ceaselessly, not only as a whole, but in each of its constituent parts.[17] The body is made up of autonomous elements, but these cannot exist on their own account — only collectively with the other organs. Hence, for the organism as a whole to survive, there must be an ordered relationship between the various components. The puzzle posed by the organic world is to discover the rule that governs this relationality of the parts: "And the view would be yet more interesting if we could sensually perceive also the incessant change of all these phenomena and the cause of the same, namely, the constant modification of the forces of all organs by means of their own actions."[18]

On the basis of what "inner principle,"[19] Reil asks, does the organic body preserve itself and perpetuate itself into the future? And what exactly are the characteristics of this principle, which inheres in organic matter and brings forth an organic world marked by "great mutability and at the same time the greatest constancy"[20] — a nature that continues its past into the future, reinventing itself again and again, yet not arbitrarily, but on the basis of what already exists?

Reil locates this cause or principle of the organism's autonomous modification in the transformations of the life force. The life force is a property of the form and composition of organic matter, and it is in the relationship between matter's form and its composition that we find "an important ground for its alternating and changeable phenomena."[21] The cause of organic life's mutability, then, is the changing relationship between matter's form and its composition. Reil calls this relationship the "temper" (Stimmung) of the life force:[22] the force changes "either in degree or in nature," and "the natural degree of the life force, as it is suited to the preservation of the individual, I will call the Stimmung (temperies)."[23] He adds: "The temper of the life force is changeable, and must be so for the health of man and the organs of which he consists."[24]

Transformation of the life force is what keeps the body alive and allows it to run "its measured arc from the point of becoming to the point of dying, along the wheel of change."²⁵ As the principle of life, the alternations of the life force are found on many different levels. For example, changes may be either transitory or more sustained; they may "occur rapidly and frequently" or "slowly and rarely," "either in all the body's organs at the same time, or locally in this or that individual organ."²⁶ Through "frequent repetitions of several actions at the same time or in a particular sequence," the life forces are set in relationship to one another or, as Reil puts it, "associated" with each other and mutually "tempered" or attuned.²⁷ Finally, "the periods of the life force's changing temperature occur in measured stretches of time or they do not; they are fixed or mutable, short or long. Some periods begin at particular times and last for the course of several years of life."²⁸ The "tempering" of the life force means that its transformations follow "certain rules": "These rules are determined by the causes that change the life force. According to these rules, the changes of the life force have a certain relationship to time and occur in particular periods, the intervals between them having a more or less measured duration. We observe this periodic alternation in all of nature."²⁹

Reil gives examples of such alternation: "every plant blossoms, the fruits ripen, animals mate" and "carry and cast their young at particular times."³⁰ What he describes as temper, periodicity, and measured intervals of time, as happening "at the same time" or "in a particular sequence," following "certain rules" and in a particular temporal relationship, is the rhythm of the life force's alternations — which occurs in the organism at different locations and different times, but always according to a rule. Although Reil does not use the term, there can be no doubt that he is thinking of rhythm when he describes the regularity of such organization.

This becomes obvious when he goes on to explain in more detail "the rules according to which the periods of change in the temper of the life force occur."³¹ First, the life force is subject to the different periods of the human being's life, from childhood to old age. Additionally, it alternates according to the times of day and the seasons of the year.³² Then there are "a) the annual changes in irritability, which depend especially on the quarters of the year, namely, at the equinoxes and the solstices," "b) monthly change in the temperature of the life force," and "c) the daily change in the temperature of the life force," which itself is not the same every day but varies rhythmically.³³ Finally, "inner irritations present in the body" and "habits and associations" also create their own rhythm.[34] The body is suffused by the rhythmic reign of nature: Reil's examples of the economy of the sick and healthy body range from teething to gonorrhea, from consumption to pregnancy, epilepsy to constipation, fever to body weight, hypochondria, and psychic suffering. As a doctor, Reil also knows that every disease has its rhythm. This makes it nonsensical, he argues, to administer medicines at the same dose every hour. Instead, "the periodic changes of the temper" must determine the timing, and "the degree of its changes" the dose itself. "We would do infinitely more good if we were able to bring the timing and dosage of medicines into a harmonious relationship with these changes of the body."[35]

The rhythm in which the transformations of the life force reorganize a living being from the inside, over and over again, is the rhythm that characterizes nature as a whole. Yet the individual organism is not completely absorbed into the great rhythm of nature. As we have seen, Reil emphasizes the autonomy of every organism as a coherent unit that produces its formations out of itself; accordingly, the periodic changes of the life force cannot "be defined solely by temporal change." Instead, the individual organism attunes the life force's alternations with its own rhythm, modifying the force's effects upon it. This process occurs not only in the organism as a whole, but also in each independently operating organ, so that "one external cause may produce a change of temper in one of the body's organs, another in a different one."[36]

Up to this point in the essay, Reil has traced the rhythmic organization of life in concrete terms, through the various rhythms of the phases of life, the seasons, or day and night. But he now also tries to capture in more general terms the "certain rule" constituted by the rhythm of the life force. It is here that Reil's social metaphors become particularly interesting. Rhythm, he writes, is the order that produces the republic of the body — a republic whose individual members are separate and independent, but that, in its unity, creates a functioning structure out of all the parts. This recalls Plato's notion of rhythm as the ordering principle that organizes the community of the polis. The order of rhythm weaves a complex web of relationships between the parts and the whole. Looking at the body through Reil's eyes, we see the rhythm of life not only where it becomes immediately obvious — breathing, the pulsing of the blood — "but everywhere in the body and in every fiber of the same."[37] Thus, for the whole body, in all its components and functions,

> in a healthy state, a change of temper must occur regularly according to all its relations, namely: a) Considering time; it must happen at the right time, not too soon, not too late. b) Considering frequency, not too often or too seldom. c) Considering intensity, not be too weak or too strong. d) Considering the right parts, occur either in all of them or in a single and particular part . . . . e) Considering the irritation, finally, it must be stimulated by the usual irritations. When these laws, which must guide the varying temper of irritability in the healthy state, are overthrown, and those changes occur according to different rules, then a man is sick.[38]

Time, frequency, intensity, part, stimulus — Reil's description of the life force's rhythm is a musical one. The tune of life must start on the right beat; it must have a regular keynote, follow a meter, and exploit the full repertoire of tones. If that is the case, argues Reil, then the organs of the body unite "not by hazard, but according to necessary laws."[39]

Physiological Times

Around 1800, perceptions of the organism shifted from the topography of the body, as described by anatomy, to the circulation of the blood, the secretions of the glands, and the organism's nutrition and growth, emergence, and development. When the physiologist contemplated the body, it was set in motion — sent out of timelessness and into temporal flux.[40] Physiologists now asked how the body's structures and forms come into being, develop, and change; how substances move within the body, react to one another, and are transformed into one other.

Physiology's point of departure was the nature of organic matter, which, at the end of the eighteenth century, was considered to be an essentially unstructured and fluid substance. In his *Theoria generationis* of 1759, Caspar Friedrich Wolff had characterized vegetable and animal matter as a viscous liquid, a mass composed "of rather incohesive little globules [*Kügelchen*] that are simply heaped together . . . transparent, mobile, and almost liquid."[41] Karl Friedrich Kielmeyer, too, in "Ideen zu einer allgemeinen Geschichte und Theorie der Entwicklungserscheinungen der Organisationen" (Ideas for a general history and theory of the developmental phenomena of organizations) of 1793–1794, described the material, "to the extent that it can really be observed," as being "liquid." It is "unformed in itself and mobile in itself; as a whole it is, precisely due to this fluidity, receptive to all forms and all movement and all life in as much as life consists in movement."[42] The treatises of the period abound in descriptions of organic tissue as mucus (*Schleim, Grundschleim*) or a gelatinous substance.[43] Kielmeyer, for example, also notes that organic matter is initially "soft and gelatinous."[44] The Tübingen professor Johann Heinrich Ferdinand von Autenrieth wrote of a "neutral formless liquid";[45] Xavier Bichat, in *General Anatomy* (1801), of a "mucous substance" and a "cellular, or mucous tissue, as Bordeu calls it."[46] In 1808, Friedrich Tiedemann called the organic mass a jelly or gelatin (*Gallerte*),[47] while Karl Friedrich Burdach observed a "grayish, translucent primal mass."[48] According to Karl Asmund Rudolphi's *Elements of Physiology* of 1821, the organic substance is "a soft and of itself shapeless substance, but which is capable of adapting itself to any form whatever."[49] Rudolphi also spoke of "cellular or mucous substance, cellular or mucous tissue (*tela cellulosa, mucosa, contextus cellulosus*)."[50]

If this was the nature of living material, the crucial question was how the organism could take shape, organize itself, preserve and renew itself, even reproduce itself, out of something so amorphous and liquid. The issue of organization lay at the heart of the new science of organic life, concerning the structure of matter not only at the beginning of a life, but for the whole of its existence. The totality of physiological processes was seen as subject to the law of constant transformation — the juices that comprise and keep alive the body, for example, are "consumed and replenished at every moment in time."[51] "The organic body," commented Autenrieth, thus exists "solely in this transitional state of neutral and formless fluidity."[52] Far from being a once-formed and then unchangeable body, the organism is a transitional phenomenon at every point: it always exists "only in the crossing from formless liquid to solid crystallization."[53]

At the turn of the century, there was no doubt that life is characterized by its constant reorganization, its continual transformation. As a contributor to *Archiv für die Physiologie* put it in 1809: "The process of formation continues constantly throughout the whole course of life and accompanies every act of life, whether of formation or of motion.... But although with every new dissolution the object

is shaped anew according to its fundamental type, it is never perfectly reproduced in its previous shape. For then the shape would
become fixed, and its gradual progression would not be possible."[54]
For Andreas Sniadecki in *Theorie der organischen Wesen* (Theory of
organic beings), published in Polish in 1804, the constant transformation of nature "from one part into another" means that "life in all of
vitalizable matter is a constant transformation of forms; in each given
form, it is a constant transformation of matter."[55]

To understand this ceaseless chain of reorganization is the task
of a new science of life: "Thus all animate beings are, while they live,
perpetually organized: or, which is the same thing, all of life is a constant and uninterrupted organic process, or a perpetual assimilation.
The most important truth at which the science of life has been able
to arrive, and which will be the whole foundation of our present-day
science."[56] A science of life, Sniadecki argues, must seek the regularity or lawfulness underlying this world of constant change; indeed,
reason itself requires as much, according to Reinhold Treviranus in
his *Biologie oder Philosophie der lebenden Natur* (Biology or philosophy
of animate nature) of 1802.[57]

But what might that law be — a rule that would make it possible
to think of the organism as organized and simultaneously in constant
dissolution, always different and yet recognizably the same, old and
new at once? In 1835, Karl Friedrich Burdach wrote in *Physiologie als
Erfahrungswissenschaft* (Physiology as an experiential science):

> All organic beings differ from inorganic things in their constant advance, that
> is, in a particular progression of their existence, in a regular transformation
> grounded in themselves and in a particular goal of their existence that is inde-
> pendent of external circumstances. They are thus characterized by an inner
> type of alternation [*Wechsel*] that, although it can be altered by outer things, is
> not imposed by them, but rather resists their changes up to a certain point.[58]

In this sense, "organic" means continually changing according to a
rule and measure, independently of external conditions and driven
by the body's internal circumstances alone. In his description of the
lawfulness of organic formation, Treviranus took a further step by
defining the law of transformation as a rule of repetition and variation. He posited a "spiral" of formation:

> This [regularity] can therefore only be relative. The series of changes under-
> gone by every material system must be constituted such that, after certain

revolutions, the system once again approaches some state in which it has already found itself in the past, yet without perfectly coinciding with that state, or it must be possible to portray the rule through the image of a spiral line in which a moving body approaches a given point again and again, only to recede ever further from it.[59]

Kielmeyer also refers to plants and animals moving in "parabolas": "the point that they have once left, they do not touch again, whereas those recurrent changes are symbolized by the movements of enclosed curves."[60] Later in the same treatise, he describes development as "regular changes"[61] whose governing law must be discovered. He finds that "law of changes"[62] in the "constancy of temporal relationships" between the changes in the organism: "The changes or the identifiable classes of changes, particularly if they are compared in respect to their temporal relationships [*Zeitverhältnisse*] with each other and with exterior circumstances, now allow us, regardless of their different natures, to note a sameness or constancy in those temporal relationships—to note, thus, the constant coexistence or succession of those changes that compose one class with those that compose another."[63]

The key term in Kielmeyer's argument is "temporal relationships," since the "constancy" of these relationships is what makes it possible to determine the law of changes.[64] Formation, for Kielmeyer, is the relation of temporality. As a result, he is not interested in a single time that determines the organism, but in a complex configuration of different times. Only if the physiologist takes account of the organic order of time—the various temporal relationships among the transformations of living material—does the organism appear as "living."

Like Reil, Kielmeyer regards the organism as an autonomous structure, in the sense that it is a single whole defined by its own inner temporal relationships, an entity bearing its own temporal order within itself. In *Grundriß der Entwicklungsgeschichte des menschlichen Körpers* (Outline of the developmental history of the human body) of 1819, Samuel Christian Luca described Kielmeyer's notion of temporal relationships as a "multifarious cycle of activities and processes harmoniously united into a whole."[65] The organism must thus be considered "something that never remains identical with itself, that never stands still, that perpetually moves forward, that passes through a particular series of changes in a kind of cycle."[66] Karl Asmund Rudolphi, in 1821, saw the distinction between organic bodies and inert matter in the fact that organic bodies exhibit "a periodicity in their action."[67] And Moritz Ernst Adolph Naumann argued in 1823 that if, despite all this change, organic life nevertheless appears as a unity, that is a property "not of the material," but of the "constant alternation [Wechsel] maintained by the highest law of nature."[68]

In short, the specific characteristic of organic life appeared to be its inherent temporality. The epistemic task facing the new science of life was to fathom the organism's time. However, understanding the time of the organism entailed far more than attending to the various states and changes of life and tracking them consecutively as a sequence or series of transformations; the temporal dimension could not simply be tacked on to the fixed order of the organism. Rather, the structure of the organism had to be thought through the prism of time — that is, in terms of its ceaseless transformation. In physiology, this meant studying not a given form, but the continual changing of shapes, perpetual transformation, the fundamentally formless. By regarding the organism not as a structure established once and for all, but as one in continual flux, the new science of life posed itself a very particular challenge: to discover the principles of transformation, the law of metamorphosis.

The spiral, the cycle, and the period were images of such a law, the rule according to which the organism organizes itself in continual change. But they also described the core elements of rhythm: the ordering of time, the coupling of repetition with variation, and the relationality of the individual components. If nature, as a contributor to Schiller's journal Die Horen put it in 1795, "uses finite means to pursue infinite ends,"[69] or in Goethe's words, "always only plays, and playfully brings forth multifarious life,"[70] then rhythm is the rule of her game. Rhythm is the law by which the finite number of organic forms can perpetuate themselves into ever new and infinite plenitude, by which the organism incessantly repeats and simultaneously transforms itself.

The Formations of Flow

The organism changes perpetually. Cycles, periods, and temporal relationships, day and night, the seasons of the year, and the temper of the vital force are all ways of describing the rhythmicity of the temporal change to which organisms are subject. The precondition for the organism's mutability is its fluid, malleable character — more precisely, organic matter's mobility is what makes its constant transformation possible. Besides the vital forces and the organization of organic matter, then, a further problem was pivotal to biology around 1800: the question of movement. The ability to move is the essential feature of organic matter, and the alternation between formless flow and solidification into form is the basis of its virtually inexhaustible malleability. But what does it mean to say that fluid substance "moves" in the body? And how can form arise from motion?

In microscopical studies published in an 1821 treatise on the circulation of the blood, the physiologist and anatomist Ignaz Döllinger (1770–1841) confirmed "that the so-called animal cell tissue is a mucoid, semifluid substance, which originally has no cavities or cells; it is therefore not actually a tissue; one would do better to call it mucus [*Schleimstoff*]."[71] To avoid confusing this mucus with the secretions of the mucous membranes, Döllinger called the basic material of "all animal parts" the *Thierstoff* or "animal substance (zoogen)."[72] This animal substance or "animal mucus" (*Thierschleim*) is the basic substance of organic life.[73] Thus far, Döllinger's microscopic observations matched his contemporaries' views of the consistency of organic matter, but in later deliberations, Döllinger went further. Based on its material composition, he argued, the "animal substance" is identical with another substance: blood. There is a crucial difference between the two substances, though, and that is their motion: "the animal mucus is nonflowing blood, the blood is flowing animal substance."[74] Within organic matter, only two states can be distinguished: "The whole mass in my animals was divided into two parts; one part flowed; the other lay still among the cheerfully flowing streamlets; this stillness, too, can only depend on the vitality of the animal substance; for when the inclination to flow awakens in it, then it becomes blood."[75]

On the face of it, there is nothing extraordinary in Döllinger's choosing to discuss the issue of motion in a treatise on the blood's circulation. However, he does not think of movement simply as a property of circulation. Döllinger's significance for the history of physiology lies in his identification of motion as the origin of every formative momentum in the organism. "Nobody doubts," writes Döllinger, "that all parts are generated and nourished by the blood";[76]

the blood and the body's other fluids are "the material that serves as the substrate for the processes of life."[77] All physiological processes are processes by which the blood has to change into an immobile animal substance or, conversely, the immobile animal substance has to change back into flowing blood. In another work, Döllinger refers to these processes as "modification" or "metamorphosis": "The blood should be regarded as a metamorphosis of the animal substance, from which it differs in two ways: α) in its more defined rounding, its sharper delineation, let us say: in the individualization of the granules [Körner] that make up the animal substance; β) in the mobility that characterizes the granules once they have become more individual."[78] Metamorphosis means not the transformation of substance or form, but solely the transformation of its *movement*. In this account, all physiological processes are processes of motion, transitions of a substance into a liquid. For Döllinger, metamorphosis is the transition from a rigid order into a temporally variable one.

Although Ignaz Döllinger was one of the early nineteenth century's most important physiologists,[79] historians have almost always mentioned him only as a teacher who introduced his students to microscopical and experimental practices,[80] initiating and supporting their research, for example the famous embryological studies of Pander and von Baer.[81] In fact von Baer himself, one of the main sources of information on Döllinger, significantly shaped this image of him as a teacher.[82] Even today, Döllinger's oeuvre has not been studied from the perspective singled out in 1841 by his first biographer, Phillipp Franz von Walther:

> Nobody before or since has grasped organic phoronomy—living movement, from its faintest, almost imperceptible beginnings in the movement of cell substance, in a constant sequence through the motion of vessels and the motion of the juices up to its perfected, most active, and freest development in the motion of muscles, in contradistinction to the rigid stasis and merely passive mobility of the skeletal system—more brilliantly *in its continuity and with such coherence*.[83]

In the following, I interpret Döllinger's works as an attempt to conceptualize physiological processes as regulated movements—more specifically, as rhythmical movements. In several important texts published between 1814 and 1824,[84] Döllinger sought to describe the organism as a *structure* conditioned by flowing change. He thought

146

of the whole physiological disposition of the human body — blood, metabolism, secretion, regeneration, healing — as a process of movement, or rather as a subtly regulated sequence of movements. The continuous transformation of organic matter proceeds, according to Döllinger, "in rapid haste, but following a strict rule."[85] Revealing physiological change as a structure in flux, an ordering of time, a ceaseless, but differentiated motion, that strict rule is rhythm. Döllinger's own research has not hitherto been discussed in relation to the embryological work of his students. Yet his intense intellectual engagement with ordered motion as the basis of physiology, I claim, laid the epistemic foundations for the pioneering ontogenetic models of his students Pander and von Baer.

Life Is Movement
Döllinger uses the term "metamorphosis" to describe the transition from the static mass of the animal substance to the flowing mass of the blood. The first metamorphosis to occur in the organism is the development of the embryo; this, too, he explains, is nothing other than the transformation of the animal substance into blood. When development begins, the hen's egg contains only animal substance, which "is by its nature present earlier than the blood, since the blood is only one of its metamorphoses. At first, therefore, the body of the animal embryo is also animal substance, and if blood is to emerge in the embryo, it must do so out of the animal substance."[86] It is when the animal substance begins to move, and thus to become blood, that the organic matter changes into a structured mass, so movement is the very first differentiation to occur in the hen's egg.

But what is the nature of that movement? Döllinger prided himself on having been the first to see how the static animal substance is mobilized and starts to stream through the body as blood:

> Near to a flowing stream of blood, a streak of the immobile animal substance begins to move; out of the mucous granules there forms, so to speak, a little mobile column, which at one end almost abuts the stream of blood at a right angle, and at the other end is turned away from it; this streak, or this little pillar, shifts now toward the stream of blood, now away from it, all at regular periods of time; the mucous granules, of which the shifting or oscillating streak consists, line up in an ordered way, and by degrees take on a more defined, less indistinct shape.[87]

In order to understand the movement of blood, Döllinger brings its flow to a stop.[88] Anyway, blood can "not be called a liquid in the strict sense"; it does not flow like water, "but like fine sand in an hourglass, by virtue of the smallness and mobility of its soft granules."[89] Döllinger here focuses not on the continuity of the flow, but on the mobility of its components. In the animal substance — the static, mucous, basic matter of the organism — "roundish, more or less distorted granules are stuck together with a semifluid material."[90] In the blood, in contrast, the granules are "in streaming motion."[91] Whereas the granules or globules in the animal substance still have a definite form, this dissolves as soon as the substance starts to move. For Döllinger, motion means change in the structure and interrelationship of the globules.

In his observations, Döllinger isolated particular globules from the mass and followed their routes through the fluid. One globule, for example, "turns rather often" or flows "at different speeds in different places," or two granules "settle next to each other."[92] Döllinger visualizes the movements of the globules in a sequence of scenes:

> Mostly the globule flows slowly, occasionally it rests; it still runs on vigorously, even if the impetus from the trunk is very weak....I have observed a single granule that remained behind, removed from the others of its kind, and seemed to have become caught at one of its blunt tips, for its other end floated to and fro, and did so precisely to the rhythm of the blood's flow.[93]

In Döllinger's vivid descriptions, the blood becomes a torrent of globules running or being swept along:

> Now, when such a globule has remained lying for a while at the base of the branch, sometimes it is suddenly swept away, but then also separately hurled along, so that it travels parallel to the trunk for approximately 10 times the length of the globule's diameter, but far enough away from the current that between the globules hurling along and the edge of the current there would probably be room for 2–3 of them.[94]

Döllinger also records the movement of the globules in an illustration (figure 6.1), in which F. 3 is glossed as follows: "*ab* ... the trunk, where the blood flows from *a* to *b*, *c* is the branch joining the trunk, *d* is a globule that has been swept away, *e* a swept-away globule that falls back into the stream, *de* is the line described by the globule in its leap."[95] The figures show the globules' direction, route, and rotation,

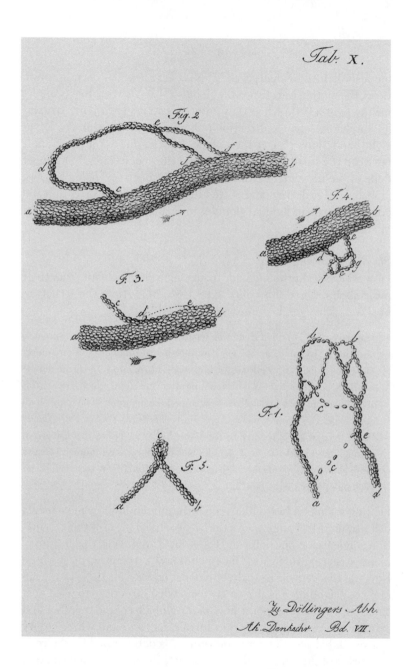

Figure 6.1 Visualization of the blood's movement. Döllinger, "Kreislaufe des Blutes," 1821.

but also their individual shape. This is because under certain conditions, the motion can change the globule's shape, so that the "granules having bent a little when they entered a strongly flowing streamlet and had to change direction quickly," or when, as in the lower right-hand figure (F. 1, *e*), the globule "buckles due to the sharp angle between the small arterial current and the venous one."[96] Döllinger depicts the movement from different perspectives, using arrows, alphabetical ordering, and dotted lines to explain its direction and comparing the individual globules in terms of their velocity, flow, or stasis.

Physiological Motion
However, blood is formed not from single globules, but from their multiplicity. Once this collectivity is set in motion, the structure breaks up, changes, begins to reorder itself:

> I saw two rows of globules, running separately, come together at one point to unite into a streamlet; at once, one row impeded the other in its onward movement, the globules now piled up on both sides into 6, into 8, and formed some small, thick pillars as the globules coming after kept adding themselves to this accumulation; both little pillars now joined the common little trunk at an acute angle; this trunk, which had previously only drawn 1 or 2 globules, finally began to draw 4-5 of them; now there was space; the pillars that had arisen through the accumulation of globules became shorter and thinner, because more globules were subtracted than were added, and finally the simple rows of globules were there again.[97]

Döllinger looks at blood like a general commanding the movements of his soldiers from high above the battle, observing how their formation condenses at one point and thins out at another so that its movements structure the space. Once an unknown streaming mass whose flow was inaccessible to analysis, blood is now as calculable as a "heap of peas."[98] Döllinger is able to say from what direction, to what destination, and at what point how many globules come together and arrange themselves, where they form structures and where empty spaces. In his account, each movement is determined by the configuration of the whole and in turn affects the whole. As a formation of globules, the flow of the blood can now be precisely qualified: its parameters are form, direction, arrangement, speed, coordination, pulse, and rhythm.

At the same time, it is not the case that blood is first formed and only then begins to move. On the contrary, the blood's motion begins "instantly, as it is formed." When the animal substance starts to move, it changes into "oscillating streaks of mucus,"[99] and the characteristic movement of the blood is "oscillating," in other words, inherently structured.[100] As a formation of globules, the oscillating motion itself creates "regular moments of time [*geregelte Zeitmomente*],"[101] and as such, it cannot be attributed to purely mechanical causes, such as the pumping of the heart or the pressure of blood in the vessels: "The oscillating motion passes over into the regular motion of the blood, the circulation, in the measure that the mucus forms into granules of blood. The oscillation, whether original or secondary, is divided into the two momentums of flowing in and flowing out and acquires its unity from the turning point that it sets itself."[102] Closer observation shows that the blood's current is not a linear onward flow, but a constant coming and going of the granules, a thickening and thinning, a converging and drifting apart. This pumping movement is ordered by the tipping point — and the turn from movement to countermovement is "set itself," that is, inherent in the organic matter. The corpuscles "have the grounds of their movement within themselves."[103] Only when the movement tips over into its contrary does motion come about. Without that tipping point, there would only be thickening (or only thinning) of the globules, and the process could not sustain itself. The flow of the blood is thus a rhythmical movement in the sense, first, that it is a single flowing motion composed of discrete states (thickening and thinning, flowing together and apart), and, second, that the movement has its very own metronome — it can, so to speak, conduct itself and define its own specific rhythm.

The oscillating movement of the globules enables the various vessels to take shape. As soon as the first streams of blood are present, more can form from them through ramification. In this process, individual globules move "sideways into the animal mucus, and . . . gradually clear themselves a path"; the first globule is followed in this channel by a second, then more and more globules, "until finally a coherent row of onward-moving granules forms."[104] Arteries and veins form different topographies, depending on the direction of the flow: "The arterial streamlets are simpler, smaller, more frugal, more sharply delimited; in their distribution, a tree form prevails: the venous streamlets are more composite, broader, frequently merging into each other, without regularity, with a distribution more netlike than treelike."[105] To illustrate the formation of a venous vessel, Döllinger gives the example of a globule that "departed from the trunk of a vein, made its way in the opposite direction, downward from the heart, then, describing an arc, turned back into the forward-flowing direction and incorporated itself back into the trunk."[106] This convoluted journey gave rise to the netlike structure characteristic of the veins.

Döllinger noted that the streams formed by the moving blood initially lack any walls or demarcations, basing this comment on his embryological observations.[107] It would be wrong, he wrote, to imagine that as soon as the blood leaves its channels it pours out into its surroundings, "forgetting all order and discipline."[108] The blood does not require an externally imposed order, because its own rhythmic motion has made it an internally structured phenomenon: the interplay of movement and countermovement, flux and coagulation creates form and order through its own rhythm. Thus, the animal mucus — congealed motion — checks the individual streamlets, "so that by the mucus itself they acquire a wall, just as the river acquires a bed from the soil and does not need to be enclosed in pipes in order to flow onward in a regulated way."[109]

The starting point of Döllinger's physiology was the distinction between animal substance and blood. However, his focus was not the quality of the organic substances, but the quality of their movements. "The essential aspect of blood is its motion,"[110] since only that enables the blood to access all parts of the body. Having arrived, it loses motion and becomes animal substance again, and as a substance at rest, it forms the structures and organs of the body. In these processes of motion, the substances themselves were of little interest to Döllinger; static animal substance was always only "the indifferent third party, either what is generated or what can be used to develop the generating forces."[111] This implies nothing less than that movement is what creates forms from formlessness. Only in motion does substance become capable of shaping, generating, creating: matter is "never more than life makes out of it."[112]

"The Grounding of Life in Itself"

Life, then, is motion. Döllinger derived the development of the embryo and the circulation of the blood from elementary processes of movement. In his vision of physiology, indeed, *all* physiological processes are founded on configurations of motion:

> A complete anatomy, sufficient on all sides . . . must solve the connection of all the organs, it must tell us how the parts grow, how they change in form and tissue as they grow, change their position, how what was separate is combined, how what was simple gradually diverges into variety, it must exactly describe the edifice in the state of completed growth, in the perfect fulfillment of the concept of the individual life.[113]

Döllinger describes the transformation of animal substance into blood as modification or metamorphosis. This transformation is in fact a setting into motion and becoming liquid or, conversely, a renewed standstill and solidification of organic matter. In his important treatise *Was ist Absonderung und wie geschieht sie?* (What is secretion and how does it occur?), Döllinger defines all the body's structures as "merely modifications of the animal substance," which, however, "are distinguished by their appearance and chemical properties; they include the muscle fibers, the bone substance, membranous formations, etc."[114]

If all physiological processes — generation, nutrition, secretion, circulation, and so on — were processes of motion, or more precisely the continuous alternation of movement and rest, then physiology as a science of life needed to consider all organic processes in their interrelatedness. It was this that distinguished physiology from other sciences, such as physics. For physicists, it was enough to bring together individual observations "according to the simple causal series that alone is valid for the intellect," but organic life could not be grasped through simple causalities of this kind. Döllinger stressed "how dangerous it is to tear apart the phenomena of life and try to study each one separately, as if it existed outside its connection with the whole."[115] The "incomprehensibility of life" was to be found, instead, in its "inner purposiveness," which should be thought of as "resting on its inner, internally coherent interaction [*Wechselwirkung*]."[116] For this reason, the knowledge to which the physiologist should aspire lay "in the unity of the form in which all alternation [*Wechsel*] takes place."[117] In order to attain such knowledge, the comparative anatomist must "assemble facts, examine in what respects they are similar and in what they are dissimilar . . . and investigate how one and the same thing is formed through a series of metamorphoses."[118] The incessant metamorphosis of life leads the organism from the original animal substance via its differentiation through movement back to the formed animal substance — and this in an incessant to and fro:

> The ground and soil in which life is rooted is the simple animal substance; this not so much *is* as *becomes*; life in general is only a becoming, and what is part of life as something that is, does not persist, it comes and passes away, only the form of becoming remains, and this form is what we perceive as constant. All the substances in the living body are always alternating and transforming themselves, because it is part of the essence of the animal substance, of which all discriminable animal parts are only modifications, to alternate and transform itself. The living transformation of matter [*Wechsel der Materie*] is mediated by the circulation of the blood. Life mediates itself, that is, one attribute of life mediates the other; the blood is animal substance in motion, but without the transformation of matter there would be no motion, and without motion no transformation of matter.[119]

To understand life, physiology would have to take the "form of becoming" as its object. The physiologist could not, like the physicist, assess the quality of substances and accumulate individual observations to form the basis for his knowledge. Quite the contrary: physiology's knowledge was to be found in motion — in what happens between two states, in the transition from one form to another, and thus in what is constituted only in reciprocity, in an inner structure or interplay, and is lost once torn from that context. Motion alone creates forms from formlessness, life from matter. But motion, as we have seen, is not mere flow; it signifies transformation.[120] Life can "mediate itself"[121] only when its movements are regulated in a constant oscillation. The physiologist envisioned by Döllinger would therefore have to examine not so much the states between which life moves, but the rule of the transitions between those states. Döllinger described this rule as alternation, oscillation, an inner interplay, a formation of globules according to regulated temporalities. Organic movement, in other words, was rhythmically structured, and it was this rhythmic motion alone in which the "grounding of life in itself" became manifest.[122] If physiology managed to "verify this nature of the life process," then it would receive the "stamp of a science," and then, too, it would have understood "the accidental as necessary."[123]

PART THREE

Serial Iconography

CHAPTER SEVEN

The Iconography of Motion

So far, I have shown how, around 1800, rhythm emerged as a new episteme for the explanation of physiological or organic developmental processes. However, the trope of rhythm was present not only in aesthetic and biological theories, but also in visual culture. This chapter looks at the history of an iconographic form that reached back into the sixteenth century: the rhythmic pictorial series. By this I mean a series of images that evokes gradual change picture by picture. Among the earliest serial visualizations of this kind is the representation of human motion in pictorial instructions; that is, graphic representations designed to explain, by visual means, how to use an unfamiliar object, how to handle something in a particular way, or how to carry out a certain movement of the body. A succession of physical movements and gestures is broken up into a sequence of images — from one picture to the next and across the whole series — that is intended to familiarize viewers with the movement and teach them to perform it themselves.[1] To an extent, this anticipates the technique of moving pictures that was later deployed by cinematic technology.[2]

In the following, I trace the history of this iconographic convention, showing that it was primarily in the domain of visual instructions for physical activities that the rhythmic pictorial series became canonized in the seventeenth and eighteenth centuries. At the same time, this study of the iconography of series lays the groundwork for the next chapter's return to the story of early embryology. Around 1800, I argue, instructional graphics on the arts of movement supplied the basic iconographical framework for one of the most important forms of visualization in the new field of biology: the representation of embryonic development as a picture series.

The Beginnings of Instructional Graphics

Representations of movement are documented from the Paleolithic onward, depictions of moving bodies on ancient Egyptian and Greek vases and reliefs being particularly well known. In the European Middle Ages, the body was not so much studied in its movements as measured in its proportions; the quest was for the exact norms of its portrayal. However, in the late fifteenth century, attention returned to the perception and depiction of the moving body in art and nature — whether in Leonardo's anatomical studies, Michelangelo's sculptures, or Botticelli's figure of the nymph.[3] The pictorial representation of the human body in motion now became coterminous with the representation of life itself, with Leonardo da Vinci (1452–1519), for example, referring to motion as "the cause of all life."[4] Motion and emotion were synthesized here. Together, movement as an inner feeling and its external expressions communicated the vitality of a form. Leonardo distinguished three kinds of motion: "motion of place," which happens when "the animal moves from one location to another" and includes "rising, falling, and going on the level"; "motion by action," the motion that "the animal makes in the same place without change of location" and is "infinite, together with all the infinite activities in which man often indulges"; and finally, compound motions "composed of motion by action and motion of place." These compound motions are infinite, as well, "because amongst them are dancing, fencing, playing, sowing, ploughing, rowing."[5]

Among the earliest texts that use pictorial resources to teach movement are fencing manuals. In the Middle Ages and Renaissance, these books did not teach only fencing in the narrow sense, but were comprehensive explanations of many different forms of combat — armed or unarmed, with or without armor, on foot or on horseback, against one or several opponents, or using several weapons.[6] The first documented cases are manuscripts dating back to the fifteenth century. These early texts include the 1409 *Flos duellatorum in armis, sine armis equester e pedester* (The flower of battle with harness or without, on horseback or on foot), by the Italian fencing master Fiore dei Liberi, the *Fechtbuch* (Fencing book) of Hans Talhoffer dated 1443, and Albrecht Dürer's *Fechtbuch* of 1512 (figure 7.1).[7]

With the rise of book printing in the sixteenth century, this literature began to grow. Andre Pauernfeindt's *Ergrundung Ritterlicher Kunst der Fechterey* (Explanation of the knightly art of fencing),

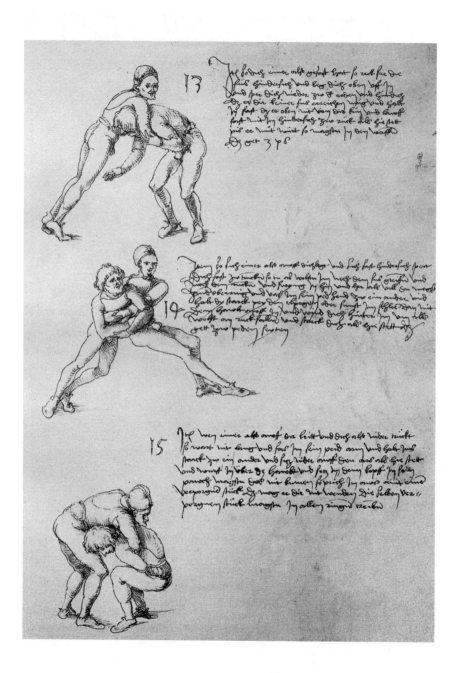

Figure 7.1 Drawings from Dürer's *Fechtbuch*, 1512.

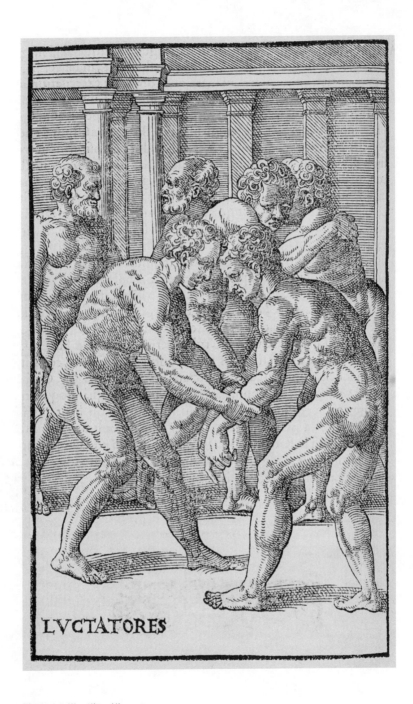

LVCTATORES

Figure 7.2 Wrestling. Hieronymus
Mercurialis, *De arte gymnastica*, 1569.

published in Vienna in 1516, is considered the first printed fencing manual to include illustrations.[8] Illustrations on physical training, gymnastic exercises, and combat techniques such as wrestling can also be found in Hieronymus Mercurialis's *De arte gymnastica* of 1569, the first postmedieval book to address not only physical exercise, but also matters of hygiene, diet, and the healing of illnesses (figure 7.2).

Until the end of the sixteenth century, however, such treatises and manuals on martial arts remained very rare.[9] That changed at the beginning of the seventeenth century. The period around 1600 saw the emergence of a field crucially concerned with the teaching of physical movement: military drill. The visual representation of forms of bodily movement now found a new area of application and a mode of its own, serial instructional graphics. No other area contributed to this genre more importantly than the military *exercitium*. In the early seventeenth century, it prompted an enormous proliferation of military manuals and a new mass production of images.

Military Drill
Starting in the sixteenth century, completely new challenges for military training arose from some radical changes in the practices of war, among the most important being the introduction and growing prevalence of firearms.[10] The new forms of combat made possible by firepower first transformed fortress construction, then, at the turn of the seventeenth century, led to a fundamental restructuring of armies.[11] They also demanded a new form of instruction for soldiers, because "skill in war was no longer the natural birthright of every gentleman," but "had to be learnt by study and experience."[12] The beginning of modern military practice was marked by the Dutch army reforms of 1590–1620 under Maurice of Nassau, Prince of Orange (1567–1625).[13] These reforms were modeled on ancient warfare, with classical Greek and Roman virtues invoked to counter the lack of military efficiency among the free mercenaries and their egregious moral decay. The essential elements of the new order — drilling, the language of commands, the structure of the army administration, and the forms of weapon use — all derived directly from classical sources. Discipline became the most important unifying principle that would enable the new army to function. Scholar Justus Lipsius, who helped pave the way for the military reforms, analyzed discipline and expanded its meaning to cover practice, order, self-discipline,

and exemplary conduct. These components were intended not only to underpin the army's reform, but also to ensure the moral rejuvenation of the soldiers.[14] What was new in terms of military practice was the systematic training of troops, which was built around the drill.[15]

Drilling molds soldiers to a norm. Accordingly, historians have often studied military drill as a technology of power, associating it with control, subjugation, and manipulation and their far-reaching consequences. Thus, for Foucault, "through drilling the soldier becomes something that can be made; out of a formless clay, an inapt body, the machine required can be constructed; posture is gradually corrected; a calculated constraint runs slowly through each part of the body, mastering it, making it pliable, ready at all times, turning silently into the automatism of habit."[16]

However, the literature of the military drill is no less intriguing from a different perspective. Besides power and control over the soldier's body, the literature highlights the necessity of standardizing the form of instructions, the representation of the movements required in order to control that body. Manuals on the art of war were produced in order to teach soldiers the motions and perfect their performance. The movements to be learned depended on the construction of the weapons deployed at any one time — the early seventeenth-century infantry, for example, used the pike, the musket as a heavy weapon and the harquebus as a light one, and a combination of sword and round shield or "buckler."[17] A glance at these weapons shows that the movements to be learned by a soldier were anything but self-explanatory. A seventeenth-century Dutch musket weighed around fourteen pounds and had a range of just 200 to 250 yards; no fewer than forty-three movements, taking up two to three minutes, were necessary to load it for a single shot.[18] Not surprisingly, the manuals of the period are full of minutely detailed descriptions of procedures such as lighting the fuse for firing. The following account is from Johann Georg Wallhausen's *Kriegskunst* (Art of war), 1615:

> When this has been done, /bring your right hand up to the left in a bow shape, /and take one end of the burning fuse, /that is, stretch out your two foremost fingers, /place them beside the next finger to mix the powder inside your hand, /that is, so that the outermost flat part of your hand is turned outward /

and the inner part is turned toward your body, /place the two fingers a finger's width apart, /grasp the fuse and put your thumb on the fuse, /as if you wanted to push the through your two fingers with your thumb, /hold it well and bring it up to your mouth, /bend your head a little toward your hand, /as if you wanted to come to the aid of your hand.[19]

With an abundance of detail that was in principle infinite, descriptions of this kind attempted to capture a movement that happened too quickly to be perceived exactly. But such instructions were impractical for military purposes — partly because of their length and partly because many soldiers were illiterate. The question, therefore, was how best to represent a movement so that it could be easily grasped and memorized, the prerequisite for soldiers being able to practice it at all.

Pose and Series
In the fencing manuals of the sixteenth century, the movements to be learned were portrayed in a characteristic pattern, with one page of a manual, or several pages in a row, illustrating one or more stances from fencing, wrestling, or combat. A brief text appended to the illustration served to explain and comment on the position portrayed (see figure 7.1). As a rule, the figures of the combatants appeared against an empty background, with no reference to the location or circumstances of the encounter. The perspective, format, and arrangement of the visual elements remained the same across the pictures of the various poses. This focused the gaze on the individual position that the viewer was to understand and reproduce. The explanation added to the image, furthermore, offering a deeper comprehension of the pose by allowing the user to compare what he saw with what he read.

In the early seventeenth century, the Dutch engraver Jacques (Jacob) de Gheyn (1565–1629) elaborated this form of representation into the iconography that would long remain the prevailing visual convention for teaching military movements.[20] The plates he engraved in 1607 for his *Wapenhandelinghe van Roers Musquetten ende Spiessen* (Handling firelocks, muskets, and pikes) aimed to help "instruct the untrained souldiers and to reinforce the minde of the expert by the sight and reading of it."[21] To teach the required move by pictorial means, de Gheyn sliced up the flow of movement into a carefully determined sequence of stationary positions. This meant dividing the action into its key elements: the components absolutely necessary for the movement had to be distinguished from less important ones, a selection made. As de Gheyn noted: "No man shall finde it strange that wee in drawing of the Pikes, only set that which for the use of the same is most necessarye, omitting diverse maners of tossing of the pike by forme of recreation, which in militarie exercise bringeth little benefite or profite."[22] Ultimately, while the portrayal of the figures had to incorporate the greatest possible amount of information, that information also had to be reduced as far as possible to avoid distracting from the movement, and it had to be communicated using simple visual resources.

De Gheyn produced 117 engravings to visualize drilling with weapons. Pierre du Moulin, a member of the governor of The Hague's bodyguard, posed for him — complete with weapons and costume — in the individual stances for each movement (figure 7.3).[23] Each pose, or elemental movement, was depicted on a copperplate, and the associated command was added. The movements portrayed, writes de Gheyn, were those that Count John himself "doth observe as the perfectest and best patterne . . . . Having to that purpose drawn all the postures that come in the holding of using of the armes by order and the same described with his reasons and wordes of command."[24] John II (1561–1623), Count of Nassau-Siegen, commissioned the book from de Gheyn sometime around the turn of the century. De Gheyn's visual approach draws on John's manuscript for the exercise of arms, in which, most likely for the first time, he combined words of command with illustrations.[25] In the manuscript, a drawing probably made by John himself shows the loading and firing of the caliver in twenty figures, labeled with short commands and arranged in a circle around the commanding officer (figure 7.4).

Segmenting the flow of a movement into different components, each with its associated command, made possible a new, controlled mobility and a myriad of potential variations — movements could be rearranged into different exercises and rehearsed in isolation.

De Gheyn engraved each pose individually onto its own copperplate. Only when several such positions were assembled did a motion sequence emerge. For example, the four illustrations in figure 7.5 show how to light the match; more precisely, they show four steps that have to take place between igniting the fuse and igniting the powder.[26]

Figure 7.3 The pikeman. De Gheyn,
*Wapenhandelinghe van Roers
Musquetten ende Spiessen,* 1607.

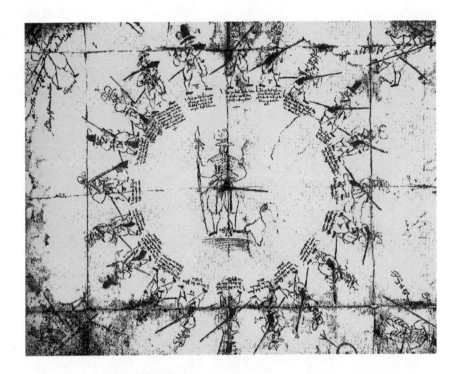

Figure 7.4 The manuscript of Count John
of Nassau-Siegen, reprinted in de Gheyn,
The Exercise of Armes, 1999.

Figure 7.5 Lighting the match in four steps. De Gheyn, *Wapenhandelinghe van Roers Musquetten ende Spiessen*, 1607.

Each engraving is a variation on this pattern. In his 117 plates, de Gheyn modifies one and the same basic figure in the full spectrum of different poses. All the plates have the same format, showing the same frame from the same perspective; the background is reduced to a simple floor line, the soldier placed centrally upon it. Alongside these constant elements are the variable ones, such as the positions of the limbs. The engravings thus represent the changing poses in the form of repetition with variation.

De Gheyn's visual model laid the foundations for all the instructional graphics of the period that followed. When his arms manual appeared in 1607, it carried a special privilege designed to protect it against plagiarism, but to no avail. By the time the original copper engravings were reissued in 1613 and 1640, other drill books were already appearing that made their own direct or indirect use of de Gheyn's illustrations.[27]

Whereas de Gheyn portrayed each posture separately, subsequent publications intensified the succession of poses into series. Military treatises of the seventeenth to nineteenth centuries typically strung together multiple postures on one page. As the figures, in their various positions, became smaller in scale and were crammed closer together, their individual detail and significance faded from view. What predominated now was the impression of a *sequence* of figures, merging the changing poses into a single fluid movement.

One of the most prolific military writers of the seventeenth century was Johann Jacob von Wallhausen (c. 1580–1627), who served Maurice of Orange-Nassau. It was through Wallhausen's writings that the Nassau military system took hold in Germany.[28] In 1615, he published *Kriegskunst zu Fuß* (The art of war for infantry), one year later, *Ritterkunst* (The art of cavalry), *Kriegskunst zu Pferdt* (The art of war for cavalry), and *Manuale militare oder Kriegsmanual* (Military manual), and finally, in 1621, *Defensio patriae oder Landrettung* (The defense of the land). These were parts of an encyclopedia that was intended to comprise eight volumes, but remained incomplete.[29] In his works, Wallhausen not only made extensive use of the instructional graphics that de Gheyn had introduced, but was also one of the first to adopt a consistent system of linking up figures into long series of positions.

The three copperplates shown in figure 7.6 are from Wallhausen's *Kriegskunst zu Fuß* and show the use of different weapons: Plates B and C show how to handle the musket, plate D the pike. The shape and poses of the individual figures are taken directly from de Gheyn, but de Gheyn furnished his figures with more artistic detail. At times he clothed them differently and changed their armor — "not that we holde it for necessarie, but that such varietye might give the fuller ornament to the pictures, and to showe to posteritie the manner of souldiers apparel used in these days." This included representing "the right maner and fashon of the arminge of his Exces. owne Garde, as it is at this tyme."[30] In later publications, the images' function as a historical source became less important than the impression of a sequence. The figures became smaller, and the series of movements unfolded across elongated horizontal formats. The densely packed rows did not invite the eye to dwell on the individual figure, but hurried it on to grasp the structure of the overall image and thus the movement as a whole. One reason for this visual design was quite pragmatic: the plates in these manuals were items of practical use. The picture series were actual instructions that the soldier could carry with him in the field and during drill and showed quickly, conveniently, and directly what he was required to be able to do. They were also affordable. Wallhausen's *Kriegskunst zu Fuß*, for example, included

> also a printed sheet in the form of a copperplate, which I have added for the assistance of every soldier who does not have the ability or sufficient money to pay for this book, in which he finds, quite perfectly described and instructed, the whole discipline and the science concerning his weapon, alongside all the steps for handling the musket and the pike, as well as the exercise or drill, which is a concentrate and summary of the 1st, 2nd, 3rd, 4th part.[31]

The pictures were intended

> for the use and benefit of all soldiers made in such a form that every soldier can buy it for a low price, and small so that he can carry it with him easily, although it would be necessary for every soldier to have the whole book, even the soldier who cannot write or read can see before him very beautifully all the steps in handling the musket and the pike in the copperplates just like life.[32]

In Switzerland, it was Hans Conrad Lavater's *Kriegs-Büchlein* (War booklet) that disseminated the Dutch army's drill techniques. Until the end of the seventeenth century, Lavater's 1644 instructions formed the basis of the Swiss training regulations. Lavater used the same type of instructional graphics as Wallhausen (figure 7.7).

Figure 7.6 Handling the musket and
the pike. Plates B, C, D in Wallhausen,
Kriegskunst zu Fuß, 1615.

The representation of the positions on a single page in Wallhausen's and Lavater's works took de Gheyn's principles to a new
level: the repetition of the poses (and their variation) was intensified into a series. The movement now unfolded before the viewer's
eyes as a whole; the series visualized it as analysis and synthesis at
once. For this relationship of images, into which the series dissolved
the individual movements, the lacuna—the space "between" the
images—was no less constitutive than the pose portrayed. The series
functioned first and foremost as a sequence of representation and
omission, image and interstice, fullness and emptiness. Only through
serialization was the pose translated into movement, form into formation, the dead body into the living one. In these images, the series
becomes the law of one pose succeeding the next. The law of the
series is the choreography of interplay, the controlled and regulated

Figure 7.7 Introduction to the use of musket
and pike. Lavater, *Kriegs-Büchlein*, 1644.

relationship within which the postures succeed one another. In order
to create picture series of this kind, graphic resources were chosen that almost completely concealed the distinctions between one
image and the next. It was the impression of movement that counted;
all the picture's other elements were pushed into the background.
The transition from one image to the next seemed to disappear, creating the illusion of a single fluid movement.

   An early example of this procedure can be found in the treatises
of Johann Georg Paschen (1628–1678) on drilling, wrestling, fencing, flag-waving, and vaulting, which appeared in numerous editions around the middle of the seventeenth century. In these works,
Paschen's artistic form closely tracks his teaching of the various
movements, with the images arranged like mathematical tables. Each
is annotated only with a brief text, so that the text is an accessory

Figure 7.8 Four successive plates teaching
fencing. Paschen, *Kurtze iedoch deutliche
Beschreibung vom Fechten*, 1664.

to the image, and not the other way around. Paschen's instructional
drawings privilege the principle of seriality over the quality of representation in the individual figure: whereas his figures are portrayed
with extreme simplicity and crudeness, their seriality is elaborated
with the greatest of care.[33] Here it is in the serial form — the image
type of the *series* — and not in the artistic quality of the picture itself
that the movement's aesthetic is to be found. Form is systematically
downgraded in favor of sequence.

Figure 7.8 shows Paschen's illustrations of fencing. Paschen represents all the various kinds of movements according to the same
principle. The plate is divided by horizontal lines into three (for
vaulting) or four compartments (for fencing and wrestling), each
compartment representing one position in the move to be learned.

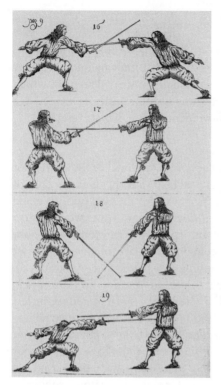

The individual positions are numbered sequentially across all the plates. Each compartment shows the figures against a neutral white background; shading merely hints at the ground on which they stand. The size, frame, perspective, and placement of the figures remain constant. From one plate to the next, there are only minor variations between the particular positions of the arms and legs (or of the body vis-à-vis the horse) that the pictures are to teach. This allows the changing movement to be tracked in detail through the sequence of images. Paschen's graphics emphasize the visual sense of seriality with particular consistency, partly due to the modest quality of his drawing. The multiplication of the individual poses on one plate, the succession of plates over many pages, the repetition of the same schema — the empty background, the focus on the figures, the elimination of anything dispensable to the demonstration of the movement — combine to generate a new visual impression, the impression that the figure is moving. Paschen's manuals are just one example of the rigorously pictorial teaching of movements by means of seriality, albeit a particularly striking one; similar motion series can be found in almost every significant military manual. Military drill books were among the most visually powerful areas in which the new iconography of the series was systematically developed as a new convention for teaching movements through pictures.

The Law of Rhythm

The sequencing of pictures in these instructional graphics did not follow the simple pattern of metrical beat, but the complex laws of rhythm. The movement concerned was not divided up into equal segments based on metrical regularity. Rather, it was woven into a rhythmic pattern as a complex interplay between individual pose and overall movement. This becomes particularly obvious in a series such as that in plate B of Wallhausen's *Kriegskunst zu Fuß* (figure 7.6). In this engraving, the individual figures are numbered 1 through 83 (continuing up to 123 on plate C). The numbers are placed to the left and right of the figure's head. In this sequence, there are considerable variations in the allocation of numbers to the particular figures. While some figures have just one number, others have four, five, or even seven, referring to the number of movements to be carried out in any one case. The series portrayed is thus not a simple sequence where one movement follows the next, separated by a gap. On the contrary, what we see is an intricately woven pattern of motion. For the detail reproduced in figure 7.9, this means that the first figure shows the preparation to fire, gathering seven movements (20–26):

Figure 7.9 Firing the musket. Detail of plate B in Wallhausen, *Kriegskunst zu Fuß*, 1615.

20. Raise your right hand up //bring it to your left shoulder like a bow. 21. Begin to stride with your left foot. 22. Let the stand in your hand sink down to the right along your body. 23. Let the musket slide down a little. 24. Turn it a little toward your right hand. 25. Grasp the musket under the pan. 26. Raise it away from your shoulder.[34]

The fourth figure, in contrast, consists of only two actions: "36. Gracefully take the match from your left hand using two fingers and your thumb. 37. Begin to stride with your right foot." The ninth figure is annotated only as: "38. Blow the match out."[35]

This complex pattern of movements attains its order from the *rhythm* of its execution. Military drilling in the early modern period was a rhythmic art of motion. The beating of the drum, the burst of the fanfare, the sound of the flute ensured the harmonious coordination of the posture's individual components into a total movement (see figure 7.10).

Rhythm was the timekeeper and cue for holding a position and changing to the next; it determined a different number of movements for each particular period or count: "When you have now finished in your posture, that is, with the first three measures [*Zeiten*], you must also make yourself ready in three measures and steps and fire, and so conduct yourself such as to do this gracefully, easily, and nimbly."[36] Instruction in military movement was musical in the sense that it followed a rule: the movements were to be repeated, they recurred, but they also varied within this recurrence according to the order of rhythm. And, importantly, they always had to be performed in unison.

In this context, special importance accrued to the illustrations' seriality, in which figure and lacuna formed a rhythmic order. That is, the gap, the unknown quantity that separated the movements, the barely perceptible flux between positions, became an interval, a regulated and comprehensible quantity. In this way, the series could also capture the practical knowledge that the soldier acquired through active and reenactive practice. Individual motion sequences were combined into different exercises following the rhythm of music. This made it possible to practice particular maneuvers separately, if necessary, or to vary the drill again and again.

This point is illustrated by figure 7.11. The recurring figure of the horseman is arranged across the engraving in lines, creating a

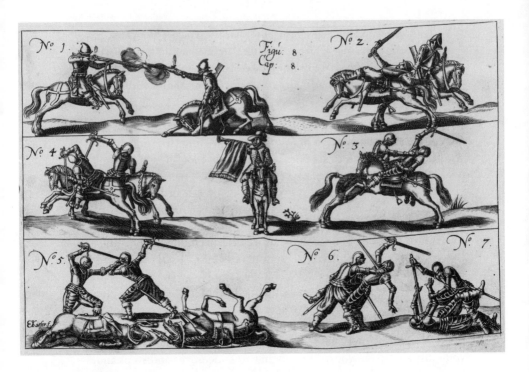

Figure 7.10 Orders are obeyed and the troops' movements coordinated by the rhythm of the music. The first illustration is seventeenth-century, from Wallhausen's *Ritterkunst*, 1616; the second, a century later, from Fleming's *Der vollkommene teutsche Soldat*, 1726.

Die vier Commando der 24. Tambours und 12. Querpfeiffer, wie es in der Königl: Residenzstadt Dresden gebräuchlich.

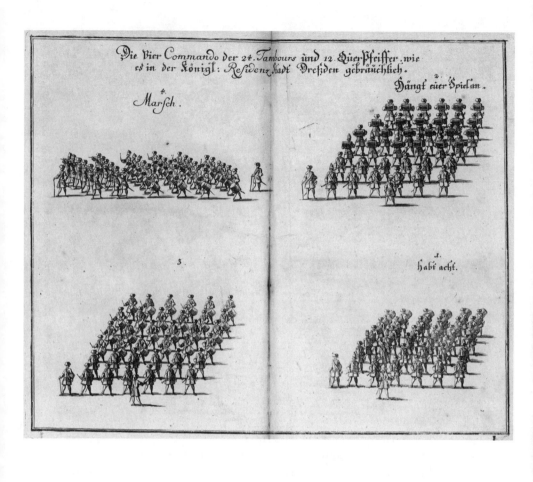

4.
Marsch.

2.
Hängt euer Spiel an.

3.

1.
habt acht.

Figure 7.11 Drill sequence.
Wallhausen, *Ritterkunst*, 1616.

division into three horizontal rows. Following the numbering of the figures, the gaze moves along the first line from left to right, changes direction at the end to travel back from right to left, and finally halts, in the lowest line, at the sight of a battle between two horsemen. The first five figures form a sequence, showing the preparation of the weapon: 1. The rider sits on his stationary horse, his gun at his side, 2. he takes the gun in his right hand, 3. the horse begins to move and the rider raises his gun with the help of his left hand, 4. he takes up his firing position and fires, 5. after firing, he reloads. In contrast, in the second line, each figure is an independent pose, one for each different way of firing: forward (6), to the left (7), straight ahead (8), or backward (9).

Clearly, instructions for movement are not simple linear sequences of motion, but rhythmic movements that are modified repeatedly and in their repetition. In figure 7.12, the rhythmicity of

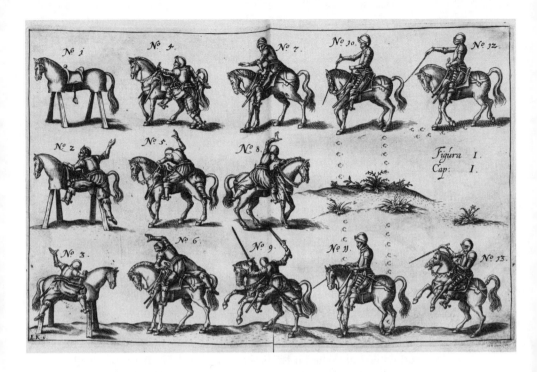

Figure 7.12 Mounting the horse.
Wallhausen, *Ritterkunst*, 1616.

the drill is also visually implemented in the combination of horizontal and vertical lines of vision, so that the gaze glides over the image like a wave — on the one hand, driving the movement onward, on the other, catching at the pose and its variations.

The first figure in this engraving shows a wooden horse, the second mounting from the left, the third mounting from the right. Nos. 4 through 7 show how to mount a real horse in a sequence of four steps:

> No. 4. Here you see firstly how you set your left foot in the left stirrup is your first step. No. 5. Instructs how to swing yourself up from the ground with your right foot. No. 6. Shows halfway into the swing with the right foot in the air. No. 7. How you betake yourself over the horse through the upward swing and put your body into the saddle and your right foot into the right stirrup.[37]

No. 8 repeats no. 5 — mounting the horse — but now from the reverse side, swinging the left leg. In no. 9, mounting is hampered by the facts that the horseman now has a sword in his left hand and a firearm in his right and the horse is no longer stationary, but moving. Nos. 10 to 13, finally, show various postures that the rider can take on the horse with his gun in his right hand. Nos. 10 and 11, arranged at the top and bottom of the image, are connected by hoofprints — they indicate a sideways movement, whereas hoofprints in front of the horse in no. 12 mark a backward movement and those in no. 13 an upward movement of the horse. The figures are evenly distributed across the picture surface; the horse stands on an imaginary ground that is merely suggested by the shadows cast by horse and rider. Three horizontal fields are formed, arranged one above the other. The numbering of the figures traverses this horizontal segmentation, running from top to bottom. As a result, a wave moves through the horizontal arrangement, generated by the sequence of labeling in the individual figures.

As these examples show, the aesthetic of the pictorial series is an aesthetic of rhythmic segmentation. Rhythm is the rule according to which the movements are performed, repeated, and varied, assembled into drills, then disassembled. The series can make a movement reproducible only when it finds the right rhythm, that is, the right rhythmicity of pose and lacuna, of fixed image and imaginary flux.

Eighteenth-Century Drill

By the mid-eighteenth century, instructional graphics had become an integral part of visual culture, especially in the military domain. The wars of the eighteenth century faced great nations such as France, Prussia, and the Habsburg Empire with new challenges. The need for instruction grew, new strategies and tactics were drawn up, and soldiers had to be trained and disciplined in ever greater numbers. Comprehensive "regulations" or "points of observation" explicated the procedures, rules of conduct, and moves that were required for warfare and that must be observed by the soldiers. An example is Ludwig Andreas Khevenhüller's *Observations-Puncten* (Observation points), published in 1749 (figure 7.13).

In 1726, Hanns Friedrich von Fleming, an officer in the service of Elector Frederick Augustus I of Saxony, published *Der vollkommene teutsche Soldat* (The perfect German soldier). Fleming's book, of similar importance to the eighteenth century as Wallhausen's had been to the seventeenth, dedicated eleven large-format plates to drills with various forms of weaponry (figure 7.14).

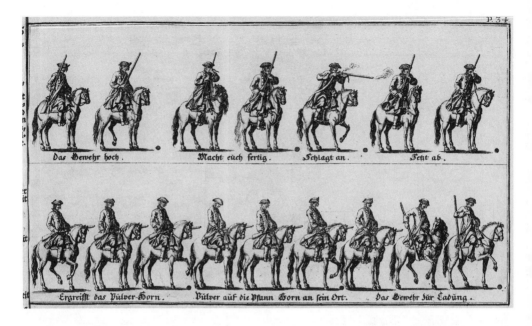

Figure 7.13 From an eighteenth-century regulation book, Ludwig Andrea Khevenhüller's *Observations-Puncten*, 1749.

Figure 7.14 Drills. Fleming, *Der vollkommene teutsche Soldat*, 1726.

The first regulation book to be made generally binding for the troops of the Habsburg Empire was published in 1749, at the decree of Empress Maria Theresa. What was special about this *Regulament und Ordnung des gesammten Kaiserlich-Königlichen Fuß-Volckes* (Regulations and order of the whole of the Imperial-Royal Infantry) was that the entirety of drills for all ranks of the army was presented in pictures, running to more than a hundred copperplates. This probably makes the *Regulament* almost unique. The use of picture instructions in such quantities was dictated by the linguistic patchwork of the multiethnic Habsburg state; in fact, it was only in the mid-eighteenth century that the Habsburg Army had settled on German as the obligatory language of commands.[38] Figure 7.15 shows the first five of the total twenty-five double plates on the weapons drill of the grenadiers and the fusiliers.

The first plate juxtaposes position, command, and rhythm with particular complexity. All the movements are executed to a beat, and the sequence of movements jumps to and fro between the figures so that the drill entwines the individual poses into a complex pattern. The numbering of the moves crosses from the first figure (no. 1) to the second (no. 2), then back to the first (no. 3), down to the lower left-hand figure (no. 4), then back to the upper left (no. 5), then to the upper right (no. 6), and so on. To keep control over the movements, they are timed. Each of the commands listed "occurs in three beats [*Tempo*], and when the last command has been spoken, one begins, without a pause, to carry out the first beat, and counts secretly to oneself, of course without drawing out or stopping, One, Two, Three, Four, with which counting to four the second beat is made, and after an equal pause the third."[39] All the beats must be "fresh, also perfectly wrought, and, as mentioned above, one well set off from the other, so that one can be well distinguished from the next."[40] For the remaining four plates in figure 7.15, the same principle applies as in the first, although the sequence of movements is now less complex — the movements take place one after the other, and for each tempo a single figure is given (thus, in no. 10 the command "Present arms" is carried out in three tempi, which are given one by one).

Instructions and their representations had not always been as complicated as this, but in the eighteenth century, there were important changes in the technology of warfare and the handling of weapons. For example, the advent of the flintlock mechanism, paper cartridges, and iron ramrods in the first decades of the eighteenth

Figure 7.15 *Regulament und Ordnung des gesammten Kaiserlich-Königlichen Fuß-Volckes*, 1749. The first drill book for the whole of the Habsburg Empire.

century reduced the number of moves required to fire a weapon.[41] Again, new strategies and tactics were required, while far larger numbers of soldiers than before had to be trained and disciplined. Especially at the end of the eighteenth century, with the French Revolution, a new era of warfare began that profoundly altered the character of the army.[42] The professional warrior was increasingly replaced by the untrained soldier, so that simpler fighting techniques had to be developed and rapid instruction of the troops ensured. Serial instructional graphics could be adapted both to the new weaponry and to changing military conditions. The example from a French military manual published during the Revolutionary Wars (figure 7.16) shows that although drill instructions were reduced to a minimum, with less detail and scantier illustration than had been the case in the mid-eighteenth century, their serial structure remained untouched. Indeed, the genre of instructional graphics proved to be one of the most stable components in the fast-moving art of warfare. In the mid-nineteenth century, in spite of the transformations in techniques, weaponry, and equipment, the use of the bayonet, for example, was taught using instructional graphics very similar to the ones that had been prevalent since the seventeenth century (figure 7.17).

N.º 10.   pag. 95.

Macht euch Fertig.

N.º 11.   N.º 12.

Schlagt an.   Setzt ab.

N.º 13. 14.   pag. 97.

Feuer.

N.º 15.   pag. 98.

Ergreift die Patron.

N.º 16.   pag. 99.

Lincks schwenkt euch zur Ladung.

N.º 17.   pag. 100.

den Ladstockh in Lauff.

N.º 18.   pag. 101.

Setzt an die Ladung.

N.º 19.   pag. 101.

Versorgt den Ladstock.

Figure 7.16 Late eighteenth-century French instructional graphics, from *Instructions militaires simples et faciles*, 1792.

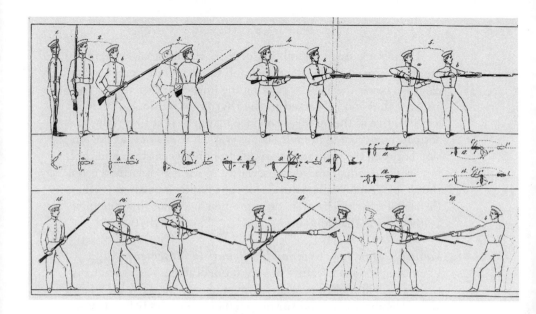

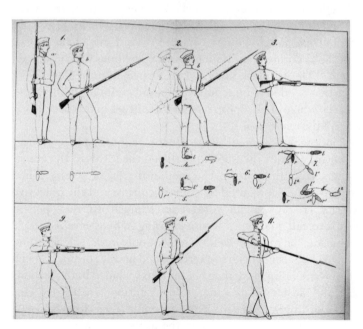

Figure 7.17 Two mid-nineteenth-century examples of instructions for bayonet fencing. The first is from Rhein's *Bajonettfechten*, 1840, the second from Rothstein's *Anleitung zum Bajonettfechten*, 1853.

The picture strips on military drill are rhythmic series: a movement is segmented into depicted and nondepicted elements that can then be synthesized by relating the images to each other. They are repeated and rehearsed to the rhythm of music. Self-evident and intuitive as the picture sequence may appear at first sight, it is actually constructed in multifarious ways. It exists only as the tension between the pose and its dissolution, between stasis and flux, between individual image and seriality. What is represented is not the simple, supposedly natural flow of a bodily movement; rather, through its rhythmicization, the movement becomes an autonomous aesthetic form that contains its own law of motion: rhythm.

Vaulting, Dancing, Gymnastics: Beauty in Movement

Exercising was not a feature of military drill alone. The characteristics of the drill applied to other forms of movement, as well. Diderot and d'Alembert's *Encyclopédie* included in its 1754 entry for "Danse" all the "regulated movements of the body, leaps, and measured steps that are made to the sound of an instrument or voice."[43] In early modern Europe, moving did not mean involuntarily changing the position of one's limbs — movement was synonymous with practiced movement.[44] The use of the body and the form of its movements were practices that needed to be acquired. Not only military drill, therefore, but the whole of European movement culture in this period was a culture of the *exercitium*.

This notion of movement as the ordered, coordinated, and measured control of one's own body paralleled the ideals of the court. Since the sixteenth century, the medieval trial of strength had increasingly been supplanted by courtly culture and the objective of dexterity. "Today's gallant world requires much of a young person whom it is to call qualified," wrote Fleming in his *Der vollkommene teutsche Soldat*: "he must be well versed not only in the sciences, and the fair arts, but also in diverse exercises and bodily routines."[45] The ideal of movement for the *galant homme* thus included both weapons exercises and mastery of dancing, riding, vaulting, fencing, or gymnastics. Accordingly, the *Encyclopédie* entry for "Exercice (manège)" covers "practice on horseback, dance, the action of drawing arms and of vaulting, all military exercises, the skill necessary for tracing and for constructing fortifications, drawing, and in general everything that is taught and everything that should be taught."[46] Movements were thus not simply carried out: they were practiced, repeated, and polished in countless exercises until they fulfilled the aesthetic criteria of the day. During the seventeenth and eighteenth centuries, beauty of form was the yardstick by which the body's movement was judged. The ideal of early modern bodily movement was an *aesthetics* of movement. It turned dancing, shooting, and physical exercise into an art of motion, an art that must be understood, studied, and diligently mastered.[47]

Physical movements, then, were to be not just precise, but also graceful, pleasurable, and beautiful to look at, regardless of whether they were performed in the ballroom or on the battlefield. The ideal of harmony, proportion, and elegance was fulfilled by the decorous, discreetly and pleasingly executed movement. Thus, the military drill — its endless repetition of moves, steps, and stances — was to give the soldier no less formal pleasure than dancing would give the nobleman at court. The angles of arms and legs and between the parts of the body were meticulously noted down in fencing manuals; riding manuals prescribed the patterns of hoof beats in circles, straight lines, or diagonals; and dancing masters drew the paths of the minuet onto the dance floor. It was no coincidence that the description of firing a weapon "gracefully, easily, and nimbly"[48] resembled descriptions of activities such as vaulting (figure 7.18):

> But this is threefold, consisting in rising, leaping, and swinging, and you must take care that everything is done with your feet rigid, and when you raise yourself at the horse that you jump with a fine softness onto your toes and then back upward, standing with a fine gracefulness and this also avoids wrenching the ankle. Your arms and your body must be rigid, and the less you touch the horse, the more gracefully you will leap.[49]

Arts of movement that were so sophisticated, based on rhythm and aesthetics, required instruction. They were taught by masters of exercise, dancing, drill, or fencing. Enjoying high social status and engaged at all the courts of Europe, these were professional instructors of movements that had developed their own traditions, elaborate enough to require specialized knowledge and therefore specific training. Little is known about these masters and their schools, but from the late fifteenth century onward, an increasing number of their treatises and disquisitions survive, texts in which they recorded and passed on their knowledge.[50]

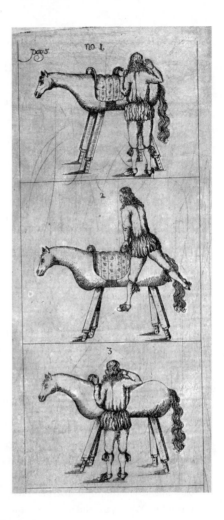

Figure 7.18 The art of movement: vaulting. Paschen, *Kurtze iedoch gründliche Beschreibung des Voltiger*, 1664.

The teachers wishing to transmit their art to their pupils faced a dual task. On the one hand, they had to build a vocabulary to analyze and describe the movements of fencing or wrestling, handling the sword or the lance, the position of the feet, the grip of the hands, attack and defense, backward and forward—in short, the coordination of all the individual parts of the body, along with any weapons or on horseback, on gymnastic equipment, with or without an opponent. Even more importantly, they had to create a visual repertoire by means of which the necessary movements could be reconstructed. The aesthetic of a movement, the complex interplay of the parts of the body it demanded, relied on pictorial representation. The "graceful" carriage

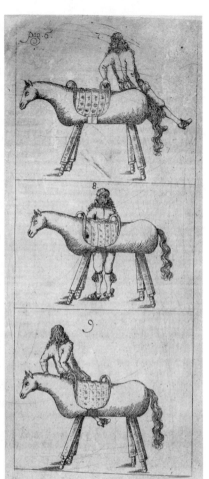

of arms and legs or the "decorous" and "harmonious" coordination of the body from head to foot had to be visualized, not left to verbal description alone—after all, these were performing arts, perfected through a living reconstruction in the grace of the movement itself.

The crux of the aesthetic conception of body movements that dominated movement behavior in the seventeenth and eighteenth centuries was rhythm. Gymnastics, vaulting, and riding, no less than the firing of guns, were rhythmic arts of movement: their motions followed the rules of rhythmic execution and were nothing other than the choreography of learned, controlled, minutely defined positions of the body.

A particularly interesting example here is the representation of fencing in Diderot and d'Alembert's *Encyclopédie*.[51] The plates and explanations on fencing, the encyclopedists note, are "drawn entirely from a treatise by M. Angelo published recently in London."[52] This was the *École des armes* by Domenico Angelo, which had appeared in 1763, just two years before the *Encyclopédie*, with carefully colored copper engravings (figure 7.19).

Angelo's illustrations stood in the tradition of fencing manuals since the sixteenth and seventeenth century: they divided the movements of fencing into a series of specified positions, of attack and defense, *garde* and *parade*. In his original edition, the plates are dispersed through the text in landscape-format folios, accompanying the passages that describe the position concerned. The *Encyclopédie*, too, begins with the analysis of movements as poses: "The fencing master begins by 'breaking in' the body, habituating the student to the different positions that he must adopt in order to make the joints move more easily and to give suppleness to his motions; next the master teaches the student to execute the movements of the arm and above all else the hand, which deliver an attack to the enemy or attempt to deflect his attacks."[53]

Above and beyond this, however, the *Encyclopédie* explains fencing as a rhythmic art of movement. The legends to the plates make it clear that attack and parry are metrically ordered, the movements of the two fencers relating to each other in a rhythmic counterplay:

> Disengagement and the attack of the foot is done in a single moment; and in the tempo [*tems*] when one disengages, one must reach one's adversary with the sword, attacking with the foot, and drawing straight and close to the body. It will be observed that this operation, although it consists of two tempi [*tems*], of which the first is the disengagement with the attack of the foot and the second is to draw, must be executed as quickly as if one said to oneself "one, two."[54]

The crucial term here is *tems*. It describes the rhythm of movement and countermovement within which the adversaries interrelate: "'Tempo' means the favorable moments that one must choose to lunge at the enemy. They are infinitely varied, and it is impossible to say anything particular about them." The rhythm of these movements cannot be analyzed; it exceeds the mechanical quantification of meter. Instead, it is "the fruit of long labors and the necessary science of arms."[55]

The feel for *tems*, or rhythm, can be practiced, sensed, acquired only by physically recapitulating the movement. Yet if the rhythm of movements cannot be treated analytically, it can be visualized. The *Encyclopédie* lines up all Angelo's illustrations in a row. Three of his plates are gathered on a page, with a total of forty-seven plates on seven successive double pages. The image plus lacuna of the series makes fencing a rhythmical order of movement; the serialization creates the impression of an ongoing fencing bout (figure 7.20).

In early modern movement culture, then, dance was not exceptional in its ordering of movement according to the rhythm of music, but rather the most important exemplar: if rhythm, agreeableness, proportion, regularity, elegance, and grace were already ingrained through dance, it was easier to implement them in all other exercises, as well.[56] The choreography of dance began with one's own body. Before a movement could be carried out in space, dancers first had to "choreograph" their own limbs. In 1725, the French dancer and

Escrime.

Escrime.

Escrime.

Escrime.

Figure 7.20 Plates on fencing from the *Encyclopédie*, 1751–1780, after Angelo's engravings.

dancing master Pierre Rameau (1674–1748) published his *Maître à dan-*
ser, one of the eighteenth century's most influential dancing manuals. Using no fewer than sixty illustrations, which he drew himself,
Rameau explained how to move, the position of the feet, pliés, steps,
and the holding and turning of the arms. For dance more than for
almost any other kind of movement, it was vital to isolate the movements of the limbs and draw them into an aesthetically heightened
performance to the rhythm of music. In Rameau's treatise on dance,
it is not the dancing body as a whole that appears, but fragments. The
body is a collection of limbs:

> We reckon three Movements of the Arms as well as the Legs, the which are
> relative one to the other; viz. that of the Wrists, that of the Elbows, and that of
> the Shoulders, which must agree with those of the Legs . . . because the Elbow
> ought not to move without being attended in its Motion by that of the Wrist
> the same as the Instep and Knee, which cannot finish its Movement without
> rising on the Toes.[57]

In the example shown in figure 7.21, Rameau deals with the movements of the hands in an analogous way. As well as the descending
sequence of the postures, he uses a circular layout in the labels to
capture the spatial rotations of the hands.

Courtly culture gave rise to the prototype of controlled, measured movement coordinated by the rhythm of music. The nineteenth century, in contrast, has generally been regarded as the era
of a completely new view of motion, marked by "improved performance, tension, speed."[58] New forms and ideals of bodily exercise
and experience arose: horse racing, swimming, and ice-skating, not
to mention the waltz, were among the physical activities practiced
with enthusiasm in the nineteenth century.[59] However, movement
schooled by aesthetics and its associated visual genre of instructional
graphics did not disappear with the emergence of this new configuration of European movement culture around 1800. An important
example is the adoption of drill-like exercises in physical education
by the Philanthropists and the gymnastics movement.

The educational reform movement of Philanthropism or Philan-
thropinism arose in the German-speaking world in the late eighteenth century, accompanied by intense media attention. Its objective was a new education of the human being both in mind and,
through physical education, in body. With this new bourgeois

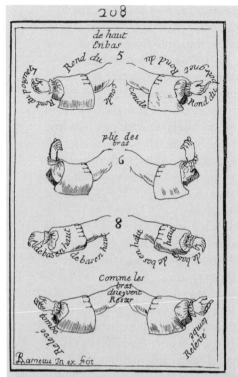

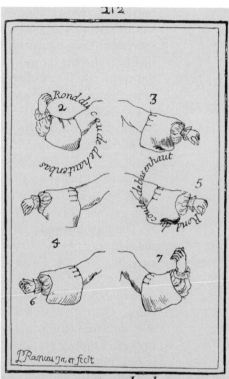

Figure 7.21 The French dancing master
Pierre Rameau combined typography
and movement to represent the twists
and turns of the body.

movement culture of physical education and gymnastics, the Philanthropists carried the eighteenth century's measured aesthetics of movement, trained in precision and rhythm, into the nineteenth century by reshaping the moves of the drill into new exercises. Similarly, Philanthropist literature continued the use of instructional graphics to teach movement. The four-volume *Elementarwerk für die Jugend und ihre Freunde* (Elementary work for youth and its friends) by Johann Bernhard Basedow (1723–1790) appeared in 1774, with one hundred copperplates by Daniel Chodowiecki (1726–1801).[60] In Chodowiecki's plates, fencing, games, ice-skating, riding, vaults, and so on were brought to life before the readers' eyes. The plate reproduced in figure 7.22 is entitled "Extraordinary arts in movement," and its first figure shows a tumbler in the eleven phases of his jump.[61]

The Philanthropist Gerhard Ulrich Anton Vieth (1763–1836) was the author of two works published at the end of the eighteenth century, *Versuch einer Enzyklopädie für Leibesübungen* (Attempt at an encyclopedia of physical exercises) of 1794 and *Oeconomische Encyclopaedie* (Economic encyclopedia) of 1796. In a continuation of the culture of drills, both works divided physical exercises up into exercise elements that could be recombined in various formations.[62] In Vieth's representations of vaulting, it is only the new degree of abstraction that shows the passage of the years, turning Paschen's horse with its saddle and tail into a blunt-headed pommel horse (figure 7.23).[63]

Little more than a hint of the equipment remains in the illustration published in Johann Christoph Friedrich GutsMuths's *Gymnastik für die Jugend* (Gymnastics for the young), a work that first appeared in 1793 and that was highly influential across Europe (figure 7.24).[64]

The objectives of the exercises propagated by the Philanthropists did not differ substantially from those of physical training since the seventeenth century.[65] Like the exercises of earlier periods, GutsMuths's exercises were designed to make the body "robust, strong, nimble, and skillful"[66] — all movements were to be "carried out without noise, on one's toes," and "the posture of the body always remains seemly."[67] The new forms of movement brought forth by the late eighteenth and early nineteenth centuries did not make instructional graphics redundant. Quite the contrary: the changes to movement culture at the turn of the century were anticipated and prepared by the rhythm of the series. While the manuals of the seventeenth

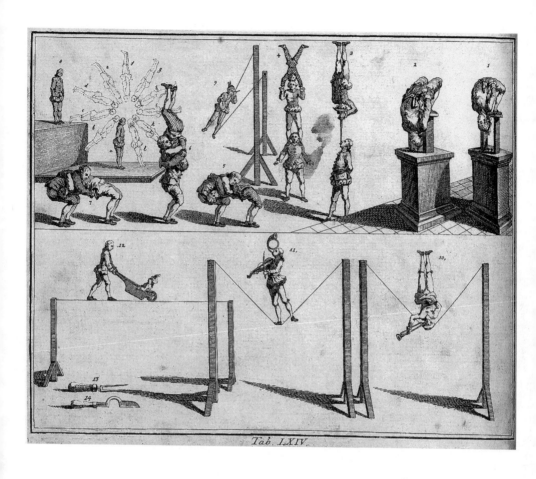

Figure 7.22 Chodowiecki's tumbler.
Basedow, *Elementarbuch*, 1774.

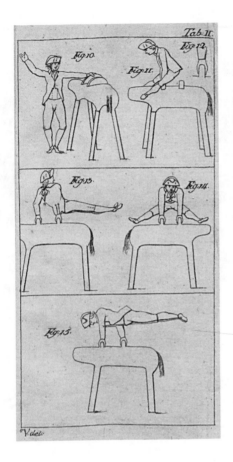

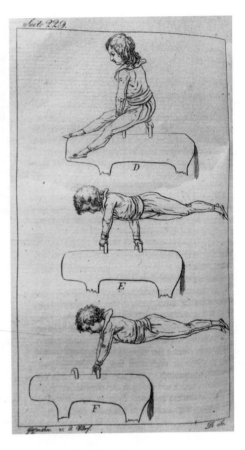

Figure 7.23 Vaulting. Vieth, *Versuch einer Enzyklopädie für Leibesübungen*, 1795.

Figure 7.24 Vaulting. GutsMuths, *Gymnastik für die Jugend*, 1804.

century had presented the figures of fencing, riding, or the minuet using circles, straight lines, diagonals, and precisely defined angles, the gymnastics books of the late eighteenth century showed artistically curving lines and flowerlike formations. In sports such as skating and swimming and dances such as the waltz, which became fashionable in the late eighteenth century, bodily rhythmics was not abolished, but received new accents. (See the portrayals of skating and swimming in figures 7.25 and 7.26.) The rhythm of the body was only read differently, shifting away from form and toward formation, away from pose and toward the flow of movement, which now found expression in the élan of the waltz, the velocity of horse racing or skating. The figures of ice-skating differed from those of dance only in the swoop of the lines, not in the bonding of movement to the body's rhythm. In the new sense of movement, rhythm as the law of motion continued to govern the pulse of its performance.

Dance, Formation, Evolution: The Choreography of Motion

The early modern aesthetics of movement subjected movement to form. In the pose, the body froze to a perfect form in which each limb stood in a defined, regulated relation to the other, and rhythm heightened the individual body's performance into aesthetic motion. But rhythm was no less important for the choreography of the ordered, carefully orchestrated movements of two or more people and their figurations in space. The movements of fencing, riding, or dancing were "spatial artworks,"[68] and the interaction of the participants, their crossing and conquest of space, was subject to aesthetic criteria just as much as was the individual body itself.

From the end of the seventeenth century, complex notation systems for dance began to be invented. Feuillet's notation, presented in his Choréographie of 1700, left its mark in the familiar, complicated geometrical figures that filled the dancing manuals of the eighteenth century (figure 7.27). Here, the patterns of the dancers' movements on the floor, with the notes of the music to which the dance was performed, did not differ in principle from the spatial configurations created by horses in a riding arena. The training of the horse likewise followed the rhythm of a careful choreography (figure 7.28).

In military contexts, the choreography of movements was a matter of life and death. With the transition from medieval cavalry to early modern infantry, marking the beginning of modern warfare,

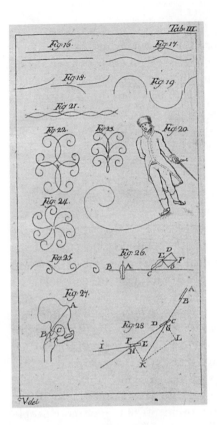

Figure 7.25 Skating in the eighteenth century. Vieth, *Versuch einer Enzyklopädie für Leibesübungen*, 1795.

Figure 7.26 Learning to swim with Heinitz's *Unterricht in der Schwimmkunst*, 1816.

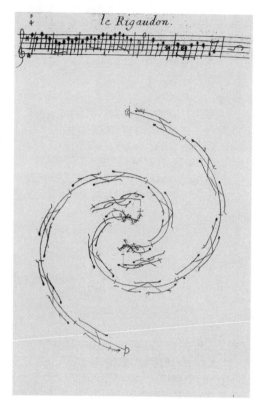
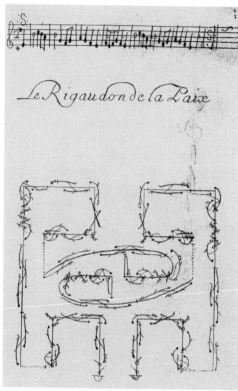

Figure 7.27 Feuillet's notation of the *rigaudon*, a seventeenth- and eighteenth-century court dance derived from an older French social dance.

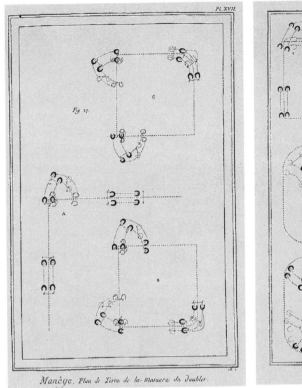

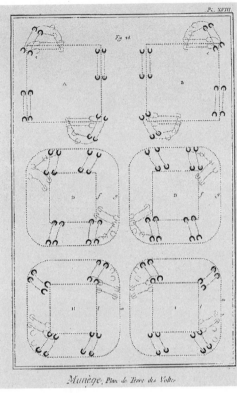

Figure 7.28 Two illustrations for "Manège,"
Encyclopédie, 1751-1780.

weapons drill had to focus on mastery of the body, control, and precision. It was not only the survival of the individual soldier that depended on these skills; the standing armies that arose from the mid-seventeenth century and the linear tactics they adopted could not function without discipline and training. The whole logic of warfare now rested on the elementary principles of bodily discipline, for the mass of soldiers could be coordinated and moved around the battlefield only if the individual fighters coalesced into a single corpus. Accordingly, the movements of troops and units were conceived of and represented in terms of movements of the body. The French language mirrors this most strikingly, bringing together man, army, and death in attributes of the body: the *corps humain*, the *corps d'armée*, and the *corps défunt*.

The Nassau military reforms not only introduced weapons drill on the basis of strictly defined movement elements attached to particular commands, but also applied this system to a battle order in which the tactical movements or "evolutions" of the troops and units followed fixed patterns of formations and commands.[69]

The countermarch was one of the crucial innovations in the Dutch battle order, linking firearms drill with the positioning of the soldiers on the battlefield.[70] What was new about the countermarch formation was that it enabled an uninterrupted hail of fire despite the lengthy process of reloading muskets. The soldiers were lined up in ranks, and when the first soldier had fired, he marched back behind the soldier in the last rank, leaving his successor to fire while he himself reloaded. In the sketch by the Count of Nassau — among the earliest images of the countermarch — this is represented as a simple sequence of dots (figure 7.29). The number of soldiers required in each rank depended on the rapidity and precision with which the individual soldier could reload his musket (figure 7.30).[71] In figure 7.31, the soldiers' changes of position in the countermarch are indicated by the sequence of letters of the alphabet and dotted lines or arrows.

The military manuals of the seventeenth and eighteenth centuries contain innumerable depictions of military strategies and moves. Drill books prescribed the battle formation of the standing army down to the smallest detail. The aim was to amalgamate the soldiers into a tactical body, forming a single wall of fire at the front line that would advance unremittingly toward the enemy. With the increased

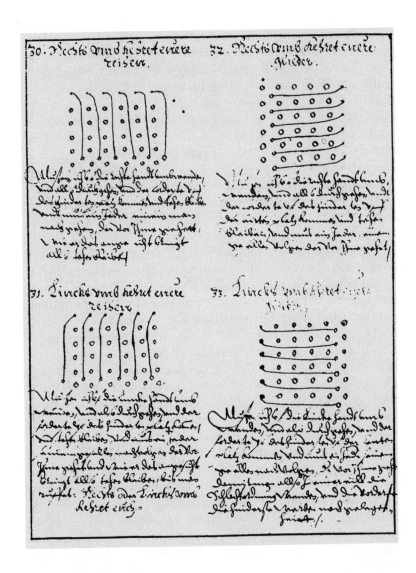

Figure 7.29 Variants of the countermarch,
from the notes of Count John of Nassau-Siegen.

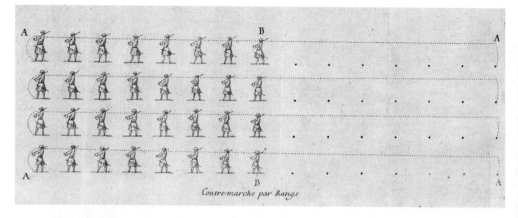

Figure 7.30 Depiction of the countermarch in a French warfare manual, Bottée, *Études militaires*, 1731.

sophistication of firearms, reducing the time required for reloading, the front line became thinner, but longer. Ever larger numbers of soldiers, spread out ever farther across the battlefield, had to be led toward the adversary in ordered movements.[72] If the movements of arms drill were organized rhythmically, this was doubly true of the organization of the army as a whole. Without the metrical beat of the drums and the commands of the fanfares, it would simply have been impossible to coordinate the units. As early as 1616, Wallhausen concluded his *Ritterkunst* with an engraving (figure 7.32) that he annotated as follows: "So I also wanted to show you this in the thirteenth figure as a general picture, in which the kindly inclined reader who has never attended such weddings and dances, may see how gaily and jointly they dance around the bride and pull each other's hair."[73]

Wallhausen's comparison of battlefield and wedding, bride and enemy, dance and combat is not simply the poetic euphemization of a bloody business. It also reflects the underlying rhythmic organization that characterized both the measured steps of the dance and the timed about-turns on the battlefield during the seventeenth and eighteenth centuries. What that meant in practice becomes clear when we descend from the heights of poetry to the depths of the regulations. In 1786, an Austrian cavalry regiment described service on the battlefield thus:

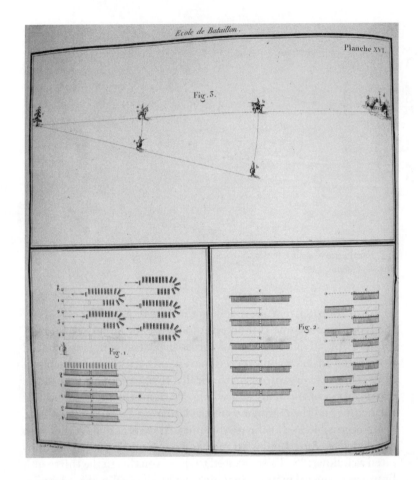

Figure 7.31 Formations in a French book of military rules from the revolutionary period, *Règlement concernant l'exercice et les manœuvres de l'infanterie*, 1791.

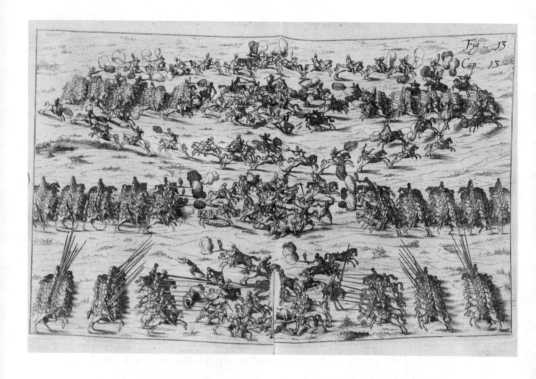

Figure 7.32 The battle as a dance.
Wallhausen, *Ritterkunst*, 1616.

The signal to attack is given from the front, by the bugler, with the second part of the sally on horseback and in procession. All the buglers from the whole attacking wing respond by playing the same signal again.... Shortly after the signal there comes from the bugler: a trumpet blast! Immediately, all the division commanders command: Forward march! All buglers of the attacking wing sound the second part of the sally on horseback and in procession and arrange themselves such that between one and the next there is an interval during which one can count to twenty. After this follow: two double blasts![74]

In the field, it was of prime importance that the soldiers advancing in ranks did so in synchrony. Only thus could the second line confront the enemy in an orderly way once the first had broken up, and so on. During drilling, therefore, the key point was not so much to integrate the soldier into a larger formation, training his posture and the use of his weapon, as to discipline him in rhythm.[75] In this process, music was more than just a signal: its rhythm had to coordinate the soldiers' movements. Rather than the speed of the troops' advance, it was the precision of their movements that would guarantee their survival in battle. The beat set by the drum bound the soldiers to march in step,[76] a pattern in which they advanced toward the enemy and that turned the paces of the individual into part of a unified formation, an evolution moving across the battlefield to the rhythm. "Evolution" here refers to tactical changes in formation, switches of position, and maneuvers for attack and defense in the field, which — just like the use of weapons — were practiced in military drill. The term is first documented around 1600 in the context of the Dutch military reforms and became standard from the eighteenth century on.[77]

Representations of these evolutions range from quasi-naturalistic battle scenes right up to abstract columns of letters and number games on paper.[78] Whereas in Khevenhüller's *Observations-Puncten* the movements of the cavalry are depicted naturalistically and the maneuvers indicated with dotted lines and shaded fields (figure 7.33), Bottée's *Études militaires*, published around the same time, reduces the moves of the troops to diagrammatic displacements of letters (figure 7.34).

Just as in the individually executed weapons drill, the most important point for evolutions in the field was the order and regularity of movement. For the soldiers to be quickly coordinated, the maneuver could not be too complex, yet at the same time, a certain

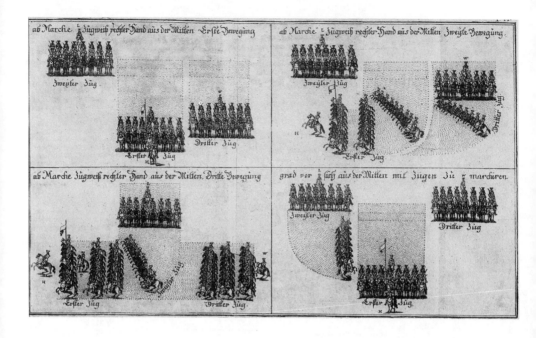

Figure 7.33 Khevenhüller,
Observations-Puncten, 1749.

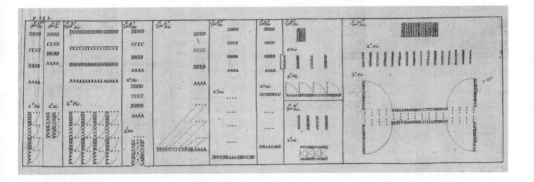

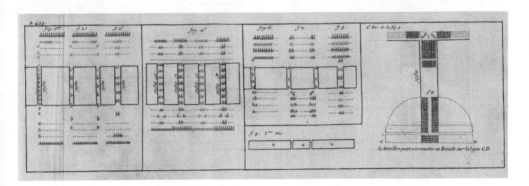

Figure 7.34 Bottée, *Études militaires*, 1731.

tactical freedom had to be preserved. The coupling of repetition and variation in the rhythm of the drill was thus the ideal order of motion for this kind of warfare. The succession of movements was visualized in various different ways: by layering information through the use of dotted lines, digits, alphabetical sequences, vectors, or arrows; by arranging the movement sequentially; and, not least, by combining these two methods. The illustration in figure 7.35 shows the crossing of a bridge in three stages (26–28), ordered horizontally.

In John Russell's *Series of Military Experiments* of 1806, the attack is arranged along a time line. The enemy line is immobile on the left of the frame, whereas the attack is visualized as a movement that advances toward it from the opposite side. The changes in tempo over the course of the attack are captured in the horse's gait: as the horse comes closer to the enemy, it goes from a walk to a trot and a canter, until it finally meets the cannon of the enemy line at full gallop (figure 7.36).

The French *Règlement concernant l'exercice et les manœuvres de l'infanterie* that appeared in 1791 shows the drawing up of six battalions on the front line as a quarter-circular movement (figure 7.37, fig. 1). The vertically arrayed battalions move diagonally toward the horizontal front line, pictured at the upper margin of the image. From the first to the last battalion — from right to left — various stages of the array are traversed. When the first battalion has completed the movement, the last is only halfway through.

In the *Encyclopédie*, numerous illustrations are also dedicated to the *évolutions militaires* — the tactical movements of the infantry on the ground, the cavalry on horseback, and the navy at sea. Here, the movements, especially those of the infantry, fuse into a parade of dots, lines, and circles. The indications of moves are reduced to a bare minimum: the dot becomes an emptied circle — "zéros ou points blancs"[79] — where the soldier has left his position and taken up a new one; dotted lines indicate the path already covered as diagonals, arcs, or vectors; and the turns of the soldiers can be read from a little flag, suggesting the weapon, that is attached to the soldier's dot-body (figure 7.38). This symbolism of movement can also be combined with serial representation, as in figures 67, 68, and 69 of the plate on the cavalry's evolutions (figure 7.39).

The plate shows the horses "in their projection perpendicular to the ground," so to speak from a "bird's-eye view," in order to "render the execution of their different movements more sensitively

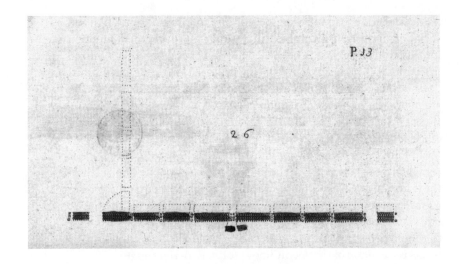

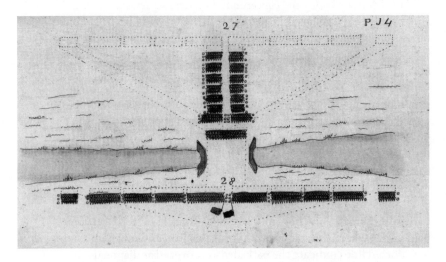

Figure 7.35 Crossing a bridge.
General Review Manoeuvres, 1779.

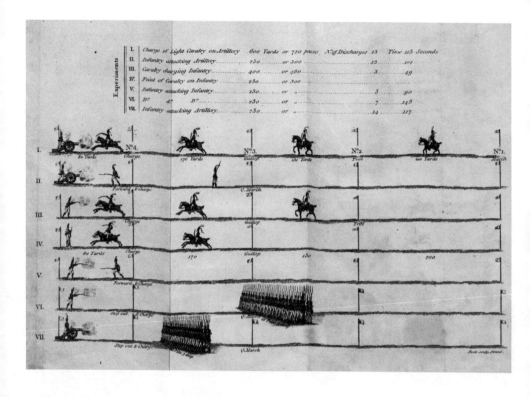

Figure 7.36 The attack as experiment. Russell, *A Series of Military Experiments*, 1806.

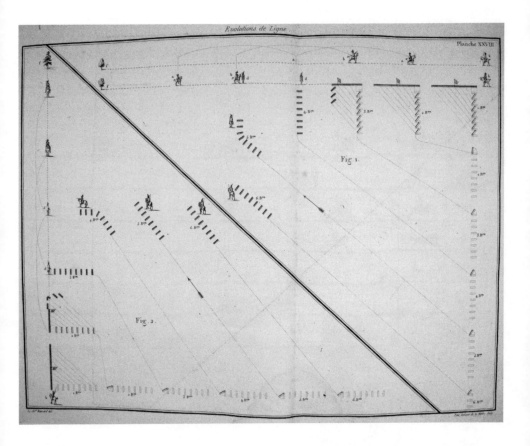

Figure 7.37 Choreography of lines.
*Règlement concernant l'exercice
et les manœuvres de l'infanterie*, 1791.

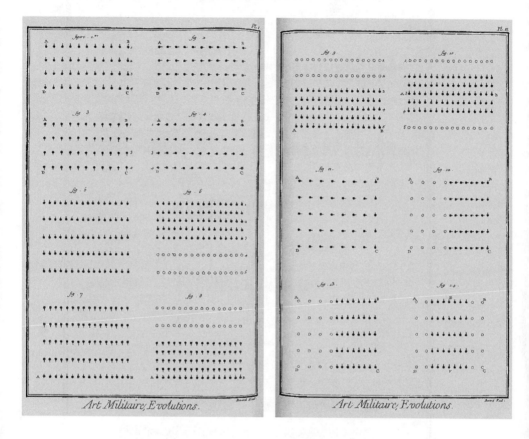

Figure 7.38 Parades of dots, lines, and circles: the infantry's *évolutions*. *Encyclopédie*, 1751-1780.

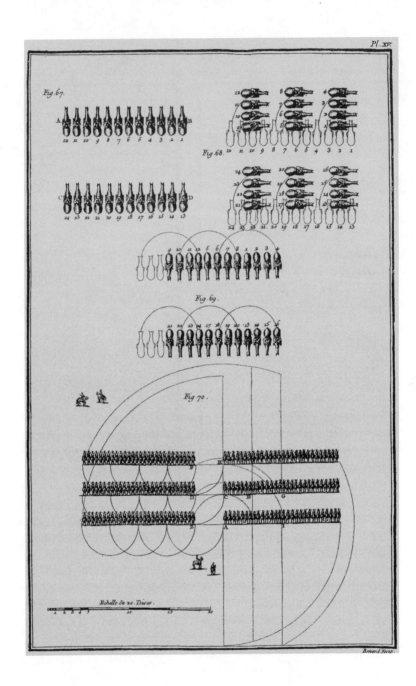

Figure 7.39 The dance of the cavalry.
Encyclopédie, 1751-1780.

and more distinctly."[80] Concretely, the movement of the horsemen consists in a 180-degree turn of two cavalry ranks, executed in two stages of one half-turn to the right. Within the unity of the troops, the movement around the axis of the horseman becomes a complex hybrid of forward and sideward motion.

In the plates for Dalrymple's *Military Essay* of 1761, finally, the maneuver has coalesced into pure line (figure 7.40). Seventy-two battalions, the explanation of the plates tells us, are being moved in six groups — first along the same path, then in parallel. Here, the movement is completely reduced from the moving body to the rising vertical line, the elegant arc, the interrupted line. The result of the series is an almost organic quality of movement.

As I will discuss in more detail in the following chapters, organic development was represented in a very similar way. Figure 7.41 comes from Johann Moritz David Herold's *Entwickelungsgeschichte der Schmetterlinge* (Developmental history of butterflies), published in 1815. The transformation of the caterpillar's nervous system during its metamorphosis consists essentially in a shortening of the nerve fiber, "due to which certain ganglia move closer together and merge into larger ganglia."[81] In this process, the ganglia, numbered in sequence, move toward the brain (here labeled "a") — like the battalion across the battlefield in Dalrymple's illustration.

Handiwork

The soldier's body was drilled, the bodies of innumerable soldiers maneuvered across battlefields to the beat of music — the movements both of the individual and of the mass had to be trained, ingrained, and their implementation choreographed in time and space. Movements were multiplied in the bodies of the many; conversely, the body was also fragmented, with movements isolated and practiced separately. Of all the body parts, none was more important than the hands. The hands, as tools, were located at the center of the active human being. Cultural techniques such as writing, drawing or molding, sewing or embroidering, the crafts and their technological development — all depend on the mastery of elementary or complex, simple or highly elaborate, painstakingly coordinated moves of the hand.

In the eighteenth century, the French encyclopedists envisioned a new transparency for the crafts. Among the central objectives of Diderot and d'Alembert's *Encyclopédie* was to enhance the status of

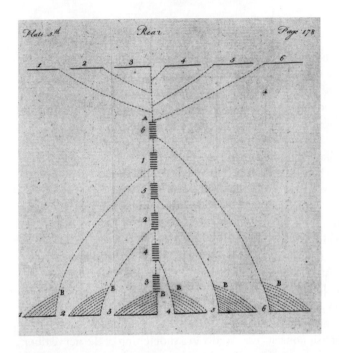

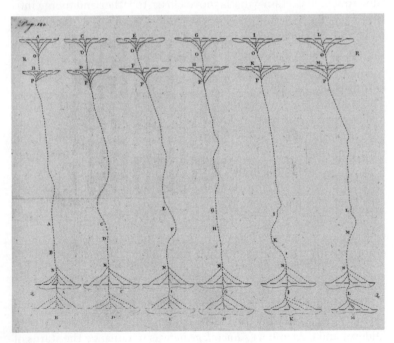

Figure 7.40 Maneuver. Dalrymple,
A Military Essay, 1761.

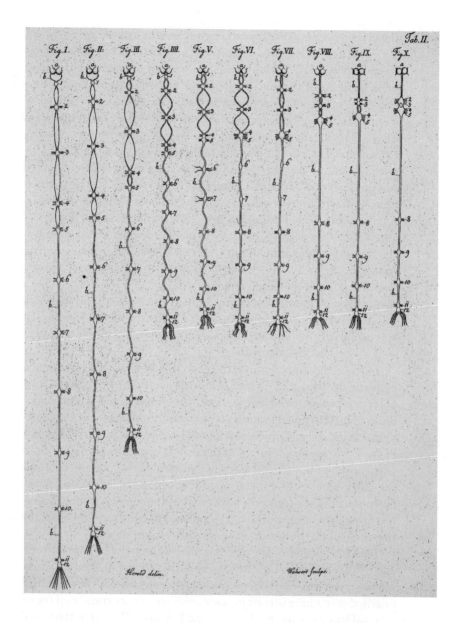

Figure 7.41 The nervous system of
the caterpillar during metamorphosis.
Herold, *Entwickelungsgeschichte
der Schmetterlinge*, 1815.

the crafts relative to the fine arts by publicizing knowledge of the *arts et métiers*, uncovering their traditions, and introducing the reader to them step by step.[82] The *Encyclopédie* was not only among the great book projects of the Enlightenment;[83] its volumes of illustrations also made up one of the eighteenth century's most extensive pictorial compendia. The visual representation of knowledge was by no means a secondary or merely ornamental element in the project of encyclopedic collection. The number of illustrations grew steadily along with the size of the encyclopedia, which had initially been planned as a translation of Ephraim Chambers's 1728 *Cyclopaedia*. The two volumes of illustrations originally envisaged became eleven large-format folio volumes, which appeared between 1762 and 1772 and contained almost three thousand *planches*, or plates. A complex system of cross-references between text, picture, and prefatory essay (the *explications* preceding the plates) wove a web of knowledge within which image accrued as much epistemic importance as text, in which the various authors supplemented or contradicted one another, and in which the closed encyclopedic universe ceded to discursive knowledge that was essentially incomplete and piecemeal, drawing on every kind of source — verbal and visual alike.[84]

In the eighteenth century, the *Encyclopédie* was a "véritable encyclopédie de l'image";[85] Ernst Gombrich even called it "the greatest enterprise in pictorial instruction."[86] Its importance as a compendium of instructional graphics lay primarily in the area of the crafts or mechanical arts. Representations of the crafts make up a large part of the *Encyclopédie*'s volumes of plates.[87] If depicting machines at all was a novelty for a general reference work in the eighteenth century, the sheer quantity of illustrations dedicated to the *arts et métiers* was far more so. They formed a panorama of the manual trades, manufacturing processes, working techniques, and hand movements at the threshold of the Industrial Revolution.

Diderot's vision of the *arts et métiers* was that of an "applied science."[88] Accordingly, the *Encyclopédie* analyzed and divided up working processes into basic operations — the mysteries that the guilds had guarded and administered for centuries were now translated into a rational language and made public, and this is reflected in the visual strategies of the *planches*. The majority of the illustrations on the *arts et métiers* are constructed on a single pattern: they show the place of work and the tools required, usually with a bisection of the image so that the

Figure 7.42 Combing cotton.
Encyclopédie, 1751-1780.

upper segment (the vignette) illustrates the workshop interior and the lower segment presents the tools and machines in turn.[89] However, the engravings also took the dissection of the work process a step further by presenting individual moves of the hand and the sequence in which these must be performed. The pictures made by Louis-Jacques Goussier (1722–1799), in particular, reveal a special interest in hand movements. Goussier was one of the main collaborators on the *Encyclopédie*, and between 1747 and 1760, he worked closely with Diderot.[90]

Goussier's illustrations on cotton manufacture show the combing of cotton in a sequence of four hand movements (figure 7.42). The flocks of cotton are first combed around a single card (originally the prickly dried flower head of the teasel, a plant of the Dipsacaceae family) (fig. 1), then divided between two cards (fig. 2), and finally transferred from the larger to the smaller card. The fourth figure shows the waste fibers removed from the cotton in the process.

The encyclopedists made use of existing images and material for the plates wherever these seemed useful,[91] merely adapting them to

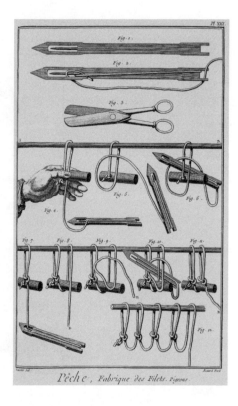

Figure 7.43 Knotting a fishing net using the system "sous le petit doigt." *Encyclopédie*, 1751-1780.

Pêche, Fabrique des Filets. Pigeons.

a greater or lesser degree, but in the case of the pictures on combing cotton, Goussier is known to have made his own studies and preliminary sketches in situ for the illustration of the various movements. A surviving preliminary drawing shows that Goussier copied the actions from the cotton workers and turned their practical, performative knowledge into a series of sequenced movements.[92]

Goussier was also responsible for the plates illustrating the entry "Pèches." For the fisherman knotting a fishing net, Goussier returned to the representation of moves through a picture sequence like the one he had used for cotton combing.[93] Here, too, the hands are staged *pars pro toto* for the craftsman at work. The sequence of individual hand motions shows the larger movement that enables the fisherman's net to continue growing.

Plate 21 (figure 7.43) shows the first knots of the net; the following plates illustrate different ways of knotting using the little finger or the thumb ("sous le petit doigt," pl. 21–23 and "sur le pouce," pl. 24 and 25 in figure 7.44; an unnamed third system is illustrated in pl. 26 and 27).

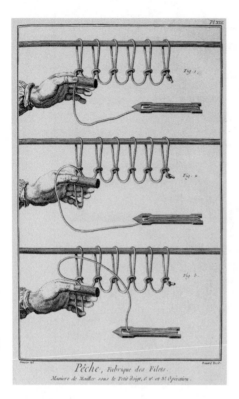
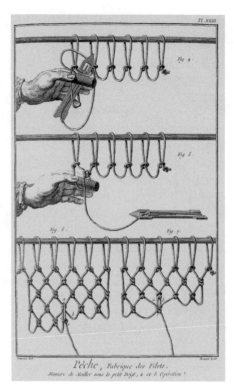

Pêche, Fabrique des Filets.
Maniere de Mailler sous le Petit doigt, 6.e 7.e et 8.e Opération.

Pêche, Fabrique des Filets.
Maniere de Mailler sous le petit Doigt, 4.e et 5.e Opération.

In this case of handiwork, again, the plates do not simply show a succession of individual steps. A fishing net can take shape only when the sequence of the hand's moves combines with the corporeality of the laborer performing them — when the worker performs the procedure in the rhythm of his bodily possibilities. Just as for the plates on fencing, this becomes clear both in the seriality of the illustrations and in the verbal explanations. These describe the sequence of work processes using the concept of *tems*:

> Fig. 1. First tempo [*tems*]; the thread must be over the *moule*, where it is stopped by the thumb of the left hand. 2. Second tempo in the formation of the net; it consists in passing the shuttle from the top of the net downward, such that the thread is behind the *moule* . . . . 3. Third tempo in the formation of the net; the thumb must be passed underneath the *moule* in order then to move the thumb as is shown in the figure.[94]

Graphic instructions for movement were a complex visual invention. On the one hand, they produced movement out of static

227

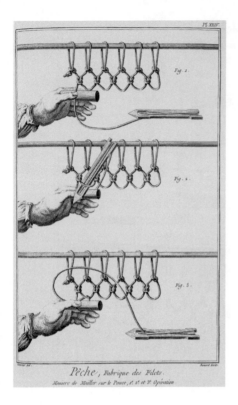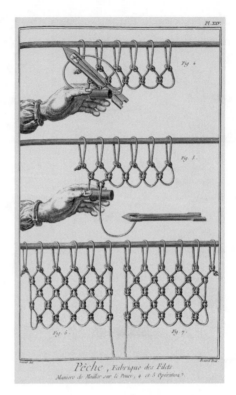

Figure 7.44 Knotting a fishing net "sur le pouce." *Encyclopédie*, 1751-1780.

moments, the poses: the motion to be learned was isolated, the continuous flux of motion segmented into a sequence of individual figures of the same size, perspective, and frame. The background was emptied out, the gaze guided to the figure. On the other hand, each pose was defined in terms of the sequence of poses before and after it: movement arose out of the relationship between images in the series. Motion was therefore something relative, something possible only in the alternation of image and lacuna, pose and formation, form and law. This interplay of analysis and synthesis, flux and discontinuity, pose and space constituted a complex ordering of time — in other words, rhythm. With rhythm, movement carried its own temporal order within itself. The motion series was not simply a linear succession, but a complex rhythmic pattern in which varying numbers of impulses were apportioned to the different individual poses.

As a rhythmic figure, movement always also signifies a coalition of constant and variable elements; the series unites the regularity of repetition with the spectrum of variation. This is precisely how Pander and von Baer described embryogenesis. And indeed, Pander's and von Baer's visual strategies resembled those of a seventeenth-century or eighteenth-century dancing or fencing master: they disengaged the idea of embryonic development from the chains of chronological time and reapprehended it through the concept of rhythm. The stages portrayed in the developmental series fulfilled the same role as the poses or positions in the instructional series on movement. Both constituted the form and its transformation simultaneously and as a visual relationship between figures. In both types of series, the grouping of variously complex changes into an ordered sequence was carried out according to the logic of rhythm.

CHAPTER EIGHT

Epigenetic Iconography

Starting in classical antiquity, chick embryos were the most important material for emerging theories of development, and from Fabricius ab Aquapendente's early illustrations in *De formatione ovi et pulli* of 1621 to von Baer's *Über Entwickelungsgeschichte der Thiere* more than two hundred years later, the hen's egg remained key to observation and the representation of developmental thinking.[1] However, although images had been present in the study of organic development long before Wolff, Pander, and von Baer, it was not until the period between 1760 and 1830 that they took on a constitutive role. This chapter will show that in the famous Haller-Wolff debate, Caspar Friedrich Wolff used the visual image as ammunition against Albrecht von Haller's preformationist approach, placing pictures at the heart of his argument that the structures of the egg could not possibly be preformed. Ignaz Döllinger's project on chick embryogenesis, too, was planned from the outset as a pictorial enterprise, though he could not realize that plan until he had assured Christian Heinrich Pander's financial assistance and Eduard d'Alton's artistic expertise. Pander's *Beiträge zur Entwickelungsgeschichte des Hühnchens im Eye* (Contributions to the developmental history of the chick in the egg) may also be read as a volume of illustrations, rather than an illustrated text. I will argue that by switching from text to image, Pander also moved from a linear, chronological order to the pictorial series as a *relational* structure of forms. Karl Ernst von Baer, finally, picked up on Pander's approach and expanded it by using schematic representations.

The serial iconography presented in this chapter supports my wider hypothesis that development around 1800 was constituted as a rhythmical law. It was the adoption of this particular visual means of representing movement, namely, as a rhythmical series of repetition and variation, that opened up a visual repertoire for conceptualizing developmental processes in nature. In other words, the picture series was not a mere epiphenomenon of the approaches that were coming to prominence in biology around 1800; it was not the visualization of the new thinking, focused on dynamism and process, that historians have regarded as marking the dawn of modern biology.[2] Serial representation had its own pedigree, going back to the Renaissance and earlier. Around 1800, the developmental series began to emerge from a tradition of pictorial instructions for movement that proved to be fruitful for natural history and especially embryology. This shift in representation at the threshold of the nineteenth century was equally an epistemic rupture. Before the mid-eighteenth century, movement had belonged to man, not to nature. The notion of epigenetic development reversed that: nature, not man, was now the protagonist of movement.[3]

Making the image integral to thinking on development entailed a shift from picture to pictorial series — to what I will call an "epigenetic iconography." What did it mean to dissect a developmental process into a series of individual images that, viewed in their sequentiality, evoked a gradual process of development from one image to the next?[4] The establishment of epigenetic iconography depended on the revision of several ideas about development that had previously held sway. First among these was a turn away from regarding chronological succession as the only possible principle for ordering change.

The chronological tradition in embryology — equating development with the chronology of changes — dominated visual representations of development until the end of the eighteenth century. It used pictures as a twofold aid to vision: on the one hand, pictures recorded observations themselves, rather than verbal descriptions of observations; on the other, pictures tied observations closely into an existing web of iconographic representation. The visual image lent both comparability and historical legitimacy to what had been seen.

Abandoning this tradition meant that although development still seemed to occur *in* time, it was no longer simply the same as chronological succession. Instead, development became an ordering *of* time. The primacy of chronology was replaced by that of the image, which alone could capture the form of the embryo at each point in time with the necessary precision and detail. In turn, the use of pictures enabled development to be located in the image itself: in the iconography of the series, development becomes a pictorial relationship. It is the relationships between the visual forms that produce both the individual stage of development and development as a whole — simultaneously and in mutual dependence.

These innovations were nothing other than the conception of development as a rhythmical ordering of time. The order of rhythm here is both analysis and synthesis; it constitutes one developmental stage in relation to the series of changes and vice versa, and the successive formation of the embryo becomes located in the interreference of the pictures. In this sense, the series is at once a mode of representation and a mode of thinking.

Malpighi

Long before Wolff set out his epigenetic theory toward the end of the eighteenth century, the gradual changes that occur in the hen's egg had been an object of observation. The development of the embryo from day to day was traced in the late sixteenth century by Ulisse Aldrovandi (1522–1605) and his pupil Volcher Coiter (1534–1576), but the first pictorial representations of the egg's development were by Hieronymus Fabricius ab Aquapendente (1537–1619) in *De formatione ovi et pulli*, published posthumously in 1621 (the images were probably made in 1604).

Five of the seven plates in *De formatione ovi et pulli* show the daily progress in the fertilized hen's egg, from the second day of incubation until hatching. Each picture is labeled with a number indicating the day when the egg had the appearance shown. One picture is dedicated to each day, the first picture in the third plate (figure 8.1) showing the egg on the second day, the last on the thirteenth day. The pictures are arranged in consecutive rows. The opened eggs are sometimes interspersed with portrayals of the isolated embryo at various stages of incubation. Fabricius's illustrations were the outcome of remarkably precise observation, which here is displaced from the text into the plate. The plates are attached to the treatise, unexplained, while Fabricius's philosophical erudition meanders into scholastical detail quite devoid of reference to his own observations.[5]

With these images, Fabricius inaugurated the tradition of chronological visualization in embryology, a tradition in which the picture refers to the time of the scientific observer. The forward movement

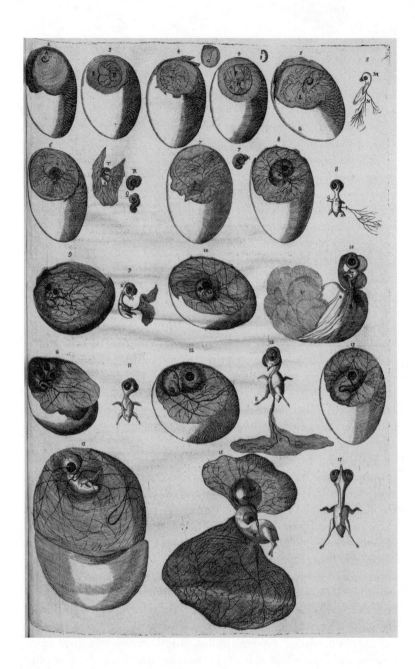

Figure 8.1 The first pictorial representation of development in the egg. Fabricius of Aquapendente, *De formatione ovi et pulli*, 1621.

of time is the reference point of the individual images, and their task is to record a particular state in the egg at a determined point in time.

In the seventeenth century, anatomists also began to use microscopes to delve into the egg's depths and make them visible. The drawings of the egg's development published by Marcello Malpighi (1628–1694) in his two treatises *De ovo incubato* (1672) and *De formatione pulli in ovo* (1673) remained seminal well into the eighteenth century. In these illustrations, Malpighi tracked the hourly progress of development inside the hen's egg, from the earliest phase until a few days after hatching.[6] He used detailed descriptions and delicate red chalk drawings to report on the various states he had observed. Malpighi's pictures were an attempt to register his observations as precisely as his eyes and his artistic talent would allow.

Each of the plates — figure 8.2 is the second plate of *De formatione pulli in ovo* — combines several separate observations. The individual figures show, for example, the formation of the heart (fig. xv), the emerging embryo (fig. xi), and the formation of blood vessels on the yolk (fig. xiv). For every observation, Malpighi specifies the time at which it was made: fig. xi after thirty-eight hours of incubation, fig. xiv after forty-eight hours, and so on. Malpighi's sketches of the chick embryo and its organs constituted an apex of embryological illustration.[7] Nearly a century later, they could still serve as a crucial point of reference for the embryological studies of the anatomist and physiologist Albrecht von Haller (1708–1777).

The Image as an Aid to Seeing
In the mid-1750s, Haller carried out a long series of microscopical studies of chick embryos, which he published in French in the two-volume *Sur la formation du cœur dans le poulet* of 1758. Supplementary observations followed in 1778, in the eighth volume of his *Elementa physiologiae corporis humani*. Haller's observations of the development of the heart were among the most precise in eighteenth-century embryological research.

In his laboratory diary, Haller made sketches of his work. These served him as an aid to vision in a dual sense. They assisted the process of observation: sketching helped him to understand what he had seen through the medium of the picture. And they were a visual aid in comparing his own observations with existing representations of the

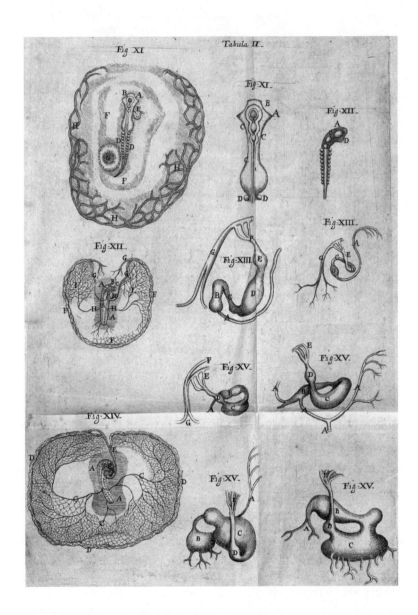

Figure 8.2 Malpighi's drawings of the embryo and the heart, 1673.

egg — especially Malpighi's illustrations, which were still canonical in Haller's day.

The objective of this comparative way of seeing with the help of the image was to record the chronology of development: at what moment during incubation the structures in the egg had what appearance. More exactly, it was to specify the chronology of sequence by knitting the scholar's observations and their representation into the fabric of existing iconography, which was thereby continually enhanced. This approach to the visual image is revealed with particular clarity in the Haller-Wolff debate on the development of the hen's egg, perhaps the most important controversy of late eighteenth-century natural history. Its crux was the development of the heart and the blood vessels in the area vasculosa and specifically the sequence in which the vessels and the heart come into being. Haller argued that the vessels are already present in the egg, preformed, but become visible only at the moment when the heart begins to beat and pumps blood through them.[8] For Wolff, in contrast, the vessels are new structures, the beginnings of which can be clearly seen before the heart has formed completely and begun to beat.

Whereas Wolff's *Theoria generationis* aimed to study the composition and emergence of these vessels in detail and to prove that they appear in the very first hours after fertilization, Haller pursued a chronology of formations, intending to demonstrate the priority of the heart's formation over that of the vessels.[9]

Examining the role of visual images in this debate, it becomes obvious that pictures played very different roles in the developmental theories of preformation and epigenesis.[10] Haller's and Wolff's extant pictures consist of engravings in their published writings, sketches made by Haller in his laboratory diaries, and hand drawings by Wolff in a 1764 letter to Haller. The first striking aspect is the asymmetry in the two men's deployment of pictures — the alternating play of absence and presence. Wolff added illustrations to his published works, whereas Haller's physiological publications contained no illustrations, though they were dependent on the pictures he had made during his research.[11] Haller kept minutes of his observations as he worked. As well as written notes, he made small sketches, dividing his pages into two columns, in the first of which he noted his observations, while the second was reserved for references to the literature. He placed his own sketches in the margins or in among the notes (figure 8.3).[12]

Many of Haller's drawings show the development of the heart. These small pictures were mainly executed in outline; he rarely used shading. Sometimes he drew with a broad nib (probably the same one he used for writing), sometimes with a fine line. The example in figure 8.4 shows the heart at fifty hours of incubation, one of several sketches that Haller made to record the organ's structure as he saw it in the egg at this point. Some hours later, the structure of the heart had already become more complex, which altered Haller's style of drawing (figure 8.5).

Haller created and discarded sketches in several cycles, trying to capture directions, angles, and curvatures in order to ascertain the structure of the heart. In the published work, *Sur la formation du cœur*, these numerous sketches are absent.[13] Using only the written notes from his protocols, with minimal revision, Haller presents the heart to the reader in purely verbal terms. After forty-eight hours, for example, it is "round or a very short oval. . . . The whole heart has somewhat the aspect of a horseshoe or a parabola with very short ordinates."[14] After fifty-four hours, it already consists of three vesicles (*Bläschen*). "The ventricle is large, it appears diagonal, and largely imitates the shape of a kidney when one looks at it from the side."[15] The image as such has now disappeared from Haller's text—yet as a reference point, its presence is unbroken. Haller refers, however, not to his own pictures, but to the famous illustrations by Marcello Malpighi, almost a century old at the time of his observations. Why did Haller need these pictures?

What is certain is that Haller wove each step of his observations into a tapestry of reference to the images made by his predecessors, especially Malpighi. When he describes an egg at twelve hours as a circle with a "whitish cloud, which one could regard as a fetus," and as "a small round body, positioned above the follicle, this was the amnion and the fetus," Haller tells us that his observation corresponds to "the fourth figure in the work of MALPIGHI, or the sixth hour."[16] Haller maintains this system in cases where Malpighi's "fifth figure" only "quite" resembles his observation,[17] and even in those where Haller has "never seen" the configuration described by Malpighi "in any of my observations."[18] Formulations of the type that Haller's observation "responds to" one by Malpighi abound in Haller's text, as do qualitative appraisals of Malpighi's figures, covering style and even coloring techniques as either "not good," "too cramped,"

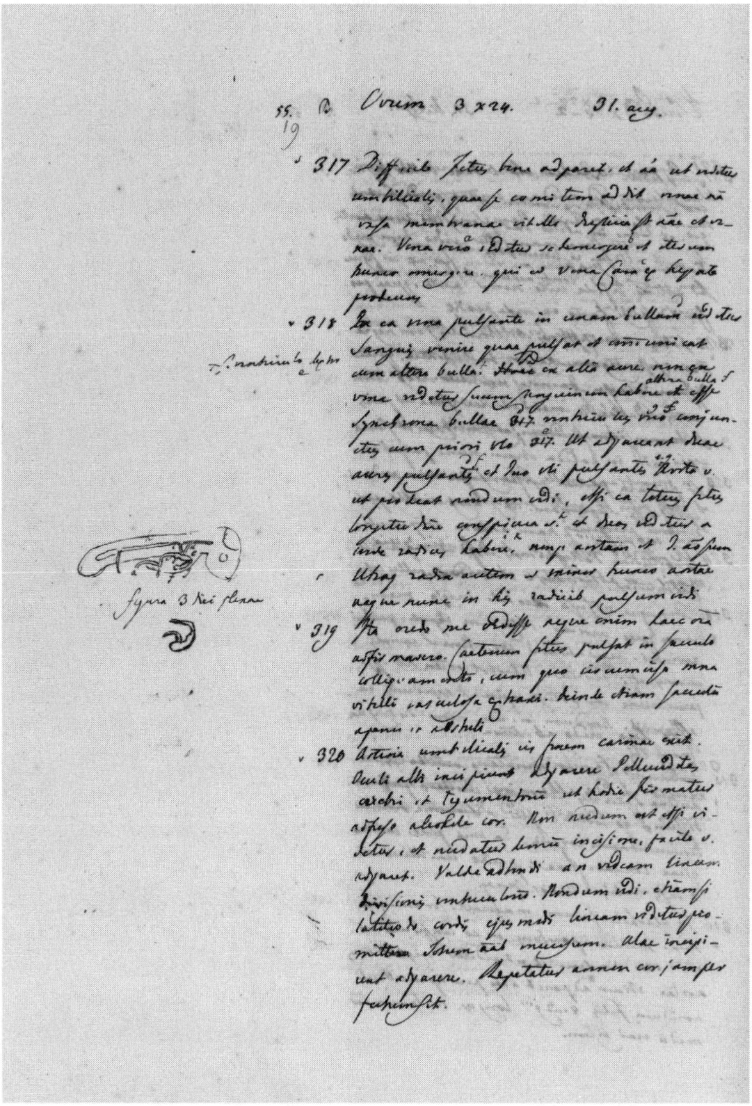

Figure 8.3 A page from Haller's notebook.

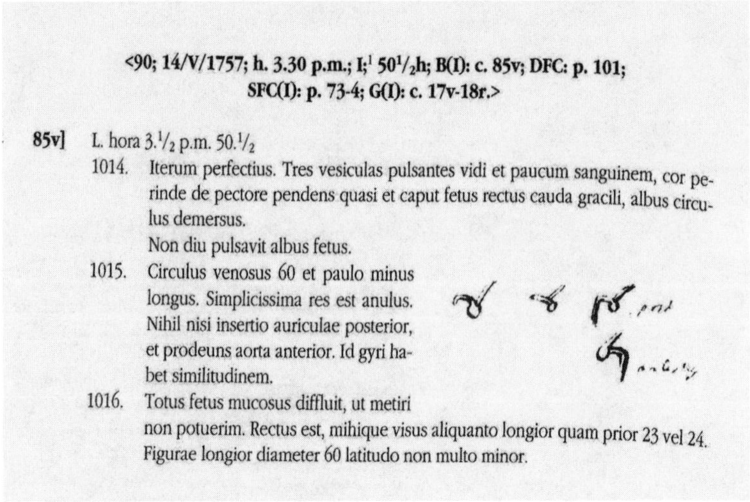

<90; 14/V/1757; h. 3.30 p.m.; I;¹ 50¹/₂h; B(I): c. 85v; DFC: p. 101; SFC(I): p. 73-4; G(I): c. 17v-18r.>

85v] L. hora 3.¹/₂ p.m. 50.¹/₂

1014. Iterum perfectius. Tres vesiculas pulsantes vidi et paucum sanguinem, cor perinde de pectore pendens quasi et caput fetus rectus cauda gracili, albus circulus demersus.

Non diu pulsavit albus fetus.

1015. Circulus venosus 60 et paulo minus longus. Simplicissima res est anulus. Nihil nisi insertio auriculae posterior, et prodeuns aorta anterior. Id gyri habet similitudinem.

1016. Totus fetus mucosus diffluit, ut metiri non potuerim. Rectus est, mihique visus aliquanto longior quam prior 23 vel 24. Figurae longior diameter 60 latitudo non multo minor.

Figure 8.4 Haller's sketches of the heart at fifty hours. Reproduced in Haller, *Commentarius*, 2000.

<160; 15/V/1757; h. 3 p.m.; L; 74h; B(I): c. 86r; DFC: p. 137; SFC(I): p. 125-6; G(I): c. 30r.>

86r] Die iterum 15. hora p.m. 3. quae est 7[7]4. L.

1024. Bene. Figura transversim latior 80 cum figurae longitudo esset 69. Fetus caudam non bene determinatam habentis longitudo fuit 30.

1025. Pulsus bene. Ventriculus pene horizontalis. Summa et pene transversa pars venae cavae est auricula pulsans, in aortam vere introrsum surgentem sanguis pellitur.

1026. Ventriculus et aorta perfecte pallescit. Et hodie et nunc distinguitur arteria umbilicalis.

Et sub bulbo aortae canalis est, et ab auricula, iste absconsus, quia posterior.

1027. Exemti animalculi cor de more angulus unice pene rectus ex auricula formatus et ex ventriculo et aorta. Omnino uti semper vidi his diebus.

Mirum quam laterales hae figurae sint, et minime in ovi vertice.

Figure 8.5 Haller's sketches of the heart at seventy-four hours. Reproduced in Haller, *Commentarius*, 2000.

or "more advanced."[19] Haller's exact and detailed references to Malpighi's work in his protocols, including page numbers, the numbers of figures, and their alphabetical legends, suggests that he had Malpighi's illustrations in front of him as he carried out his own observations.[20]

Haller's complex play with the image — its presence at certain points, its dispensability at others — shows how he used images as a visual aid as he worked. The pictures helped him to perceive structures in the egg — in other words, to discover at which point which structures in the egg are *visible*. The pictures record the curvature of the embryo at a particular time, the relative positions of the vessels, or the size of the heart when it begins to beat. It was only for this purpose that he made his own small sketches during observation, and for the same purpose that he used Malpighi's illustrations. As regards his position on developmental history, however, the images had no further epistemic value. This is why they vanished from all his physiological writings, in contrast to his studies in anatomy and botany, where Haller activated his social network in search of the best artists and spared neither trouble nor expense in the production of illustrative plates.[21] In Haller's embryological oeuvre, the task of images was restricted to helping him to see during his practical working process. As soon as that task was fulfilled and the results were certified by comparisons with existing knowledge, the images became superfluous. To construct his preformationist hypothesis, Haller did not require a single picture.

Painted Tables

As early as the beginning of the seventeenth century, Fabricius ab Aquapendente's anatomical plates had presented the daily progress of development in the hen's egg in rows and columns. In 1758, the Nuremberg naturalist, microscopist, and copper engraver Johann Rösel von Rosenhof (1705–1759) depicted the tadpole's development in his *Historia naturalis ranarum*. In rows to be read from left to right and columns from top to bottom, the viewer of the plate could retrace the path from egg to completed frog (figure 8.6).

Probably the best-known illustration in this form, however, is the first of the two plates published by the Frankfurt physician Samuel Thomas Soemmerring (1755–1830) in his *Icones embryonum humanorum* of 1799.[22] Soemmerring may be credited with having

Tab. X.

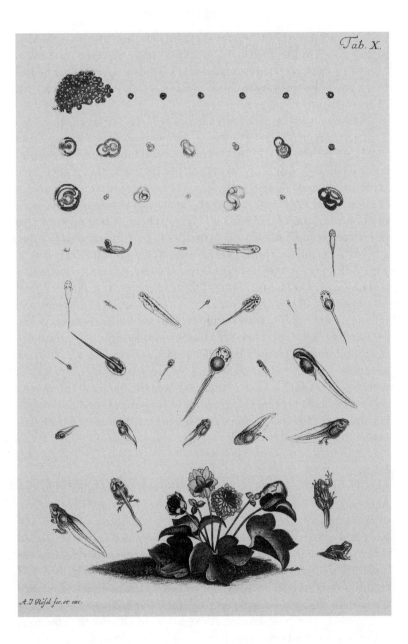

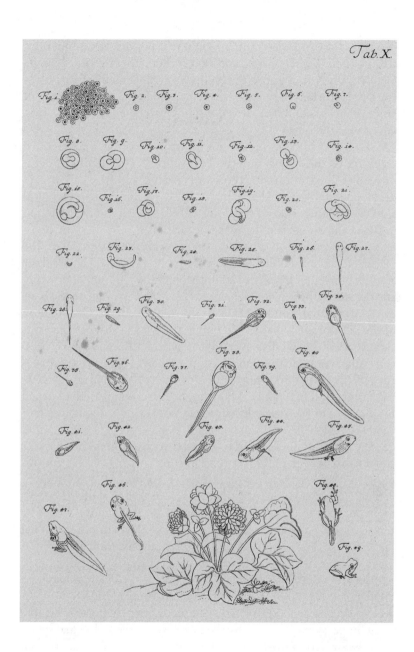

Figure 8.6 From tadpole to frog. Rösel von Rosenhof, *Historia naturalis ranarum*, 1758.

been the first to isolate human embryos, picture them individually in their actual shape, and bring them together in a single plate (figure 8.7). Because of the linear arrangement of this illustration, historians of science have regarded it as an icon of nascent embryology around 1800. It is considered to be embryology's first developmental series.[23]

The copperplate certainly shows human embryos from the third week up to the fourth month of gestation[24]—but neither Soemmerring's illustrations nor those of Fabricius and Rösel are in fact developmental series. They remain firmly within the chronological tradition of embryology: the point of reference in the series found in Rösel and Soemmerring is the linear passing of time. The chronology of observation organizes the form of the image. Here, the picture acts as an aid to vision. It records observations. The observer assures himself of his observation through the medium of drawing and organizes the various elements of his contemplation on the basis of the particular point in time at which he made it. The purpose of the picture is comparative looking. It serves to sharpen observation, to identify and display similarities and differences. Precision and comparison, distinction and similarity here organize the structure of the image; its world is that of the tableau, naturalist classifications, and taxonomy. Accordingly, Rösel von Rosenhof called his pictures of the frog's development not series or sequences, but "painted tables."[25]

The illustrations supplied by scholars such as Camper or Lavater should also be seen in this context. Johann Caspar Lavater's continuum from the head of a frog to the head of Apollo, or Pieter Camper's geometrical calculations of the facial angle in various skulls, proceeding from apes to the sculpted faces of classical antiquity, would be more properly characterized as ordering schemata or observational patterns than as series.[26] In these representations, comparing forms was a way of reaching conclusions on similarities, demarcations, proportions, or commensurability.[27] The same is true of Soemmerring's copperplate. It stands in a tradition of anatomical illustration that was organized around isolated specimens, precise portrayals, formal comparisons. This is why the plate orders the individual specimens by size, from smaller to larger; it proposes a tableau of prenatal beauty in natural proportions, and not a series of successive formation like the ones posited by epigenetic theory.[28]

Soemmerring's Icones embryonum humanorum

Soemmerring's *Icones* contains twenty images of embryos on two plates. Seventeen of the images are on the first plate (figure 8.7). Drawn life-size, they are chronologically ordered from the third week to the fourth month of gestation. The second plate (figure 8.8) shows a fetus in the fifth month, inside and outside the uterus. The engravings are large in format, measuring approximately eighteen and a half by twenty-five inches. *Icones* is essentially a volume of plates, with the accompanying text dedicated to explaining them. The "Praefatio" begins by listing previous embryological works, from those of Fabricius ab Aquapendente in the early seventeenth century up to those of Georg Ferdinand Danz, whose two-volume *Grundriss der Zergliederungskunde des ungebohrnen Kindes* (Outline of the anatomy of the unborn child) had been published in 1792–1793 with notes by Soemmerring. After the preface, the "Explicatio figurarum" gives detailed explanations of each of the twenty images of embryos and the figure in the title-page vignette.

A Collection of Specimens

The object of Soemmerring's investigations was the anatomical specimen. Dissected, preserved, submerged in alcohol, and sealed in containers, or as skeletonized heads and bodies, embryos were demonstration material kept in cabinets of curiosities and anatomical collections. It was on the basis of this kind of collection that Soemmerring's doctoral adviser in Göttingen, the anatomist and teacher of midwifery Heinrich August Wrisberg (1739–1808), wrote his *Descriptio anatomica embryonis observationibus illustrata* in 1764, a work whose anatomical skill Soemmerring praises in the preface to *Icones*. Soemmerring probably began to build a collection of his own as early as 1779.[29] In 1780, on behalf of his university in Kassel, he purchased the collection of the Jena anatomist Karl Friedrich Kaltschmied (1706–1769). This contained malformed embryos along with thirty-two normally developed ones of both sexes.[30] Soemmerring also prepared embryo specimens himself.

The anatomical collection expanded on random principles — the sources of the embryos ranged from Scotland to Poland, and local doctors and midwives also brought Soemmerring recently dead embryos. An impression of the collection's coverage is given by the *Catalogus musei anatomici quod collegit Samuel Thomas de Soemmerring,*

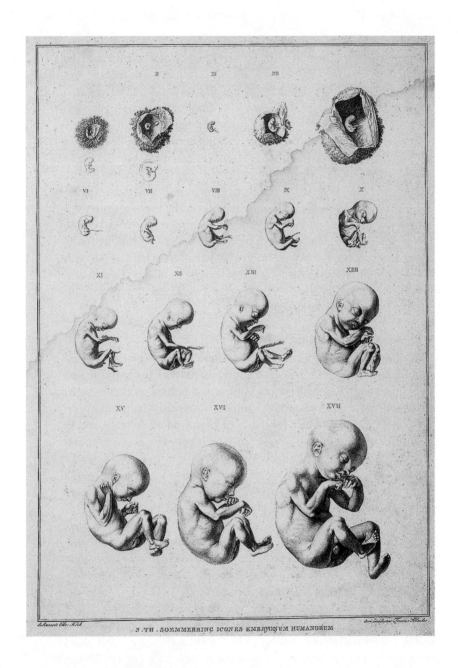

Figures 8.7 & 8.8 Samuel Thomas Soemmerring's illustration of human embryos. Soemmerring, *Icones embryonum humanorum*, 1799.

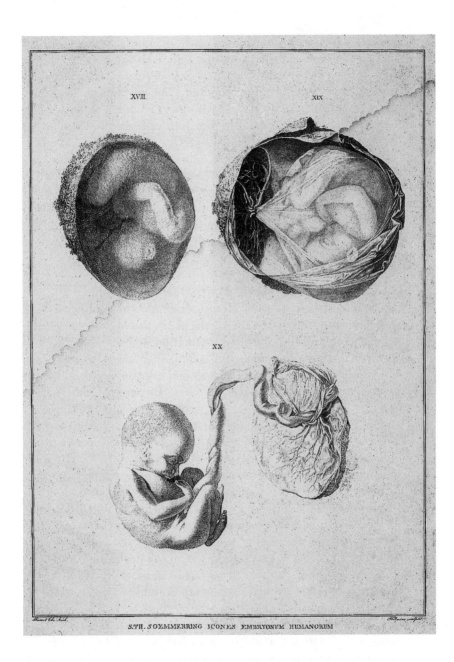

XVIII

XIX

XX

S.TH. SOEMMERRING ICONES EMBRYONUM HUMANORUM

published by his son Detmar Wilhelm Soemmerring in 1830, which lists eighty-seven embryological specimens. Ulrike Enke has shown that some of the embryos pictured in *Icones* can be traced back to particular specimens in the collection. The first figure, according to Enke, was produced on the basis of specimen no. 39, the second from specimen no. 41, the title-page figures from specimen no. 55, fig. 14 from no. 82, fig. 15 from no. 92, fig. 16 from no. 103, and figs. 18–20 from specimen no. 107 in Soemmerring's collection.[31] Soemmerring expressly notes that figs. 3 and 7 are specimens preserved in alcohol, whereas he received the embryos illustrated in the first and second figure as fresh abortions; the fetus in figs. 18, 19, and 20 and the title vignette died by violence, and Soemmerring acquired the mother's corpse soon after her death. The origins of the embryos in the remaining figures, 4–6 and 8–17, is unknown.

The Tradition of Anatomical Illustration

No less than livers, spleens, or lungs, Soemmerring's embryos were anatomical specimens. Likewise, the plates that show them belong to an anatomical tradition of representation, and Soemmerring says as much. Introducing *Icones*, he explains that he regards his plates as complementing William Hunter's famous *Anatomia uteri humani gravidi* of 1774, which illustrated the development of the embryo from the fourth month in opulent copperplates. Soemmerring wished to extend the reach of Hunter's anatomical atlas by addressing the missing stretch of time earlier in pregnancy, so that "my two plates may be added to these plates quite in the manner of a supplement." Accordingly, his chosen format itself cites Hunter's illustrations: "These two, therefore, in association with Hunter's plates, which are exactly the same size and the same format, form a complete sequence of human fetuses from the first origins to complete maturity."[32] It is striking that Soemmerring describes his own and Hunter's plates together as a sequence of embryos from the beginning to the end of gestation, since Hunter's plates in fact show the anatomy of the uterus, with only marginal attention to the embryo, whereas Soemmerring isolates the embryo from the mother's body, which is barely seen at all. His reference to a sequence cannot mean the sequence of images, since Hunter's plates begin with the state of the uterus in the ninth month and follow the progress of its dissection. Neither can Soemmerring be referring to the embryo's gradual change, since the different focal points of the two sets of images — in one case the embryo, in the other the uterus — are mutually exclusive, rather than complementary. By "sequence," Soemmerring means nothing more than the complete duration of the pregnancy.

In other words, Soemmerring cites Hunter not out of an interest in developmental history, but on the contrary, in order to position his own embryological plates within an existing tradition of anatomical representation. It is no coincidence that one of the main aims of Soemmerring's textual apparatus is to synchronize his gaze with that of his predecessors. In this sense, his approach parallels Haller's. Soemmerring compares his own illustrations "with the most excellent pictures of Trioen, Albinus, Wrisberg, Hunter, and Denman, the better to judge what one might correct, add, or more generally better depict."[33] Almost every explanation of the figures additionally contains a precise reference to particular images in other works, which are minutely compared with Soemmerring's own.[34] In this way, Soemmerring, like Haller before him, weaves each of his pictures into the fabric of existing anatomical iconography and furnishes it with the authority of tradition.

No less than the author's own gaze, the choice of draftsman and engraver ensured this integration into the visual world of anatomy. Because illustrations were pivotal to Soemmerring's anatomical studies, the artist Christian Koeck (1758–1828), who created the drawings for *Icones*, was among his most important collaborators.[35] They worked together for the first time on Soemmerring's 1793 treatise *Über die Wirkung der Schnürbrüste* (On the effects of corsetry); the drawing of the female skeleton in *Tabula sceleti feminini* (1796) was also Koeck's work, as were the numerous drawings for Soemmerring's treatises on the sensory organs published between 1801 and 1809. Soemmerring valued Koeck's skills very highly, despite complaining of his unreliability. The son of a stuccoist, Koeck had trained in Paris at the Académie Royale and with the sculptor Jean-Antoine Houdon. His knowledge of sculpture came to the fore when he was commissioned to make the wax models for Soemmerring's essay on the human ear.[36]

The Precision of Beauty
In the process of calibration between things seen with one's own eyes and the illustrations in the existing anatomical literature, the

precise recording of structures was central. Soemmerring considered for inclusion only those specimens that "recommend themselves by their precision, arrangement, and refinement."[37] However, "precision" here meant faithfulness not to the individual detail, but to standardized types. More than that, the visual representation aspired to an aesthetic ideal of universal beauty. Given that many people viewed them "with the greatest disgust, as something malformed,"[38] embryos first had to be invested with an aesthetic quality. Soemmerring constituted the embryo as its own object of beauty, arguing that every period of life enjoys "its own peculiar shapeliness and beauty, which is very different from the beauty that we admire in the age preceding or succeeding it." This was reason enough not to call embryos ugly, "for why should every harmonious form and beauty of the parts of the body be generally absent at that very tender age?"[39]

Soemmerring's pictures of the embryo, then, went in search of perfect aesthetic representation. Correspondingly, the plates show only the collection's "best pieces," those that were not only "free of serious disfigurement, but also distinguished by harmony of the limbs and, in consideration of the age, by beauty, and that had suffered no damage during the preparation."[40] The representation revolves around "shapely [wohlgestaltete] embryos,"[41] chosen for the "charm of the very lovely face," for example, or the "symmetry of the exquisite limbs."[42] Wrinkled skin or the traces of conservation are omitted and the orientation or position of the fetus corrected.[43]

It was not least this attention to the aesthetics of the bodily form that made Soemmerring the first to notice sex-specific differences in the structure of the embryos, appearing early in development. Male and female fetuses differ, he claimed for example, in the construction of the rib cage, the swell of the abdomen, the shape of the head — the limbs of the male fetus, according to Soemmerring, being "longer" and "nobler" than those of the female.[44]

Diagnosis and Comparison
The individual object was at the heart of Soemmerring's images. This has to do with a further task attributed to his pictures: comparison. The sequence of embryos on the plate is defined by their increasing size. To render the size of the embryos faithfully onto the two-dimensional surface of the paper, Soemmerring adopted a projection technique pioneered by the Dutch anatomist Pieter Camper, whom

Soemmerring had visited in 1778 and with whom he continued to maintain close contact. In this technique, "taught by Albinus and Camper, almost in the manner of the architects," "not only the complete length and breadth of the body, the head, the trunk, and the limbs, but also that of the individual parts... are measured with the compasses and portrayed."[45] Camper used the method in *Verhandeling over het natuurlijk verschil der wezenstrekken in menschen van onterscheidene Landaart en Ouderdom* (On the natural difference of features in persons of different countries and periods of life), published posthumously in Dutch and French in 1791 and translated into German by Soemmerring himself in 1792, to compare the skulls of apes, human beings, and classical sculptures; Soemmerring applied it to the depiction of embryo size, finding that it facilitated a "very faithful comparison of the growth and gradual augmentation of the embryos."[46] In Camper's technique, stereometric projection is used to transfer all parts of the body onto paper to scale and thus offers a visual method for the "comparison of all limbs."[47] In fact, it is a method of comparing purely by size. The measuring tape makes all the body's parts equal; emphasizing or suppressing individual structures according to other criteria, such as their significance, is not possible.

Soemmerring's special focus on the size of the specimen and its representation also had a practical aspect. Because the pictures demonstrated the exact relative size of the embryos, they could also serve as the basis for three-dimensional models. As mentioned, Koeck created not only the drawings, but also wax models of the ear for Soemmerring. It seems that in Soemmerring's work on monstrosities, too, the drawings were intended as the basis for models in wax.[48] However, the drawings were not simply a substitute for the actual specimens. They were also an instrument of diagnosis, and *Icones* was a manual for the use of doctors. Its life-size pictures were designed to help the practicing physician to estimate the age of an embryo and thus to make a correct medical diagnosis.[49] Comparison, in short, was the crux of Soemmerring's representation of the embryos: the comparison of size on paper, and the comparison between the specimen as a physical body and the figure as a two-dimensional representation.

Development, Preformation, Epigenesis
Schooling its viewers in the comparative gaze, Soemmerring's engraving shows a sequence of separate, standardized specimens

arranged in increasing order of size. This size order does not in itself imply a relationship of development. Far from it: Soemmerring's plate does not constitute a developmental series, or claim to show a single embryo continuously developing from one picture to the next into the next specimen in a series, for the simple reason that the embryos portrayed are sometimes male, sometimes female. Figs. IX, X, XI, XV, and XVI show male embryos, figs. VI, VII, VIII, XII, XIII, XIV, and XVII female ones. These, Soemmerring stresses, can be distinguished "at first glance."[50] If he had really been interested in representing an unbroken developmental series, Soemmerring could easily have had two separate sequences drawn.

Clearly, Soemmerring was not approaching his object from the perspective of development.[51] His interest was in the individual specimen, the aesthetically autonomous figure, and not the relationship between them. This is shown not only by the alternation of male and female embryos, but also by a marked asymmetry in his explanations of the figures. At no point does he address in any detail how he chose to arrange the embryos. In this respect, *Icones* reveals a striking inconsistency. Soemmerring takes pains to explain to his readers the criteria for selecting the embryos, their aesthetic preparation, and the technique of projection. When it comes to their ordering, in contrast, he notes laconically: "Then I put the selected pictures into a certain natural order and divided them between two plates." Regarding this "natural order," he adds only that he chose those specimens "that in a certain measure were removed from one another in the same gradations or, so to speak, with the same progress of growth, so that, ordered according to a certain rule, they led on directly from one to the next by age and by size."[52]

The natural rules that guided Soemmerring's ordering of the images, therefore, were age and size. In practice, size alone was the rule, given that for some specimens he could only estimate the age—based on the size. Soemmerring set great store by size as the only criterion able to guide the arrangement of the embryos. It is size that Soemmerring means when he writes of the "sequence of the individual stages"; he refers to the "growth and development" of the human body synonymously.[53] The increasing size of the embryos is visually conspicuous. It is an ordering feature that requires no further explanation and seems to brook no contradiction. But as a sequence based purely on magnitude, Soemmerring's plate is not a developmental series. Different as the epigenetic and preformation-ist notions of development were, they agreed on this point: prefor-mationism did not deny that the embryo becomes larger as it pro-gresses in age or the temporal dimension that turns a germ into a complete living being within nine months, and neither did the theory of epigenesis. However, at the heart of the epigenetic theory of devel-opment is the assumption that structures form *gradually* out of the homogeneous, fluid matter of the germ. Soemmerring found him-self unable to champion either theory, and instead declared himself undecided. In his essay on deformities, he wrote that he could "avow neither of the existing theories of generation." Both the theory of preformed germs and that of epigenetic development "contain, in my opinion, truths that may very well and easily be combined with the truths of the other; yet neither of them alone seems to me exclusively true and satisfactory."[54]

In *Icones*, Soemmerring makes no further mention of theories of generation, but an example given in the preface suggests that he did not follow an epigenetic line. Instead, Soemmerring appears to be primarily concerned with the aesthetics of unborn life. To demon-strate the beauty of embryos, he adduces the beauty of a bud. If we find the full-blown rose beautiful, he writes, then the bud hidden within it must also be called beautiful.[55] It is only because Soem-merring takes for granted the continuity between the forms of bud and blossom that he can advocate the beauty of the unborn. From an epigenetic point of view, in contrast, nature's beauty lies not in the form of the complete organism, but rather in the specific form of its becoming. For Caspar Friedrich Wolff and Karl Philipp Moritz alike, the perfection of art and nature — its autonomy and indepen-dence as perfect or complete in itself — was a property not of the finished product, but of its genesis.

The Image as Argument
The Haller-Wolff debate shows that the chronological vision of development, however self-evident it might at first appear, was not the only possible way of thinking about development at this time. In the work of Haller's adversary, Wolff, we find a new concept of development — and a treatment of the visual image that is completely different from Haller's. The retreat from chronology as the order-ing principle of development brought with it a new primacy of the

picture. Wolff no longer used the image as Haller did, to assure himself of what he had seen. Instead, the picture became an argument supporting his epigenetic theory. More than that, development was conceptualized by means of pictures; the pictorial representation took the place of verbal description and, as purely visual evidence, became a proof of epigenetic, successive development.

Whereas for Haller development meant the chronology of his observations, quite separate from the pictures, Wolff constituted development by connecting his pictures. For the first time in the history of embryology, Wolff worked out a method of interrelating his observations through pictorial means. This displaced the formation he had observed in the egg into the picture itself: the representation of the change observed now also revealed how it had come into being. The new form of representation that Wolff invented for this purpose became an indispensable element of his new epigenetic theory of development.

The centrality of visual images in Wolff's oeuvre is indicated by his use of a picture to introduce the second part of his dissertation, on the generation of animals. The fourth figure of this engraving (figure 8.9) shows an embryo at twenty-eight hours: "It presents a kind of mass that is characterized only by its outer shape and its position, and otherwise consists merely of rather incohesive little globules [*Kügelchen*] that are simply heaped together, and is transparent, mobile, and almost liquid, and shows neither heart nor vessels, nor traces of red blood."[56] The *Kügelchen*, or sometimes "vesicles" (*Bläschen*) — in the Latin version, Wolff used the term *globuli* — make up the organizational principle of Wolff's drawings. All the structures of embryogenesis are represented using these little spheres. However, Wolff's *Kügelchen* are not cells in the modern sense,[57] and neither are they defined by a single shared function. The globules are a form of graphic organization, a recurrent figurative element that embeds into the picture the process by which the structures observed "become form." In this sense, Wolff's drawings are themselves an instrument of investigation. The sequence of images is deployed as a kind of zoom, enabling Wolff to move deeper and deeper into the structure of the tissue he was observing. By tracking the development in the egg chronologically, distilling it in individual details, and varying the scale, Wolff could visually circumscribe what he had seen (figure 8.10). A hand drawing by Wolff in a letter to Haller of

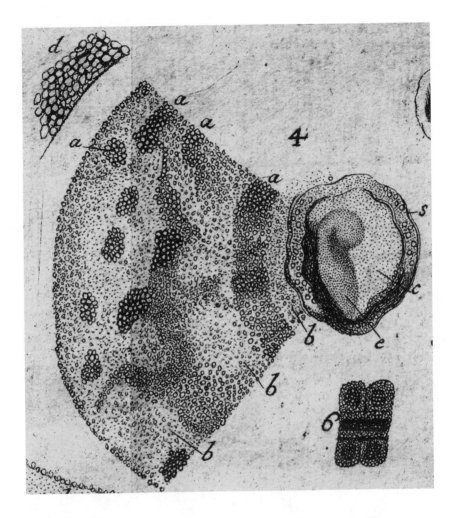

Figure 8.9 An embryo at twenty-eight hours. Wolff, *Theoria generationis*, 1759, plate II, fig. 4.

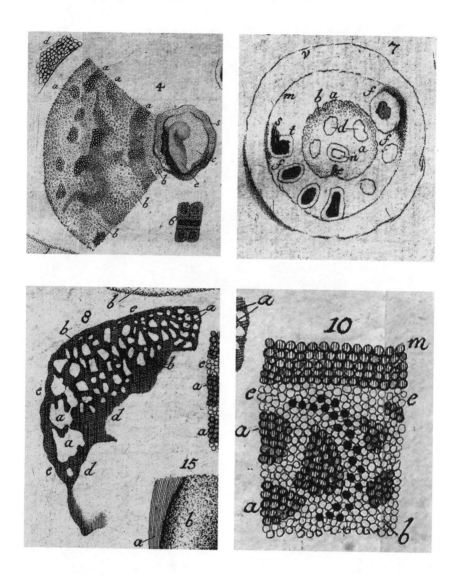

Figure 8.10 With every drawing, the gaze delves further into the tissue. Wolff, *Theoria generationis*, 1759, plate II, figs. 4, 7, 8, 10.

December 20, 1764, shows that print technology was not the defining point here: he organized his pictures in the same way when working in pencil (figure 8.11).[58]

Wolff's great achievement was to have constructed all his images from a single recurrent element, the *Kügelchen* or vesicle, thus reducing them to a common visual denominator. This homogeneous visual construction was what allowed him and every other viewer to trace the coming into being of organic structures in pictorial form. Wolff asked his readers to look in this way when he set his theory of epigenetic, continuous development out of formless mass against Haller's theory of preformed germs. By depicting the relationships between the individual pictures, Wolff made it possible to produce visual connections, draw analogies between different structural changes, and identify regularities. Image 4 at the top left of figure 8.10 shows a detail of the germinal disk at twenty-eight hours with a low magnification; image 7 shows a germinal disk at sixty-four hours seen with the naked eye; image 8 is a detail of the upper part of the previous image; and finally, with image 10, Wolff has arrived at an egg incubated for seventy-two hours, of which he shows the edge of the "vitelline vessel area" (*Dottergefäßzone*). Development begins out of the homogeneous fluid of the egg and consists in change to the arrangement and density of the globules — the change being *solidescibilitas*, the capacity of fluid to become solid. The globules form islands of more densely packed formations (see figure 8.9 and image 4 of figure 8.10) that are separated by the looser, "mobile, finer, and sparser substance (b)."[59] Thus, Wolff writes, the important image 8 shows that what initially appears to be a vessel turns out, under the microscope, to be "nothing other than the islands that fill the interstices of the netlike vessels; one can easily see that these are the same little heaps as in fig. 4 (aa)."[60] How the vessels take shape can "be discerned sufficiently by comparing the fourth, eighth, and tenth figure," and in view of all this, "no error can occur, any more than it can in any truth in the world. These are all immediate observations."[61]

In the dispute with Haller, Wolff's illustrations bore the chief burden of proof. His references to the pictures are not merely a rhetorical device; he makes them "evidence and documents," precisely the quality that Haller denied them.[62] Accordingly, Wolff opens his attack on Haller with the comment: "To show my proof, I refer to the observations that I . . . have presented, and drawn in the fourth,

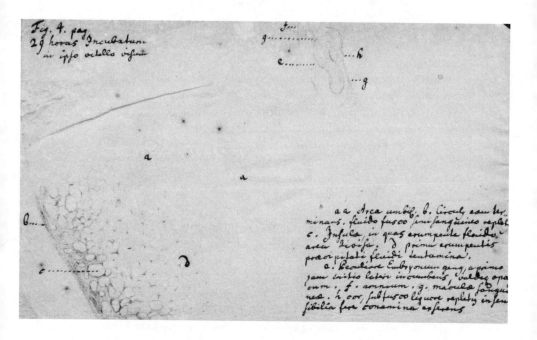

Figure 8.11 Drawing by Wolff in a letter to Haller, 1764.

seventh, eighth, and tenth figure but particularly in the fourth figure."[63] "Fig. 8 alone," he writes elsewhere, "may suffice to convince the reader of the entire theory of the formation of vessels."[64]

Wolff's portrayals of the various formations of the globules are not, then, simply a crutch for the viewer's imagination. They *are* Wolff's theory of development. Wolff drew up his epigenetic theory of development in and with the image.

Haller copied nature at selected points; Wolff conceptualized the actual forms of development vessel by vessel, organ by organ, through the structures of reciprocal reference among his pictures. Unlike Haller's, Wolff's model of development no longer aimed to reproduce mimetically what had been seen at a precisely determined moment in time. Wolff subjected his endeavors to a new observational regime by interrelating observations made at different points in time and doing so by pictorial means. Whereas Haller used images as an aid to vision, comparing his observations at one moment of development with those at earlier moments, Wolff focused on the internal relationships between his pictures. Through his choice of the globule as a unitary structural element, the law of development revealed itself from one image to the next.

Outline and Series: Tredern and Herold

Wolff's new handling of visualization, deploying the referential relationships between images and using the globule as a visual means to constitute those connections, marked the beginning of a new epistemic significance for the image in the epigenetic theory of development. As a further compositional resource, from the mid-eighteenth century, naturalist treatises increasingly moved from lifelike depictions to schematic line drawings or outline duplications of the lifelike images. In his 1758 study of frogs, *Historia naturalis ranarum*, for instance, Johann Rösel von Rosenhof drew each plate twice: once as a hand-colored picture "painted from life" and once as a schematic outline (see figure 8.6). The task of the "outline of all figures, labeled with letters" was to show "everything very clearly."[65]

Rösel von Rosenhof used the combination of picture and diagrammatic outline systematically in *Historia naturalis ranarum*. However, to my knowledge, the first use of outline drawings alone to depict embryogenesis is found in a work largely unknown in the history of embryology: Sebastian von Tredern's dissertation, *Ovi avium historiae et incubationis prodromum,* submitted at the University of Jena in 1808.[66] This text may be regarded as a kind of table of contents for a projected work that never appeared. Even in his own day, Tredern's peers—especially Oken and von Baer—were unsure what had happened to him. A member of the Russian Navy in St. Petersburg, Tredern enrolled at the University of Würzburg in 1804 and studied there in 1807 (possibly with Döllinger), visited Blumenbach once or twice in Göttingen, and completed his doctorate at Jena in 1808. His trail leads to Paris in 1811 and finally ends in Guadeloupe, where he died as a navy doctor.[67] Although so little is known about Tredern and his embryological work, the single plate attached to his dissertation (figure 8.12) is certainly of great significance. Von Baer regarded it as the "crown" of the dissertation:

> It is very simple, executed almost exclusively in outline, but with a correctness in the drawing, a precision, and a richness of detail that seem quite dispro-portionate to the brevity of the work itself. I cannot describe this plate better than by saying that — with the exception of the earliest period of embryonic life, which is here absent — it is richer than anything else in this domain of the literature. . . . It is as if the author mischievously wished to show what he could do and now left to posterity the puzzle of his identity and whereabouts.[68]

The plate shows the development of individual organs in numerous illustrations, such as the development of the beak in figs. 3–10, of the feet in figs. 12–18, and especially of the intestines in figs. 19–32. Nevertheless, the plate makes it very clear that the individual figures are not distributed according to the logic of a developmental series but on the basis of the available space, even if they are positioned relatively close together where possible. Economic constraints may have limited Tredern to a single engraving, resulting in the cramped format. But Tredern's use of line drawings proved to be highly sig-nificant for the tradition of representing ontogenesis. That tradition required new visual forms in order to conceptualize development as a pictorial relationship. The schematic outline — initially only duplicating more naturalistic portrayals — opened the door to an expanded visual repertoire. As we will see, the outline created a transition into Pander's "simulated sections"[69] and Karl Ernst von Baer's "ideal illustrations."[70]

Likewise starting from the diagrammatic drawing, the anato-mist Johann Moritz David Herold (1790–1862) created a novel rep-

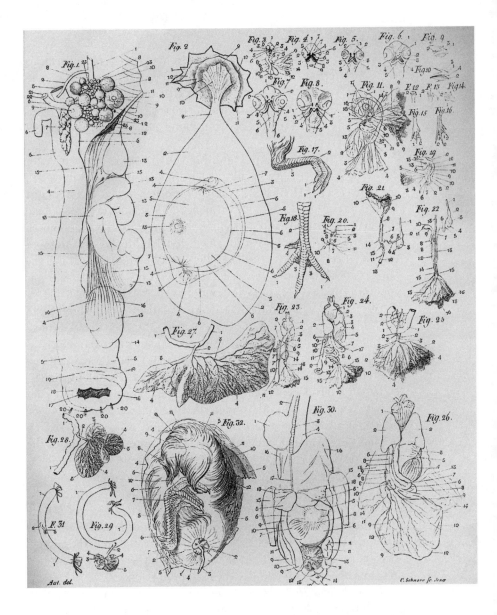

Figure 8.12 Tredern's portrayal of the chick's development. Tredern, *Ovi avium historiae et incubationis prodromum*, 1808.

resentation of insect metamorphosis in his *Entwickelungsgeschichte der Schmetterlinge* (Developmental history of butterflies) of 1815. Not only was Herold's work the first that included a reference to "developmental history" in its title, it also trod completely new paths in the study of metamorphosis and found a new visual language of transformation.

Herold is a little-known figure in the history of science. After enrolling in the medical faculty at Jena in 1806, in 1807, he moved to the University of Helmstedt, where his brother-in-law, Ernst August Daniel Bartels (who strongly influenced his scientific career), held the chair of anatomy and midwifery. In 1809, Herold was appointed prosector to the anatomist Johann Friedrich Meckel the Younger (1781–1833) at the University of Halle; in 1811, he moved again to the University of Marburg, where he remained until his death. He was appointed professor extraordinarius of medicine there in 1816, in recognition of his work on the developmental history of butterflies. In 1822, Herold became a full professor of medicine, then in 1824 professor of natural history and director of the Zoology Department.

Taught by Meckel how to dissect insects, Herold remained fascinated by them all his life. After finishing his work on butterflies, he began a developmental history of invertebrates, the first part of which appeared in 1824 under the title *Untersuchungen über die Bildungsgeschichte der wirbellosen Thiere im Eie. Erster Theil: Von der Erzeugung der Spinnen im Eie* (Investigations into the history of the formation of invertebrate animals in the egg. Part 1: On the Generation of spiders in the egg). It was the first study of spider embryology. The remainder of the work was published only posthumously, when Arnold Gerstenacker presented it in 1835, 1838, and 1876 as *Untersuchungen über die Bildungsgeschichte der wirbellosen Thiere im Eie.* Although Herold's studies made a crucial contribution to the embryology of the lower animals, none of them has hitherto been examined in terms of its significance for the history of science.[71]

Entwickelungsgeschichte der Schmetterlinge of 1815 may certainly be considered Herold's magnum opus. In his day, insect metamorphosis was one of the great unresolved mysteries of natural history. The coming into being of insects was particularly inexplicable: are they a kind of animal that is born not once, but several times—first as an egg, then as a larva, and finally as a perfected butterfly, beetle, or spider—or are they always one and the same animal in different guises? Studies of "metamorphosis" had become increasingly popular from the seventeenth century, as demonstrated by the works of William Harvey (who used the term in 1651 in contradistinction to "epigenesis"), Jan Swammerdam, Marcello Malpighi, Francesco Redi, and Maria Sibylla Merian.[72] But despite an abundance of anatomical dissections, at the beginning of the nineteenth century little was yet known about what goes on inside the insect during metamorphosis.[73] Herold noted that Malpighi, Swammerdam, Rösel von Rosenhof, and Lyonet had all supplied "only fragments" on the transformation of insects, because they had "said almost nothing at all about the changes that occur in the organization's interior during the transformation of caterpillars into pupae and of pupae into butterflies."[74]

The object of Herold's study was the caterpillar of the large cabbage white butterfly (*Papilio brassicae*, Linn.). He restricted himself to the caterpillar, leaving aside the embryonic stage due to the small size of the eggs and the consequent difficulty of observing them. In 1811, Herold discovered that the rudiments of the future sexual organs for male and female butterflies could already be distinguished in the caterpillar.[75] Based on this, Herold thought the transformation of insects afresh, representing the caterpillar's metamorphosis by concentrating on the genesis of a single structure, the sexual organs. To that end, he deployed a new visual realization of metamorphosis: the pictorial series.

Herold's work on the butterfly contains a total of thirty-three hand-colored copper engravings. He drew the pictures himself and had them engraved by the Nuremberg draftsman, engraver, and miniaturist Jakob Samuel Walwert (1750–1815).[76] The pictures show metamorphosis from a new perspective: Herold empties out each individual picture, but multiplies their number. Thirteen plates are devoted to the transformation of the male sexual organs (figure 8.13), fifteen to that of the female sexual organs. The binding of the plates in the book alternates illustrations of the male and female sexual organs; the juxtaposition makes it possible to isolate the distinctions in male and female development that Herold was the first to recognize and that he himself regarded as his most important discovery. However, the visual design of the plates also permitted a different order, allowing readers to reconstruct the caterpillar's "type of development." To trace the emergence of the female reproductive organs, for example, the viewer was to follow the sequence "pl. XXI, pl. XXIII, pl. XXV,

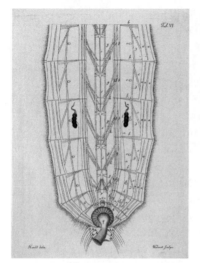
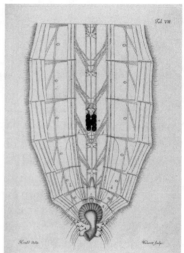
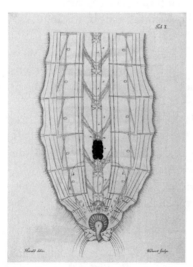
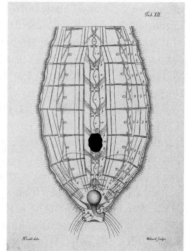

Figure 8.13 Herold uses thirteen plates to
show the development of the male sexual
organs of the cabbage white butterfly larva.
Herold, *Entwickelungsgeschichte der
Schmetterlinge*, 1815, plates VI, VIII, X, XII,
XIV, XVI, XVIII, XX, XXII, XXIV, XXVI, XXVIII, XXXII.

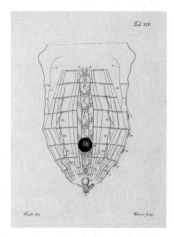

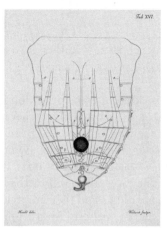

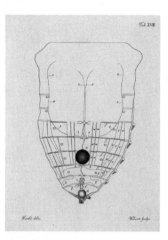

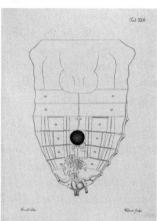

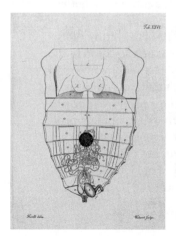

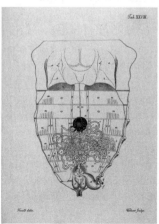

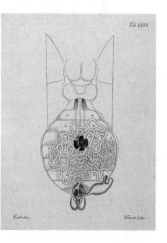

pl. xxvii, pl. xxix, pl. xxx."[77] Rather than alternately, in this case, the plates were to be read in sequence only for either the male or the female caterpillar.

This visual tracking of transformation is made possible by the format of the plates, the most striking characteristic of which is the selection and concentration of visual resources. Maximum attention is guaranteed by the use of color (which we also find later, in von Baer's *Entwickelungsgeschichte* of 1828). Red highlights the male sexual organs of the caterpillar, yellow the female ones—or rather, their changing appearance and their migration within the caterpillar's body during metamorphosis. Another important visual device is the uncoupling of foreground from background. All the tissues of the larva's body that are not involved in metamorphosis are drawn only schematically, in outline; the events in the foreground are completely disengaged from those in the background. The third key feature is the multiplication of images. These pictorial means enabled Herold to isolate a single organ and thus to follow the trail of change from one picture to the next. Taken together, the series of images produces an impression of the continuous development of two small red points into the complex structure of the sexual organs. Color emphasis, the combination of schematic and detailed views, the separation of foreground from background, and the isolation of an individual structure were techniques deliberately deployed by Herold to give primacy to the gradual changing of the sexual organs. The reproductive organs are pivotal to understanding transformation, writes Herold, because "the little caterpillar is furnished with the capacity to perform *all the functions* of an insect *apart from* the *reproductive capacity*."[78] That is, the sexual organs are the only ones available to the adult insect, but not to the caterpillar.

Herold observed not only how the rudiments of the future reproductive organs arise, but also how they change their location in the course of metamorphosis, moving farther and farther downward in the caterpillar's body. As such, he showed that metamorphosis is an epigenetic transformation. The formation of the reproductive organs out of rudiments and their displacement during development proffered a dual argument for successive new emergence, because they showed that the organs are not already present, but take the place of other structures in the course of development. For Herold, caterpillar and butterfly no longer had any resemblance:

This strange animal, formed from the fluid of the egg and consisting of the various parts I have noted, has not the slightest similarity with a butterfly. Because of the peculiar constitution of the organs of which it is composed during its emergence, the butterfly is quite dissimilar to the state of its reproductive capacity, and it appears thus *in the shape of the caterpillar as a separate, independent insect of a peculiar kind.*[79]

In Herold's account, caterpillar and butterfly were two completely different animals. They were nonetheless connected, but the connection no longer consisted in an identity of structures; it lay in a shared process of development. Whereas structural similarity was easy to convey in pictures and offered a visually accessible argument for the animals' congruence, representing a developmental relationship in which the starting and end point looked radically different required new visual tools.

For Herold, the new tool was the pictorial series. In serial representation, change itself became visible for the first time. Development was located no longer in a single image, but in the relationships between images. Organic change was an element that could not itself be observed; it could be made manifest only through the visual arrangement of a series, taking place both within the pictures and between them. The sequence itself—the alternation of representation and interval, fullness and emptiness, shown and not shown—constituted development. It was only the accumulation of changes across the series that evoked the sense of continuous forward movement. But the pictorial series did not merely make development "visible" to the viewer. Far from being just a visual technique, it allowed Herold to rethink metamorphosis—as a process in which organs come into being and replace one another.

Döllinger's Circle
Found in Herold's studies and then in the works by Pander and von Baer that ushered in modern embryology after 1800, the series is a pictorial form that even today remains paradigmatic of the interreferential structure of images and canonical in the visual representation of development. Without the series as a visual form, Pander's and von Baer's developmental theory of the folding of membranes would have been unthinkable—it was the succession of images that gave visibility to the formation happening in the egg. By entwining the

represented and the nonrepresented into an indissoluble unity, the series came to constitute thinking on embryological development.

Research on hens' eggs, which recommenced in Germany after 1800, was driven by Ignaz Döllinger.[80] At this time, Döllinger was probably the most famous physiologist in the German lands, and his reputation as an academic teacher reached much farther. His laboratory brought together students from all over the world, among the most prominent of them being Christian Heinrich Pander (1794–1865) and Karl Ernst von Baer (1792–1876).[81]

Von Baer came to Würzburg in 1815 to study with Döllinger.[82] He knew Pander from his days as a student at the University of Dorpat (today, Tartú in Estonia), and encouraged him to come to Würzburg as well in 1816, after sojourns in Berlin and Göttingen.[83] The story goes that a stroll in southern Germany in early summer 1816 was what sparked the events culminating in the establishment of embryology. Von Baer's autobiography tells of this memorable walk in the village of Sickershausen near Kitzingen am Main, taken by himself, Pander, and Döllinger to visit Döllinger's friend, the botanist Christian Gottfried Daniel Nees von Esenbeck (1776–1858). In the course of the stroll, the three men decided to begin new research on chick embryos. According to von Baer, Döllinger had a strong personal interest in chick ontogenesis, but lacked both time and financial resources. Money troubles had already forced him to give up his own studies of development.[84] Pander, the eldest son of a Riga merchant, could contribute not only an intellectual interest in embryology, but adequate funding for the experimental studies that Döllinger envisaged.[85]

Under Döllinger's aegis, Pander, d'Alton, and von Baer began work on chick embryos in Würzburg. Döllinger provided technical and scientific guidance and supervision for the incubation experiments, which lasted more than a year. Pander commented in his Latin dissertation that Döllinger had allowed him to take over and continue his own unfinished studies. He also trained Pander in the requisite techniques, instruments, and methods.[86] However, no less formative for nascent embryology than Döllinger's technical expertise was his view of art and science. For Döllinger, science and the arts were not two separate spheres, and science was by no means superior to art in its explanation of the world. On the contrary, both worked solely at the level of intuition (*Anschauung*). For this reason, he believed, science and art could make the most progress if they

worked in association—especially when their aim was to compre-
hend the living world: "Nothing, it seems, do men find more difficult
to understand than that true science knows nothing of explanation,
and that she unites with her sister, art, in this very point: that science, like art, intuits, but especially for the theory of life and its
mysterious manifestation, how can this prevail against the mania for
explanation without being woefully distorted?"[87]

Seeing art and science as an ensemble, Döllinger placed the image
at the focus of not only artistic, but also scientific practice, and especially of research on embryogenesis. Döllinger was well acquainted
with the iconographical tradition of embryology.[88] Von Baer's mem-
oirs tell us that the graphic presentation of the results had been
decided far in advance:

> Döllinger had been turning the whole project over in his mind for some time,
> as he now declared that to achieve full understanding, it was necessary for the
> different developmental stages discovered to be drawn not only accurately,
> but also artistically, so that they could serve for copper engravings to be pre-
> pared later. Thus, it would be most desirable if a draftsman and engraver could
> be found united in one person.[89]

The result of this collaboration, the treatise *Beiträge zur Entwickelungsgeschichte des Hühnchens im Eye* published by Pander, was
designed from the very outset to be a pictorial work, and accordingly, the artist and copper engraver Joseph Wilhelm Eduard d'Alton
(1772–1840) played a crucial role in the embryological work that began
in Döllinger's rooms in the early summer of 1816.[90]

Döllinger was the one who made contact with d'Alton, a "man
extremely learned both in the study of natural things and no less
so in pictures, as well as a marvelously expert engraver," as Pander
wrote in his dissertation, and he insisted on the artist's participation
as a precondition of the project.[91] D'Alton combined the skills of a
draftsman, engraver, and naturalist. He arrived in Würzburg in July
1816, and the embryological studies continued for the whole summer
and fall of that year.[92] D'Alton knew Lorenz Oken and Goethe[93] and
also the botanist Nees von Esenbeck, through whom he probably
met Döllinger. In 1810, he came to prominence with the anatomical study *Naturgeschichte des Pferdes* (Natural history of the horse).[94]
Although we do not know whether d'Alton was interested in embryological themes before meeting Döllinger and Pander in Würzburg,

his collaboration on the embryological project led to a close working relationship with Pander. The two men traveled through Europe together, and d'Alton drew and engraved the illustrations for the osteological compendium that resulted from this journey, as well as for Pander's embryological research.[95] Little more is known about the biography of d'Alton, whose papers have been lost, than that in 1818 he was appointed professor of archaeology and art history at the University of Bonn, where he taught natural history and the history of art.[96] It is unclear what contact he had with artists and naturalists and whether any further works arose from this—apart from a short report on a substantial collection of art (now lost) that d'Alton is said to have possessed. The report was written by August Wilhelm Schlegel, who speaks of "an immortalized friend," a connoisseur of art and nature in equal measure.[97]

Here we have come full circle, as the movement of knowledge between science, art, literature, and aesthetics once again becomes apparent. It may be assumed that even if they were not all personally acquainted, August Wilhelm Schlegel, Goethe, d'Alton, Pander, Döllinger, and von Baer were at the very least aware of one another's work.

Karl Ernst von Baer soon left the circle around Döllinger, and he resumed his own research only later, in Königsberg. For von Baer, too, the image was at the heart of his embryological research. "In the whole of nature, no point is as important as the formation of an organism out of a fundamental mass; it is here that the key to all physiology and biology must lie," he wrote to his friend Waldemar von Ditmar in July 1816, referring to the beginnings of Pander's embryological work. However, "without plates, the whole matter will hardly be understood by the reader. Pander has now resolved to study the developmental history of the incubated egg and to publish it with copperplates."[98] Von Baer placed great value on visual representation in his own works, as well. His *Über Entwickelungsgeschichte der Thiere* (On the developmental history of the animals), the first volume of which was published in 1828, was an initial step in a more ambitious project, a "large volume of engravings on the developmental history of all classes of animals."[99] In his correspondence with the Prussian minister of instruction, Karl von Stein zum Altenstein (1770–1840), von Baer described the difficulties of rendering his research in pictures and of finding artists and engravers suited to that task. He reported that he required a copper engraver to execute the illustrations, but "the naturalists of Königsberg suffer very much from the unfortunate circumstance that there is no copper engraver resident in within a radius of almost eighty miles."[100] Minister Altenstein promised funding, but in 1830, the dearth of suitable artists forced von Baer to travel to Berlin in search of an engraver. In a later missive to Altenstein, he wrote:

> In order to understand the history of development, illustrations are indispensable. Not only does their preparation require considerable artistic talent, but it is also necessary for the draftsman to be constantly at hand, so that he can immediately portray especially the eggs and embryos at an earlier stage, which spoil extraordinarily fast. . . . For the publication, however, it is also necessary for the engraver to work under the observer's supervision, because corrections are almost impossible to carry out adequately at a distance, as I have realized only too painfully after many irksome experiences. It would be most advantageous if draftsman and engraver could be united in a single professor.[101]

Convinced that development could not be understood without the right visual representations, von Baer thought very carefully about the preparation of drawings and plates. In Berlin, three engravers were recommended to him by d'Alton and others, but only one was available — Bernhard Wienker, who had trained at the Royal Institute of Technology in Berlin.[102] Although, under von Baer's guidance, Wienker successfully mastered the work and ultimately prepared both the engravings and the drawings themselves, he spent only one year in Königsberg, leaving the city in 1832. Von Baer would have liked to retain Wienker for three years, since otherwise he could not make "any firm plans for the course of the investigations on the first form of the mammalian egg."[103]

Von Baer's attempts to arrange for his developmental history to be illustrated in Königsberg had now reached a dead end. This appears to have contributed to his decision to leave for St. Petersburg, where he hoped to find more expertise and financial resources, in 1834.[104] At any rate, the planned volume of plates on developmental history never appeared. *Über Entwickelungsgeschichte der Thiere* of 1828 includes only three plates, and its second volume, of 1837, another four.

CHAPTER NINE

Folding into Being:

Christian Heinrich Pander

Pander's *Beiträge zur Entwickelungsgeschichte des Hühnchens im Eye*
(Contributions to the developmental history of the chick in the egg)
of 1817, the outcome of his collaboration with d'Alton, Döllinger, and
von Baer in Würzburg, was a milestone in the history of embryology.[1] While Caspar Friedrich Wolff's work on the intestinal tube had
tracked the development of a single organ from its very beginning
to its complete formation for the first time, Pander was the first to
describe the gradual formation of *all* the organs in the egg, step by
step. He focused on the first five days of development, when the
emergence of initial structures in the egg is most difficult to observe.
Yet this is the crucial phase, the one in which the foundations for the
organs' formation are laid.

To Form Is to Fold
Eighteenth-century and early nineteenth-century physiology
described the basic substance of organic life as "mucus" or "jelly,"
as liquid and amorphous, but also as a tissue, made up of granules
or globules that are constantly changing their shape. We have seen
that Wolff regarded the first structures within the liquid substance
of the hen's egg as emerging from an alternation between flux and
solidification of the globular or granular substance. Later, Döllinger
depicted all the changes in organic animal matter as ordered formations of the movements of globules. Pander, too, began from
the movements of the granular substance. What was new was his
observation of how, as incubation begins, the globules rearrange
themselves to form membranes or layers. In the unincubated egg,
he wrote, the germ layer consists "of a simple layer of conjoined
granules,"[2] but in the early hours of incubation, the globules re-form

to create two membranes. This differentiation of the germ layer is an alteration in the configuration of the globules, which previously had formed a homogeneous, unstructured substance. In the germ layer, "there forms on its outer surface, which is turned toward the shell or at first toward the yolk membrane lying over it, a new, very delicate, but dense layer that consists of less distinct granules and is more uniform."[3] After twelve hours, the globules had separated out sufficiently for Pander to be able to identify two layers: "the germ layer now consists of two quite different lamellae: an inner, thicker, more granular, less transparent one and an outer, thinner, smoother, transparent one; the latter, for the sake of a more precise description and because of its development, we will call the serous layer [*seröses Blatt*], just as we call the former the mucous layer [*Schleimblatt*]."[4]

The two layers into which the blastoderm differentiates,[5] what Pander calls the serous layer and the mucous layer, are distinguished by the different structures of their globules—a coarse, grainy structure on the inside and a more homogeneous structure on the outside. After about twenty hours of incubation, a further metamorphosis of the blastoderm commences and from the twentieth to the twenty-fourth hour leads to the formation of a third layer between the two existing ones. This new layer, which Pander calls the "vascular layer" (*Gefäßhaut*), also originates in the formations of the granular mass of organic matter. First, little islands arise, "formed from very small globules."[6] In Pander's vision of the membrane's formation, the layer of globules covering the serous layer first dissolves into a netlike tissue, torn in places. The globules divided by these tears assemble into islands of blood that gradually take on a red color. The islands then begin to interlock, forming a reddish network of islands and spaces between them. Wavelike streams of globules arise, and their flow gradually, after around twenty-four hours, gives rise to the vascular layer between the serous and the mucous layer.[7] After twenty-four hours, therefore, the blastoderm consists of three membranes, the basis for all other forms in the egg to emerge: "Whatever remarkable things may occur subsequently, it is only ever for the sake of a metamorphosis of this membrane and its layers [*Blätter*], endowed with an inexhaustible abundance of the formative drive [*Bildungstrieb*]. It is from the membrane that life radiates in all directions and back into the membrane that life retreats in a concentrating form."[8]

By deriving the emergence of all the egg's structures from the metamorphosis of elementary membranes, Pander set the scene for modern embryology. The germ-layer theory formulated by Pander and later by von Baer remains valid in outline even today. It holds that the entirety of embryonic development occurs as a gradual differentiation of originally elementary membranes, the "germ layers" or *Keimblätter*. But whereas historians of biology have frequently paid tribute to Pander and von Baer for discovering the germ layers, another and epistemically crucial facet of Pander's work has gone almost unnoticed: the idea of the fold and its role as a rhythmical figure.[9] In this chapter, I show that Pander made the fold the law of metamorphosis. For him, development, or metamorphosis, signified nothing other than folding, with all the changes in the egg taking place as a series of folds in the germ layers. The concept of the fold, I argue, gives order to physiological movement in both the temporal and the spatial dimension. In other words, the fold is the rhythm of ordered motion in space.

Gilles Deleuze saw the fold as the characteristic feature of the Baroque,[10] placing it at the core of his analysis of Baroque art, architecture, mathematics, music, and Leibnizian philosophy. For Deleuze, the fold is a figure of difference. Its movement differentiates and is differentiated, includes and excludes, embraces and delimits, is inside and outside, a surface and a form. It divides and connects matter and soul, the two "floors" of the monad, and it underlies Baroque thinking on infinity and processuality. The principle of folding, folding in, folding out, refolding, and unfolding casts new light on the object—which is now located on a continuum of variations, a perpetual metamorphosis of tipping, continued motion, flux from one state to the next, then to the next, and so on.

It is this temporal and spatial ordering, the choreography of space-time change, that defines folds for Pander, as well. Pander's fold produces a form along its line; every warp, swelling, bulge, or indentation, at every location and at every moment, directly changes the space-time coordinates of the entire embryo. Every form that closes off to the inside simultaneously opens a new space to the outside, so that every outside is also an inside. The distinction between formed and unformed matter is always only provisional; their relationship is reversed repeatedly in the course of embryogenesis. What is formed is destined not to persist as it is, but to dissolve, to change places, switch sides, transform its exterior into an interior, its surface into a body. Starting again and again, distinct yet entwined, differentiating and varying with each repetition, the folds gradually make manifest the shape of the embryo.

The Metamorphosis of Membranes

Once the germ layers had appeared, Pander no longer focused, as Wolff or Döllinger had done, on the formations of the individual globules, but on their layering. He saw the germ layers as formations of globules whose movements in space could be described as units. By gathering globules into the larger complex of the membrane, Pander was able first to trace developmental processes as the motion of these units. He observed that all the structures in the egg arise from the spatial shifts of the germ layers. In a second step, Pander derived from the membranes' movements in space—their warps, turns, approaches, fusion—a law of folding. The blastoderm itself

> forms the body and viscera of the animal solely through the simple mechanism of folding. A delicate thread attaches itself to the blastoderm as the spinal cord, and no sooner has this happened than it forms the first folds, which had to show the spinal cord its location, as an envelope [Hülle] over the precious little thread, thus forming the first foundation of the body. Then it moves on to a new fold, which, unlike the first, fills the abdominal and thoracic cavities with content. And it sends out folds for the third time to wrap the fetus formed from it and by it in suitable envelopes. So it is hardly amazing that there is so much talk of folding and enveloping in the course of our account.[11]

Like a wave, the movement of folding traverses the body in a regular rhythm, giving the emerging embryo more form with each iteration.[12] The motion of wavelike flux is constant, yet also variable, in its form. Pander is here using the concept of the fold to give a regular and spatial order to his complex observations of the movements within the egg. On the level of description, then, Pander's fold is the figure of rhythmical order.

Pander distinguishes three foldings. The first begins after eighteen hours of incubation, in the mucous layer and the serous layer, and gives rise to the primitive streak with its two primitive lateral folds (Primitivfalten), the beginning of the embryo.[13] The primitive folds take shape in the germinal area as "two faint, parallel little streaks running lengthways."[14] As development proceeds, "they merge at the wide end of the germinal area [Keimhof] by uniting in a little arch, and so the distinction between above and below comes about; because the folds at the opposite end remain separate. At the archlike, closed end of the two folds, the head of the chick develops, at the open end its tail."[15] Between the primitive folds, the beginnings of what Pander thinks is the spinal cord emerge: "a delicate streak, roundish at the top, broader and lancet-shaped at the bottom—the rudiments of the spinal cord."[16] Having initially run in parallel, the primitive folds, "as the arch that joins them now faces downward, turn over the spinal cord that is between them, serving to protect and envelop it."[17]

Underlying this succinct description is a far more complicated process of multiple bending and curving. Thus, the "approach and unification" of the primitive folds

> does not occur abruptly, suddenly hiding the spinal cord under a sheath; instead, first the primitive folds are bent in waves along their entire length, such that every expansion corresponds to the space between each of the spinal rudiments, and every constriction corresponds to its adjacent, roundish-square corpuscles. Then the edges of these folds join in the middle, and by growing together, they form a seam that covers the part of the spinal cord below it as a whitish strip. The two folds do not unite so quickly at the top and bottom; at the bottom, they separate at an acute angle, and between them there immediately appears the thread of the spinal cord with its lancet-shaped end; at the top, they also diverge at a sharp angle, but run in parallel and separately as far as the sickle-shaped fold, curve like waves, and because their edges are not folded, but stand straight up, between them there emerges a row of three to four spaces or cells, increasing in size toward the head end, apparently enclosed on each side by two lines, because the two layers [Blätter] of which each fold consists can just be seen through their edge as lines.[18]

The degree of detail in this description shows how difficult it was for Pander to see the processes in the egg as movements and make them intelligible as such. He quite rightly remarks: "If any investigation of the incubated egg poses difficulties, then it is the one by which one may reach the results I have mentioned here."[19]

A second folding, this time of the primitive streak, produces two further embryonic structures, the beginnings of the esophagus and heart. It is important to note that each movement of one membrane entails a displacement of the next, so that "when the primitive folds turn downward at the head end, forming the transverse fold . . . they not only pull along the membrane between the folds, but also the adjacent membrane above them."[20] The motion of the primitive folds stops at this point, but the membrane between them moves "from the edge even further down toward the tail, and, when it has moved down approximately one to two lines, it turns back again over itself, and runs on a single plane over the embryo's head." The folding occurs both from above and from the sides. Because the lateral folds later move closer together and finally fuse at the center, "it is natural that the space formed by the descending part of the membrane separates into two pipe-shaped tubes, each of which forms a sac closed at the top and the sides and open at the bottom." The first sac marks the beginnings of the esophagus, the second the beginnings of the heart.[21] As metamorphosis continues, the channel of the heart, formed by the membrane's folding, undergoes further folding, "sometimes binding [*Verschnürung*], contraction and expansion of its walls," until it finally turns "into the completely formed heart."[22]

While the second folding occurred at the head end of the primitive folds, the third — a "similar folding, but different in size and form" — occurs at the tail end a few hours later.[23] This third folding closely resembles the preceding two, "only now the entire development is turned outward and proceeds along the embryo's back," whereas the previous foldings had "formed on its ventral side and the folds were turned toward the yolk." The resemblance between the three foldings lies in all three layers initially working together, "until the serous layer reaches a stage when it is able to lead the process on its own."[24]

Thus, the third folding occurs at the lower, rather than the upper end of the primitive streak, on the dorsal, rather than the ventral side of the embryo, and turns outward, rather than inward. Because it starts at a different location and moves in a different direction, what it produces is not the tubular beginning of the digestive system and the heart, but the extraembryonic structures. The result of this folding is the amnion, "which at its very beginning consisted of a fold of two layers."[25] The two lateral folds gradually grow together along a dorsal seam from the head to the tail of the embryo. This produces two envelopes, which Pander calls the "true amnion" — "a sac filled with water" or a "bladder" (*Blase*) formed from the inner layer, which "wraps . . . the fetus down to its abdomen"[26] — and the "false amnion," formed from the outer layer. The false amnion, barely distinguishable as a separate membrane, replaces the vitelline membrane: "The point where the seam has closed moves . . . away from the true amnion below it, and spreads out thinly over the fetus enclosed in its true amnion, and extends over the germinal membrane up to its outermost edge."[27] Another extraembryonic membrane, the chorion, is likewise produced at the lower end of the fetus, but this happens only once the amnion has reached completion and the intestine has developed sufficiently for the esophagus and the rectum to be distinguished. The chorion "springs from the anterior wall of the rectum, at its transition into the cloaca, as an elongated bladder [*Blase*] that narrows in the middle and is thereby divided into two unequal halves."[28]

The formation of the primitive folds, the intestinal tract, and the heart, along with the extraembryonic structures of the amnion and the chorion, completes the elementary organ systems of the early embryo. These systems are produced by three foldings of the membranes, which are rhythmical in that they are repeated at different points in time, occur in different sections of the egg, and are fundamentally similar, but not identical. Like rhythmical undulations, they bring forth the forms of the future organism. The rhythmical fold is the abiding formal principle that directs the emergence of a future life, yet without subjugating it to an unbending rule—such rigidity would explain the recurrence of the motion, but not its variability.

Pander's findings met with little enthusiasm among most of his peers. Von Baer, who had left the Würzburg project early on, declared that he found the work incomprehensible and added that this impression seemed to be widely shared.[29] Other contemporaries, however, recognized the importance of the concept of folding. In 1818, the Bavarian physician and astronomer Franz von Paula Gruithuisen wrote that Pander's work "institutes a kind of folding system," even though Pander had not been able to illuminate exactly how the germ layers "conjure all the organs from among themselves by folding" or "how this magic occurs."[30]

A New Observational Regime

Pander first recounted the folding of membranes in 1817, in his Latin *Dissertatio inauguralis sistens historiam metamorphoseos, quam ovum incubatum prioribus quinque diebus subit.* His attempts to describe the folds indicate the difficulties he faced in understanding the movements of the membranes and the formation of the individual organs.

Pander was aware of the limitations of trying to capture the concept of the fold in words, and accordingly the dissertation, itself published without illustrations, promised to provide drawings in the future.[31] *Beiträge zur Entwickelungsgeschichte des Hühnchens im Eye*, which appeared the same year, contained not only the results of the dissertation in German, but also ten copper engravings—in fact, the text was merely "added to the plates."[32] This accompanying text of slightly over forty pages, twelve of them directly describing the pictures, was to be considered an introduction to the plates, as opposed to the plates being illustrations of the text. For this reason, in the following, I will treat Pander's *Beiträge* as a visual work. It reveals a transition from text to image, from description to drawing, from the law of rhythm to the pictorial series. Pander used the visual medium to tackle a task that had largely eluded him in words: portraying the temporal order of folding.[33] He now shifted the temporal order of development into the relationships between images. Development became both the subject matter of the pictures and a completely new pictorial form.

The developmental series as a novel form of visual representation involved not only the emergence of a new conceptual framework, the rhythmic episteme, but also new observational techniques and experimental practices. Among the skills that Pander had learned from Döllinger was the preparation of the egg,[34] and von Baer attributed the successful observation of the very early stages of development to a special technique that Döllinger had invented:

> First, the section of shell above the air space is knocked in, causing the entire yolk to sink. The germ layer [*Keimhaut*], including the embryo, which, after several days of incubation, has moved close to the shell, is thereby moved away from the shell. After that, a larger section of the shell above the embryo is opened and the entire yolk poured into a dish containing water. A disc-shaped section is then cut from the vitelline membrane and, provided development has not progressed too far, the germ layer is carefully detached from the vitelline membrane in order to place the former, with the embryo—floating on some water or not—under the microscope.[35]

This procedure made it possible to detach the developing embryo from the shell without damaging it, then to separate the blastoderm from the vitelline membrane intact for observation. If this technique had been known before Pander's day, wrote von Baer, the first days of the embryo's development would doubtlessly "have been completely understood at a much earlier time." Probably Pander's predecessors, "with the possible exception of Malpighi," in fact applied this procedure: "Kaspar Friedrich Wolff may indeed have used the same method, but since he did not describe it, it had to be rediscovered."[36]

Among the pioneering aspects of the collaboration between Döllinger and Pander was the experimental system that the two men used. They had special incubating machines made, enabling them to incubate not just a single egg, but dozens at once. The incubator generated a constant temperature between 28°c and 32°c. In all, around two thousand eggs were incubated in this way.[37] The natural warmth of the hatching hen was replaced by an unvarying, standardized incubation, controlled externally through the regulated temperatures and cycles of the incubator. Furthermore, forty eggs could be brought to maturity in the machine and their development tracked simultaneously.

Although this technical advance allowed the chronology of changes to be far more precisely captured, chronology was rejected as an ordering principle. This was the revolutionary contribution of Pander and Döllinger: rather than thinking of development as a course of events that follow the chronology of observation, Pander considered himself no longer "in the least bound to particular periods of time," but able to follow "each individual phenomenon up to its complete formation or disappearance."[38] Gruithuisen called this a "discursive manner" of explaining development.[39] Certainly, it was a breach with chronological sequence, which was now replaced by the logic of selection.

"So We Selected": Constructing the Developmental Series
The objective of Pander's *Beiträge* was to portray "the gradual development of the chick in the egg in its entirety": "The emergence of all the individual parts from earlier ones, and those earlier parts from the very first germs, viewed clearly and purely, was to be presented here to the friends of natural history, so that the formation of the completely developed animal, traced back to its first origins, could become intelligible."[40] The following passage explains what Pander meant by clear and pure vision. To see meant

> merely, without worrying about every possible physiological interpretation, to present the phenomena . . . just as they showed themselves to us at every moment of observation, so as subsequently, even without knowing in advance

what the observation would grant us at any one moment, *after multiply repeated investigations and comparisons to draw out what is most significant, and to be able to derive the individual phenomena from earlier, already known forms of development.* This is why the most faithful possible representation of each individual perception was our chief intent, and this is why, having for the time being refrained from making our own judgments, we could never attribute more importance to one part than another, or even wish to emphasize it at the expense of others by a greater firmness in the drawing, so as not to entangle ourselves in our own nets.[41]

It would be hard to overstate the importance of Pander's point here. On the one hand, he argues that development cannot be traced by looking at a single individual: the large number of simultaneously incubated eggs supplied by the incubator permits an abundance of observations, and that abundance is necessary because the whole problem of understanding the course of development lies in distinguishing what is significant. Development is the story of emerging forms—which "have not yet achieved their firm, lasting form, but are still continually changing," as Wolff had put it.[42] Because forms are in a state of continual change, the observer initially faces an undifferentiated mass of material, and as a result, development cannot be *observed*, but is *constructed* by selecting all those states and forms in the embryo that can be set in relationship to earlier and later forms by means of looking and comparing. The medium for such looking and comparison is the picture.[43]

Selection also forms the basis for the innovative strategy of representation chosen by Pander in 1817:

So, out of the mass of our store, we selected a cycle of depictions—regardless of the time periods, so variable in their products, in which they revealed themselves to us—and kept in sight especially those organs whose development we were able to present from their first emergence to their perfect formation with the greatest completeness. Classification based on period of incubation was bound to appear quite futile, because we realized that there was no single order in the relative time sequence of the development of different eggs; instead, eggs incubated for a shorter time often proved to be further advanced than ones which had been exposed to warmth for longer.[44]

Different eggs incubated for a comparable length of time under comparable conditions do not necessarily reach the same stage of development. In view of this temporal variability, measuring the passing of time is a useful starting point, but by no means itself an explanation of embryonal development. Pander was interested not in the chronology of development, but in the sequence of changes marking a specific embryonic structure at various, arbitrarily selected moments in development. His innovation thus lay in the arbitrary and constructed choice of figures for portrayal. Pander selected the developmental stages to be depicted not according to particular points in time, but according to possible relationships with other developmental stages in the picture sequence.

It was this criterion that enabled Pander to make the embryo, and not time, the point of reference for his study. The relationship between the pictures guided his experimental procedure: of the many hundreds of eggs examined, he selected only those that lent themselves to being set in relationship to one another and lined up serially from picture to picture—quite independently of the incubation time. Morphological comparison alone determined the selection of the forms that would be significant for development, and it was the relationship of morphological states, not the chronological sequence of observations, that permitted Pander "to derive the individual phenomena from earlier, already known forms of development."[45] As a consequence, the states described in Pander's treatise were not necessarily privileged stages of development.

What was new here was the constitution of development as a relationship of morphological forms and of their advancing change as variation relative to what had been shown before. Here, the picture was constitutive not only because it was the medium in which forms could be apprehended with the necessary precision; more than that, the pictorial series was the visual resource that enabled the relationship of forms—or, put another way, forms as relationships—to be depicted. Its reference point was no longer a time line, but a relationship of reciprocal allusion. Serial representation had the quality of literally showing both single moments and the whole picture of development simultaneously and directly. In this way, it could deliver what Döllinger had lamented as missing from his work on the circulation of the blood: "One surveys so many single moments that, taken together, yield a whole picture, but these single moments are so delicate that it is difficult to put them into words, and when recounted individually, they do not create any effect, any conviction; one simply has to see the whole itself."[46]

Serial representation achieved the effect that Döllinger envisaged—that of convincing the viewer—so brilliantly that the series is taken for granted today as the self-evident way to represent development. Yet developmental series are not the "natural" portrayals of development that they may appear to be. On the contrary, the series constructs change, in many respects: it constructs the change from one form to the next, and it constructs each individual form, which can be isolated as a single form only through its relationship to the sequence of other forms. Behind Pander's ingenuous formulation "so we selected," then, stood nothing less than a revolutionary epistemological operation.

An "Assemblage of Embryos": Pander's Plates

Beiträge zur Entwickelungsgeschichte des Hühnchens im Eye contains ten copperplates, which have never been analyzed in detail. They are preceded by an introductory text and the explanations of the plates. Unusually, bound between the pages of engravings there are line drawings printed on thin, semitransparent sheets, smaller than those used for the copperplates. These drawings are schematic representations of the plates that follow them (or rather, that can be seen through them). The only exception is the final plate x, where the added sheet shows the "table of sections," while the plate contains microscopic images. Opening the book, therefore, readers are not offered an immediate view of the engraving. Instead, their gaze falls first on the schematic drawing, through which the engraving can be glimpsed faintly. Although the drawings do not cover the engravings completely because of the smaller format, the impression nevertheless arises of an almost direct relationship of copy to original: the filigree lines of the schematic image outline the copperplate illustrations as if through frosted glass. Oken, with his characteristic attention to detail, found the plates to be "engraved so finely and transparently that one feels one could puff the figures away with a single breath."[47]

The explanations accompanying the plates refer to the lettered segments of the schematic drawings. In all the plates, the figures are arranged into rows and columns—except for the eighth plate, which shows a complete view of the embryo embedded in the blood vessels of the yolk. The plates represent various processes in the egg.

Plates I, II, III, and VII show folding processes; plates IV, VI, and VII the formation of the vascular system; and the final plate presents development through folding in the form of schematic sections. As for the processes of folding, the first plate shows the development of the germ layer up to the emergence of the primitive folds; the second, the bending of the primitive folds; the third, the formation of the embryo's ventral and vascular cavities, also through the folding of the primitive streak; the seventh, the development of the heart out of the folds at the top end of the embryo.

Plate I (figure 9.1) portrays the initial thickening of the germ layer in nine figures arranged in three columns and three rows, numbered from right to left as fig. I to fig. IX. The figures follow the chronological development from an unincubated egg (fig. I), through eight (fig. II), twelve (fig. III), sixteen (figs. IV and V), and twenty hours, up to the third day (fig. VII). The concluding figures, VIII and IX, show the state of the unincubated egg.[48]

Development begins with the thickening of the blastoderm, indicated in the transition from fig. I to fig. II. In fig. III, we see the blastoderm's first flexure, which gives rise to the primitive folds in figs. IV, V, and VI. The vascular layer is shown in fig. VII. The last two figures in the plate join up with the first, both showing the egg in its initial state — but in one case capable of development, in the other not. The intervals of time between the individual figures are irregular. There are eight hours between the first and second figure; fig. III follows after another four hours, as does fig. IV; from fig. IV to fig. V there is no interval; no comment is made regarding fig. VI, and fig. VII jumps to the third day of incubation. However, the serial representation visually standardizes these intervals. The pictures are ordered in rows and columns of strict regularity, so that the gap between the images becomes a regular space. In other words, although the periods of time between the individual states vary, the series produces an ordered, law-governed relationship between those states. The images refer to each other, rather than referring to a chronological interval that divides them. In the subsequent plates, it becomes even more obvious that chronology is irrelevant as a criterion for the egg's development: in them, Pander makes no mention of time at all.

The second plate (figure 9.2), ordered in the same way as the first, traces the folding of the primitive streak in nine consecutive figures. Here, folding takes place not only from figure to figure, but within the figures from top to bottom, that is, from the head end (the folds bend at points "c" and "d" in figs. i through v), via the center of the body, to the tail end, where the spinal cord is formed in the convergence and fusion of the folds (points "d" and "e" in figs. vi through ix). Pander and his engraver d'Alton realized the spatial dimension of the folding process by switching between front and back views of the embryo; thus, from fig. ii to fig. v, the view changes from the dorsal to the ventral aspect. This alternation requires the viewer to relate the pictures to one another in order to obtain a physical idea of the embryo, without which the complex three-dimensional movement of folding cannot be understood. In the switch from one aspect to another, the eye of the viewer reiterates the folding of the membrane. The relationality of the pictures is doubly constituted—through the alternation of perspective and through the absence of any chronological information on the individual figures.

The third plate (figure 9.3) also presents the process of folding through the pictorial resource of turning the embryo from a ventral to a dorsal position and sequencing the figures. Plate iii shows how the ventral and thoracic cavities of the embryo are formed by the folding of the primitive streak.[49] The heart exemplifies this visualization. The rudiments of the heart are shown in fig. ii in the structures labeled "i," "k," and "l." In fig. iii, the primitive folds have closed up, and the advanced outward curvature of the heart can now be seen from the back. Fig. iv portrays a further advance as the heart "tortuously begins to position itself on the left-hand side."[50]

The seventh plate (figure 9.4), in figs. i through v, shows the formation of the heart.[51] The letter "f" marks the heart, or more precisely first the head fold out of which the heart develops (except in fig. ii, where the head fold is labeled "a"). Here, folding means the contraction of the folds into a canal. The contraction takes place only two-dimensionally, which is why Pander does not switch between perspectives in this plate. However, after the third figure, the viewpoint changes, and the embryo is shown from the side. This rotation indicates an actual sideways turn that the embryo completes in the course of development. The last two enlargements of the embryo's head fold and the heart (figs. vii and viii, "f") are explained only briefly, and their purpose remains unclear.

Common to all the plates is their organization as an "assemblage of embryos."[52] This includes plate v (figure 9.5), yet that plate is highly

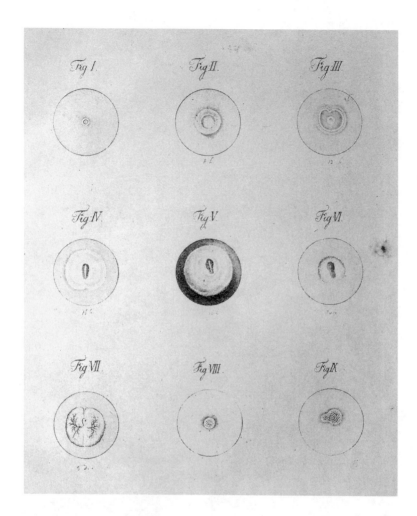

Figure 9.1 Plate I, Pander, *Beiträge zur Entwickelungsgeschichte des Hühnchens im Eye*, 1817.

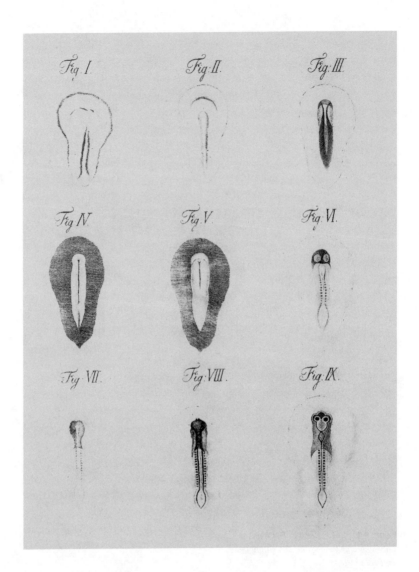

Figure 9.2 Plate II and its schematic outline, Pander, *Beiträge*, 1817.

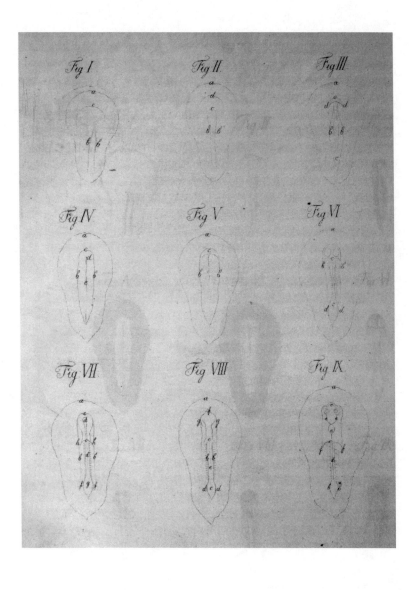

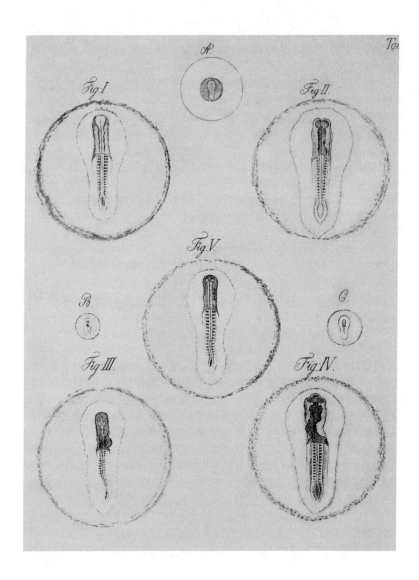

Figure 9.3 Plate III and its schematic
outline, Pander, *Beiträge*, 1817.

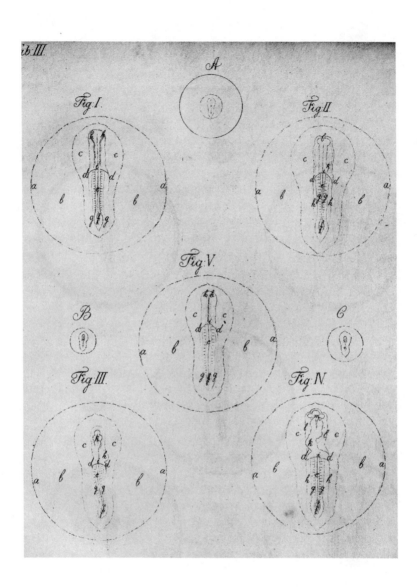

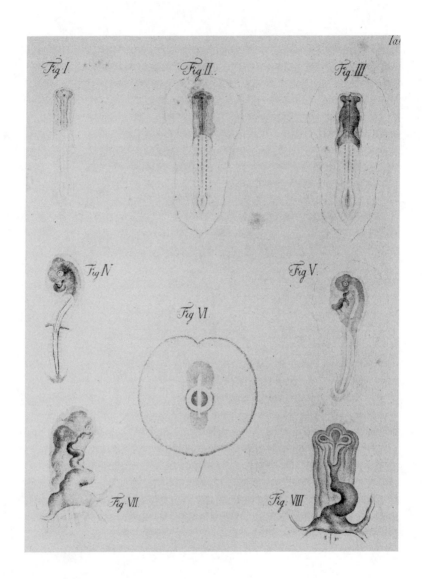

Figure 9.4 Plate VII and its schematic outline, Pander, *Beiträge*, 1817.

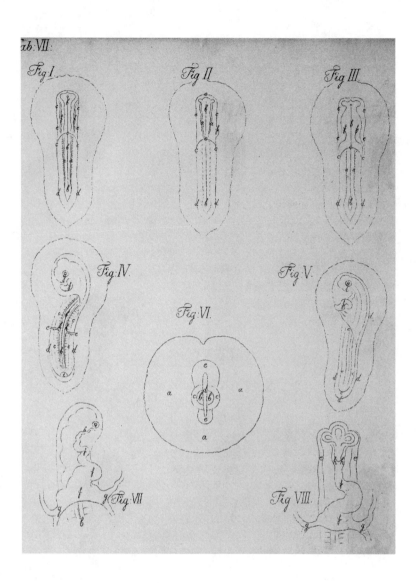

Fig.I. Fig.II. Fig.III.

Fig.IV. Fig.V.

Fig:VI.

Fig.VII Fig.VIII.

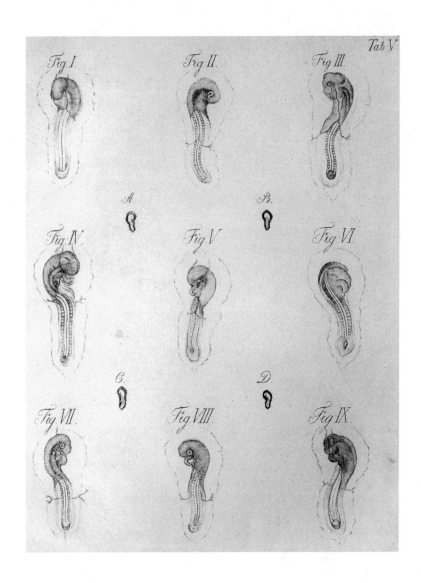

Figure 9.5 Plate V, Pander, *Beiträge*, 1817.

unusual and may be the most important one in Pander's volume. Although at first sight it seems to show a series of embryos (nine of them, numbered consecutively and arranged in rows and columns), this is actually not a developmental series. On the contrary, all the nine embryos pictured represent the same developmental stage. Why do these "Contributions to the Developmental History of the Chick in the Egg" include a plate that does not show development, but nevertheless follows the iconography of the series? In fact, one might say that Pander and d'Alton very much were portraying development in it — development as nondevelopment. If we think of the series on the basis of chronology, it is problematic for the embryo to appear in completely different forms at the same moment of incubation, but for Pander, the diversity of forms was no obstacle to his concept of development: "The selection seen here was made not so much for the sake of completeness, or in order to present an explanatory series, but rather with the intention of leaving no objectionable doubts in the mind of a future observer regarding the multifariousness of the forms in which an embryo at the same stage of development is accustomed to appear."[53]

Something that poses a constitutive problem for a chronological notion of development, the diversity of forms, is purely a factor of variation in Pander's developmental concept of the pictorial series. There, every form exists exclusively as a formal relationship and thus independently of the moment in time at which it occurs. The developmental series requires the various forms observed to be placed in relation to one another; the significance of each form for the development of the embryo can be evaluated only in relation to the forms preceding and following it. The developmental stage is determined not by the individual form itself, but by the form as a link within a concatenation of forms. In the developmental series, therefore, the individual form is constituted only by the collectivity of the series—the relationship constitutes the individual form and vice versa.

Pander's plates might be compared to a series of musical variations, in which theme and rhythm produce development. Time becomes merely a question of speed (that is, how rapidly each embryo develops). Development, in other words, is a variation in *tempo*. If the tempo slows to zero, the "multifariousness of forms" becomes visible, whereas if the forms are distributed over time, formal diversity fans out into a developmental sequence. The precondition for this simultaneous embedment in and dissolution from the flux of time is that change is constituted through the relationality of the picture series. The series is a synthesis and analysis of development at once. Development is both the individual form and the series of forms—it is stasis just as much as flow. This inherent order of time is the rhythm of the pictorial series.

Simulated Sections

Johann Rösel von Rosenhof and Sebastian von Tredern had shown the frog and chick in both lifelike representations and outlines. In Pander's *Beiträge*, too, the engravings are complemented by the schematic outline drawings in which letters label the various structures of the egg and embryo as explained in the commentary. The purpose of the outline is, on the one hand, to liberate the developmental series from all extraneous, distracting information. By shifting the system of interreference between text and image into the schematic drawing, Pander and d'Alton intensified the impact of the forms themselves and the impression of a continuous series. On the other hand, the drawings distill the information contained in the full picture by extracting the particular structures in which development is occurring. This can be seen very clearly in the transverse and longitudinal sections that Pander added to his *Beiträge* and used to explain his findings in an 1818 issue of Oken's journal *Isis*.[54] They offer a particular graphic representation of development, in which the outline acquires the status of a schema.

The "simulated sections"[55] in figure 9.6 describe the principle of folding in the abstraction of line. Pander shows the folding of the primitive streak (labeled "a"), the increasing curvature of which can be followed in longitudinal and cross sections. This folding separates the upper and lower part of the embryo (the head fold and the tail fold) and gives rise to the beginnings of the first organs, the heart and intestines.

Pander tracks the folding of all three layers—the mucous, vascular, and serous layer, labeled fig. 1, "c," "d," "f" in figure 9.6—in a further plate, which appeared with the *Isis* essay of 1818 (figure 9.7). There, figs. 1 through 4 show the developmental schema of folding. The first figure shows the undivided germ layer of the unincubated egg. On the second day (fig. 2), the germ layer has separated into the

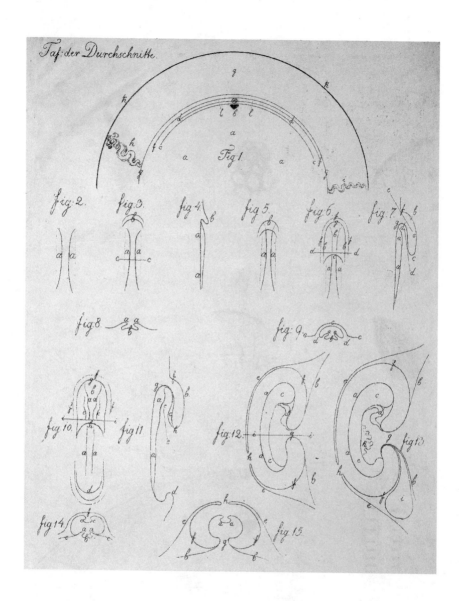

Figure 9.6 The development of the egg in longitudinal and cross sections. Pander, *Beiträge*, 1817.

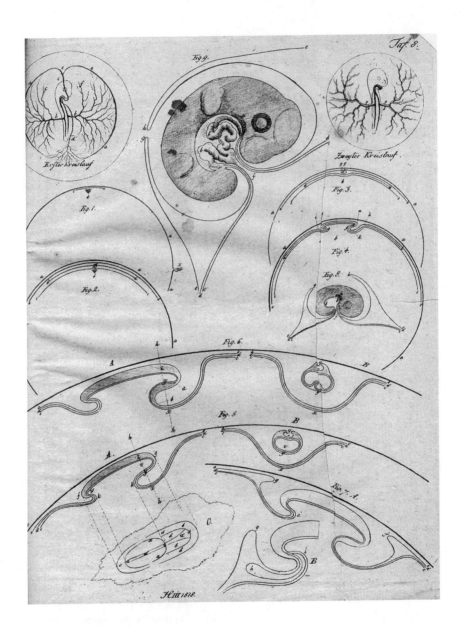

Figure 9.7 The development of the egg in longitudinal and cross sections. Pander, "Entwicklung des Küchels," 1818.

mucous layer "c," the vascular layer "d," and the serous layer "e." The third figure shows the emergence of the primitive streak (labeled "g"). The changes in all three layers can be seen in the fourth figure. In these images, Pander now also identifies which layer is responsible for which structures' emergence in the embryo. Thus, the serous layer forms the primitive folds, the sides of the fetal body, and the amnion, while the vascular layer forms the heart and the arteries (and also, Pander thought, the chorion and allantois). The mucous layer, finally, gives rise to the digestive system.[56] The next figures (5, 6, and 7) once again show the development of the various layers: the serous layer forming the primitive folds in fig. 5 and the amnion in fig. 7, and the vascular layer preparing the pericardial cavity (*Höhlung des Herzens*) in fig. 6. Each of these figures is divided into two parts, with A showing the longitudinal section and B the cross section. The principle of folding, which Pander and d'Alton sought to illuminate by combining an alternation between views of the embryo with the movement from one picture to the next, becomes completely clear only in the abstraction of the line drawings. The schematic image concentrates the information, and the form reduced to line is iconographically defined as the arena of the rhythmic movement.

Discussing Pander's work in *Isis*, Lorenz Oken's judgment was succinct: "Newly seen, newly thought, newly drawn, newly engraved."[57] For Oken, Pander's innovation was to be found in his pictures, which offered not only new representations of development, but a completely new way of thinking about it. Oken regarded images as indispensable to any understanding of the principle of folding:

> But we must wish that not drawings from nature alone (for these teach nothing), but ideal drawings will be made. More generally, every anatomical draftsman must make it his rule not to draw things as they appear, but as they are. So-called drawing from nature is always only an illusion; the eye of the painter does not see, and should not see, but the eye of the physiologist does. The illustrations that we have (for we lacked the ideal ones) do not give us understanding. When observers are able to explain and to draw the emergence of the chief parts of the body from this lengthwise folding (for that is what it is), so that the lines cohere, then they will have triumphed, and *Isis* will have the drawing engraved on a memorial coin in their honor.[58]

Oken's distinction between drawings "from nature" and "ideal" drawings, his demarcation of the "eye of the painter" from that of the physiologist, was driven by the belief that development could not be captured as a picture chronicle of the various states of the egg. For development to be conceptualized, it was not only the successive changes that had to be registered in visual relationships, but also the conceptualization of the form itself, in other words, the abstraction of the form as line. At this point, continued Oken, "once one has the folds, one also has their bending and curving, and of that we desire only line drawings."[59]

Ten years later, Oken's wish was fulfilled in Karl Ernst von Baer's *Über Entwickelungsgeschichte der Thiere* (1828 and 1837) in which the folding of the germ layers was portrayed in schematic developmental series. The first volume of this work contained two copperplates, the second volume four. Von Baer's schemata continued along the path opened up by Tredern, Herold, and Pander[60] — not only constructing development as a relationship of forms within the series, but constructing the form itself, which crystallized into the abstraction of line, the play of color, to become a vector in the space of emerging life.

CHAPTER TEN

Karl Ernst von Baer and the

Choreography of Development

Karl Ernst von Baer—like Pander, a native of the Baltic—left Würzburg in 1817 for Königsberg at the invitation of Karl Friedrich Burdach. There, he became professor extraordinarius in 1819 and a full
professor of natural history and zoology three years later. In 1834, he
moved to the Russian Academy of Sciences in St. Petersburg, where
he remained until his retirement. Unlike Pander, von Baer has been
much praised for his contributions to embryology. Particularly well
known are his discovery of the mammalian egg, described in the
1827 publication *De ovi mammalium et homini genesi*, and von Baer's
law. This criticized Meckel's and Serres's theory of recapitulation,
according to which the embryos of higher organisms pass through
the adult forms of less complex organisms during their development. Von Baer's law of increasing differentiation countered that
the embryos of one species might resemble the embryos of other
species, but not the adult forms, and that the earlier in ontogeny,
the greater the resemblance. Finally, von Baer developed a theory
of types at almost the same time as the French anatomist Georges
Cuvier. He distinguished four different types of animals on the basis
of their symmetry: peripheric, massive, longitudinal, and doubly
symmetrical.[1] Von Baer is also regarded as the founder of comparative embryology. The first volume of his *Über Entwickelungsgeschichte
der Thiere: Beobachtung und Reflexion* (On the developmental history
of the animals: Observation and reflection), which appeared in 1828,
was devoted to the embryonic development of a single animal, the
chick. A second, unfinished part appeared in 1837, as an introduction
to embryology in the form of lectures addressing physicians and students. This second volume included comparative anatomical observations of the development of other animals, such as frogs and fish.

Von Baer's studies of the chick embryo explored in far greater detail the issues that Wolff had sensed and Pander had sketched out. In von Baer's view, the emergence of structures in the egg takes place through differentiation—von Baer calls it "segregation" or "separation" (*Sonderung*)—or more precisely, through a sequence of differentiations. The three fundamental differentiations are the "primary separation" or differentiation into four germ layers, "histological separation" within these layers, and "morphological separation" into primitive organs.

Folding Layers into Tubes

At the beginning of incubation, "in the first few days of the embryo's life, no texture at all can be identified."[2] Like Döllinger and Pander before him, von Baer described the "formative tissue" (*Bildungsgewebe*), or "basic mass of all animal parts," as an albumenlike primitive mucus consisting of "incompletely isolated globules," that is, accumulations of globules distributed irregularly across the mass.[3] We have seen in Pander's work how development appears to begin when the globular substance forms into the germ layer, which subsequently separates into distinct membranes. Von Baer called this process "primary separation." He differed from Pander, however, in finding that the separation results in not three, but four layers.

For von Baer, the blastoderm first divides into two layers, a superficial and a deeper one: "From the upper layer there form the animal parts of the embryo, from the lower layer the vegetative or plastic. On this account let us call the two main layers [*Blätter*] the *animal* and the *vegetative* layer."[4] The difference between the animal and the vegetative layer consists in the arrangement of the globules within them, as a thin upper membrane that is firmly positioned like an epidermis and a lower, thicker, but less densely cohesive membrane.[5] Each of these two layers now differentiates again, although the new layers do not separate completely: the animal layer divides into a skin layer (*Hautschicht*) and a muscle layer (*Fleischschicht*), the vegetative layer into a vascular layer and a mucous layer (*Gefäßblatt* and *Schleimblatt*, corresponding to Pander's layers of the same names).[6] In the further course of development, the skin layer gives rise to the embryo's skin and the amnion, and through further differentiation, to the nervous system and the sensory organs. The lower, fleshy layer is the source of muscles and bones. The vascular layer gives rise to the main blood vessels; finally, the innermost, mucous layer forms the digestive system.[7] Thus, what von Baer calls "primary separation" is in fact a repeated process of differentiation: two main layers first form, each of which then divides into two, and further divisions will take place as development proceeds.

A second step in primary separation is the folding of the membranes to form tubes or cylinders. Von Baer calls these tubes "fundamental organs," because "it is out of them that the specific organs gradually take shape."[8] He lists five tubes, working from the inside out: first is the intestinal tract, "a tube for ingesting and transforming matter taken in from the outside world"; then the mesentery, a "tube for the movements of the newly ingested material" that surrounds the intestinal tract and is "extended to the outside"; the muscle layer, "a twin tube for the animal's movements"; the nerve tube "for its inner life, its desires and sensibility"; and the skin, "a tube to demarcate the organism from the outside world."[9] The formation of tubes is a process of differentiation that is reiterated within the individual layers—from the outermost tube, the skin, to the innermost one, the beginnings of the intestinal canal: "thus, two of the layers transformed into tubes become general systems as the differentiation that divided them is repeated in the other layers, and the original tubes are only the central components of these systems."[10]

The formation of the layers and their folding over into tubes concludes the primary separation, resulting in the fundamental organs as rudiments of the various future organs.

From Fundamental Organs to Tissues

Von Baer calls the next differentiation, which takes place in the outer shape of the layers, "morphological separation." The fundamental organs or tubes specialize first into primitive organs, then further into the definitive organs of the embryo. Morphological separation divides "the homogeneous primitive organ into heterogeneous forms,"[11] as individual segments of the tubes begin to differentiate: "the nerve tube separates into sensory organs, brain, and spinal cord; the mucous tube into the oral cavity, esophagus, stomach, intestines, respiratory system, liver, allantois, &c."[12] This process of separation follows "certain general rules,"[13] key among which is that the formation of organs occurs through the increased or diminished growth of the layers.[14] Von Baer also calls this "modified growth" over a "greater or smaller expanse."[15] Here, the diversity of the organs in terms of position and form is attributed to spatial differences in the type of growth. When it is spread across larger areas, increased growth leads to a separating of parts (*Abgrenzung*); when it is more locally concentrated, it leads to a kind of bulging out (*Hervorstül-pung*).[16] If a tube segment stretches outward, the brain or skull will form; if the stretching is lengthwise, the tube thins and gives rise to organs such as the esophagus or the spinal cord. Growth restricted to a small section of the tube results in an "outgrowth" or "bulging" that typifies the spinous processes.[17] The location, direction, manner, and extent of growth are the factors that most importantly determine the morphological structure of the later organs.

Von Baer devotes less attention to the third form of differentiation, histological separation, saying merely that it is a differentiation in the interior of the layers in which the texture of the tissues is formed "by the mass of cartilage, muscle, and nerves separating out, while a part of the mass becomes liquid and passes into the course of the blood."[18] More generally, histological separation organizes "the division into multifarious tissues that occurs within the embryo."[19]

Analytically, von Baer distinguishes three forms of differentiation, but in practice, they are intertwined. For one thing, the separations do not take place strictly sequentially. At no point in time do the tubes exist as fully formed structures, because morphological or histological separation commences before the primary separation is complete.[20] Second, the three forms of separation do not necessarily work upon all the membranes in equal measure. The formation of the individual organs out of the tubes, especially, is locally restricted to particular segments of the folded membranes. Third, similar structures are formed by different separations. For example, the circulatory system arises through primary separation in the vascular layer, but blood and blood vessels also emerge in the other layers due to histological separation. The formation of structures in the body thus occurs through the multiple repetition of differentiation—or rather, through differentiation that varies temporally and topographically: "the primary, morphological, and histological separation repeat the same differentiations, the first ensuing one above another, the second one after another, and the third one inside another."[21]

Differentiation as proposed by von Baer is a highly complex temporal and spatial process, but one that is far from haphazard. First the blastoderm separates into four layers through spatial differentiations; then the organs form out of the differentiations of the blastoderm, but now along the length of the various membranes. The formation of tissue, finally, is a set of differentiations within each of the membranes. The transition from the homogeneous mass of the egg to a heterogeneous structure, then, is an iterative process of differentiation within the germ layers. These reiterations are rhythmically coordinated — that is, they proceed within a complex spatiotemporal order, whether simultaneously or consecutively, in different layers, at different locations on the membranes, locally restricted or with a certain spatial expanse, independently, yet always referring to one another.

Because the organs emerge from the repeated differentiation of the layers, they come into being by transformation, not new formation. Each organ is a "modified part of a more general organ," being "already contained, in its entire magnitude, in the fundamental organs."[22] This summarizes von Baer's principal law of development: "there is nowhere new formation, only transformation."[23] The implication is that "the division between these differentiations in the body is not absolute, only relative."[24] Becoming form is not repetition alone, but rather the *relationship* of the repetitions that course through the germ layers like waves, bringing about differentiation into heterogeneous structures. Von Baer called the relation of repetition the "series of metamorphoses of the individual," and considered this "such a striking truth . . . that I cannot but hope it will be acknowledged as such very soon."[25]

With his three separations, von Baer made differentiation the fundamental principle of all organic formation, but this did not resolve the question of how such differentiations—entwined and interrelated with such complexity—could be qualified more precisely. What exactly is a "series of metamorphoses"? What is the nature of the "relation" between the differentiations? And what did von Baer mean by saying that the differentiations occur "one above another, one after another, and one inside another"?

From Line to Surface

Pander described the blastoderm's differentiation into three layers and the subsequent formation of the organs as foldings of the various layers. Von Baer, too, placed the germ layers at the core of his observations and of his concept of differentiation. However, he considered the movements of the layers to be so complex that he could accept Pander's notion of the fold only with reservations — the expression was "really appropriate only for the outermost layer." The process looked like a fold only "in a schematic drawing."[26] Whereas Pander had thought of the fold as a line, von Baer expanded it into a surface. Membranes were now whole planes, whose extent, orientation, movements, and spatial displacements were to be described.

This expansion makes it more difficult to describe the movements traversing the membranes in spatial terms, but not only that. Because the changes in the membrane are also temporally distinct, each fold requires highly complex coordination in time, as well. Folding becomes a rhythmical pattern according to which — simultaneously or consecutively, in one or another membrane, at one or the other end, at a single point or across a whole segment, but always choreographed as an ensemble — movements course through the various planes and axes of the future body, giving it more definition with every pulse. Von Baer replaces Pander's fold by a kind of layered, point-elastic surface. It is no longer only a rhythmically synchronized sequence of movements that folds up connected membranes, giving rise to different but similar structures in each. The process must be imagined in far greater multiplicity: von Baer sees each membrane as itself encapsulating a rhythmical pattern of motion. The series of metamorphoses within the membranes results in differentiation not only in all directions, from top to bottom, back to front, and inside to out, but also at different points and segments of the membrane in different ways.

Thus, von Baer also describes the primary separation that divides the blastoderm into four layers as a differentiation in depth.[27] Then the surface of each layer is articulated into three segments, which — each in a different layer, from outermost to innermost — now mark the beginning of the metamorphosis of the embryo's head, torso, and tail.[28] In the case of the serous layer, further metamorphosis occurs in the anterior region of each membrane, in the case of "the vascular layer in the mid region, and of the mucous layer in the posterior region."[29] Finally, the various segments of the layers, out of which the head, torso, and tail of the embryo emerge, differ in length. This is point "c" in von Baer's schematic summary of the three dimensions:

> We thus have the same sequence of distinctions: I) in the germ and the germ layer, namely a) in the dimension of depth as 1) serous layer, 2) vascular layer,

3) mucous layer. b) in the dimension of area as 1) area germinativa [*Fruchthof*], 2) area vasculosa [*Gefäßhof*], 3) area vitellina [*Dotterhof*]. c) in the dimension of length, in as much as at the front the area germinativa is widest, the area vasculosa less so, with a foremost recess, but the area vitellina predominates at the back.[30]

For the structures of the embryo, the corresponding schema is as follows: "a) in the dimension of depth as 1) animal part, 2) vascular layer, 3) mucous layer. b) in the dimension of breadth as 1) body of the embryo, 2) area vasculosa, 3) area vitellina. c) in the dimension of length as 1) brain and skull, 2) heart, 3) digestive apparatus."[31] In line with this scheme, the serous (outer) layer, for example, forms the area germinativa in the egg and the animal (nerve-controlled and will-controlled) part of the embryo, which at the end of development is manifested in the formation of the brain and the skull.

Up to this point, von Baer has distinguished the movements of the membrane into its depth, area, and length. He now goes on to distinguish them by direction, from above to below, from back to front, and from inside to outside:

> The transformation thus continues from above to below. But it also proceeds from the front to the back, for the head acquires its limits earlier than does the posterior end, and at the same time from the center to the periphery, for in the chick, the peripheral limitation occurs only on the second day, when the edges of the ventral plates are formed, the center having long since become an embryo.[32]

However, the movements of the membranes can also occur at isolated points. This is due to the globular structure of organic matter. The motions and bending of the membranes result from organic growth at particular, individual locations on the membranes. If globules accumulate in one layer, for example, and the membrane thickens, that change triggers a cascade of warps, folds, and generally of reciprocal displacements among all the structures. The area that folds is organic in the sense that folding—like every other physiological process in the body—is grounded in the movement of globules, in other words, of organic material. Every inward or outward turn of the membranes is "connected to an organic growth" and as such cannot "be thought of in a mechanical way."[33] A similar point was made by Pander, who reminded his readers that when we speak of the foldings of these layers, they must not imagine lifeless membranes with mechanically formed folds that necessarily spread across the whole area without restriction to a certain space; that idea would inevitably lead to error. Rather, the folds that are essential to the metamorphosis of the membranes are themselves of organic origin, and they form at the appropriate place, whether it is through an enlargement of the globules already present, or through new globules arriving, without the remainder of the membranes thereby being changed.[34]

For this reason, Pander argued, one must not think of organic folding as resembling the folding of a sheet of paper, in which the whole sheet participates in the fold. Instead, folding happens at isolated points "through the enlargement or accumulation of the globules that make up the membrane to be folded."[35] It is not the whole membrane that folds, only parts. Von Baer, too, regarded unequal organic growth as triggering the membranes' formation. In both the blastoderm and its subsequent layers,[36] "the upper surface develops more quickly than the lower one, the center more quickly than the edges, the anterior end more quickly than the posterior end."[37]

Rhythmical Choreography
With his schematic description of differentiation during the formation of structures within the hen's egg, von Baer implied an orderliness that was actually much more difficult to identify than the schema suggests. Most of the lengthy first volume of *Über Entwickelungsgeschichte der Thiere* consists of detailed descriptions of the changes he observed. The complex movements, displacements, and folds explained over so many pages are often baffling, but von Baer finds a narrative capable of ordering these events. He describes the formation of the embryo through the layering and relayering, bending and folding of membranes in the same way that seventeenth-century and eighteenth-century dancing masters and masters of arms had described the execution of a movement on the dancing and fencing floor or that military writers had depicted the formations and evolutions of the troops in the field. The brain, for instance, forms by the "covering of the colliculi" rising up into three oblique folds, "just as if the front part of the colliculi had had to slide over the rear part in an accelerated retreat."[38] This formulation applies the same principle that must be obeyed by a well-trained army advancing toward the

enemy in an ordered mass: whatever moves left at one point forces another point to move right. If one plate slides upward, the other slides down. If one part shifts forward, the next is pushed out and itself pulls and pushes, opening or closing a membrane, and so on.

The heart presents a particularly clear example. Its formation resembles a movement in which the emerging organ retreats into the middle of the body, simultaneously contracting and executing a turn to the left, then to the right, and differentiates into the individual substructures as the mass of future cardiac muscle grows. Von Baer describes the metamorphoses of the heart like the rhythmical movements of a dance:

> *First* the heart with its appendages retreats farther and farther back. Because at the same time the parts located above the notochord [*Rückensaite*] push forward, the relative position of the heart to the brain changes entirely.... *Second* the heart draws its various parts together as it retreats, so that the anterior parts retreat farther than the posterior ones.... *Third*, while the body is closing more and more and twisting to the left, the receptive end of the heart draws leftward.... A consequence of this is that the bend originally made by the heart to the left (§ 2. *s.*) soon ceases, and the movement now goes quite to the right. It goes so far that the curvature of the bend protrudes not only downward, but also very strongly to the right, but changing constantly, so that at first it turns more to the right, later more downward and slightly backward.[39]

The sequence of metamorphoses recalls a choreography of dancers or soldiers. The dancers or soldiers can move in formations—they can structure space in different figurations, reordering it at any point through their movements. Yet individual dancers can step out of the ranks, and starting from this point, a cascade of new movements will create a structure divergent from the one that arises from the larger formations. In both cases, the movements are synchronized, ordered by a rhythm, and attuned to one another. The same applies to the membranes. It is only through their rhythmic coordination that individual structures move not separately or arbitrarily, but as parts of a formation, of an ordered sequence of repetition and variation.

From Word to Image
In von Baer's mind, development through increasing differentiation proceeds as a careful choreography of movements that periodically

course through the germ layers and give sharper contours to the embryo with every pulse, by determining its axes, orienting it in space, shaping its body, and forming its organs. But this order of rhythm could not be identified as "easily" and "regularly"[40] as von Baer claimed. The choreography of movements and the various spatial and temporal levels of change were difficult to capture in words, as is indicated by the construction of von Baer's study. The first and longer part of volume 1 is devoted to reporting his observations, an account made almost impenetrable by its extent and detail. In the second part, entitled "Scholia and Corollaries," von Baer abstracts from these descriptions to deduce general laws of development.[41] This structure forces asunder observation and comprehension, vision and analysis, because von Baer's principal objective—to depict the relationality of all the changes in the space of the egg—remains unfulfilled by the verbal description. Neither does his law of formation become completely clear: the prosaic word cannot conjure up the rhythm of repetitions traversing the membranes, which in turn move in all directions to a subtly coordinated beat and, pulsing, progressively differentiate and shape the body.

In his introduction to *Über Entwickelungsgeschichte der Thiere*, von Baer regrets that it is "not possible at this moment" to add "a complete series of illustrations" to the book. He explains that he is "a still rather unpracticed draftsman," and "engravings that one is unable to have made under one's own eyes are seldom satisfactory," not to speak of the expense that they entail.[42] So as not to leave the book completely devoid of illustrations, von Baer published two plates of "ideal illustrations," with the aim of "discovering by this means what degree of correctness" could be achieved by engravings made externally—in other words, when observation, drawing, and the engraving of the copperplate were not made in a single move by just one person, but in distinct phases of work.[43] But the plates had more to offer than a test of the Königsberg craftsmen's artistic skill. In the "ideal" outline drawings, von Baer sets out the core of his notion of development: the individual structures in the egg arise from complex formations of movement governed by the law of the rhythm of form and formal relationships. It is the developmental series—the deployment of longitudinal and cross sections, colors and numbers, lines and points to visualize spatial relationships and their displacements during embryogenesis—that gives perceptual immediacy to the musical, aesthetic, and corporeal dimension of rhythm, present in the written text only as an abstract scaffolding.

Periodicity Replaces Chronology

Von Baer began his studies on developmental history, he tells us, as a "commentary" on the work of his predecessor, Pander, after taking up the professorship in Königsberg.[44] When studying hens' eggs with Pander, he had noted that individual eggs differed from one another significantly: "The disparities in the periodicity of development are dual in nature: 1) dissimilarity in the manifestations coexisting at any one point, 2) dissimilarity in the progress of the whole of development."[45]

The latter disparity, in particular, fluctuations in the chronological course of development, he considered "truly vexatious for the observer."[46] This acknowledgment of the "unlikeness in the time taken for the eggs to develop" was nothing new. "All observers" who had undertaken to portray "the history of development chronologically" had complained of it. That did not make it superfluous to discuss the point again, but all the more necessary "in order to present the principles according to which I have determined the various periods of development."[47]

The principles to which von Baer here refers are the replacement of chronology by a developmental course constituted not in relation to time, but through the relationships between its individual segments: "And indeed, in view of this mutability, the determination of time is something inessential, although it can unfortunately not be avoided in the description in order to show the concomitance of the phenomena. Exactitude is important only for the relative and not for the absolute measure of time."[48] Chronology, in other words, is merely a way to facilitate the *understanding* of what is portrayed, whereas development is *constituted* not through chronology, but by relating the individual forms and states to one another. Accordingly, von Baer's plates make no concrete references to time, only indicative notes such as "from the middle," "the second half," or "the beginning" of a day. In some cases, he directly names the relationships between the images, for example "6′ shows an earlier moment of formation, 6″ a later one."[49] Von Baer describes this procedure as a search for "normal development."[50] Although such a search could not completely forego all reference to the points in time at which his observations

were made, its chief interest was in the embryo at the "most instructive moments." These could not be discovered by listing the various
moments of observation; what was required was "a clear appreciation
of what one is seeking."[51] Development was to be constituted on the
basis of what sound judgment determined to be "usual." From this
perspective, governed by the primacy of form, intervals were judged
"approximately, by estimation," and, even more radically, the entire
process had to be "traced backward in formation."[52] The images in
between, the gap between two moments of formation, was thus variable, and as a relationship of images, development could be read both
forward and backward—with rhythm, it carried its own temporal
structure within itself, independently of the direction of time. As
we have seen, von Baer regarded the developmental or metamorphic
series as a "striking truth,"[53] hoping "indeed that the consideration of
the same in a continual series will vividly lay the most essential point
of developmental history, the bringing forth of the embryo from a
layered part, before the viewer's soul."[54]

To reinforce their interconnection, the images are numbered
consecutively, "so that the description in one figure can easily be
transferred to the next." In particular, the serial arrangement of the
figures means that "they easily elucidate each other."[55] The becom-
ing-form of the embryo is relocated into the images themselves: the
developmental series explains itself through their relationships as
each picture supplies a commentary and explanation of the others.
By moving through the series of images, viewers can appropriate for
themselves the concept of development.

Constructing Images, Guiding Vision
Von Baer intended his account to give a sensual insight into embryogenesis—he favors the verb *versinnlichen*, "to make sensible or sensually
perceptible."[56] In the introduction to *Über Entwickelungsgeschichte der
Thiere*, he writes: "Much as I have endeavored in the drawings of the first
two plates to combine the greatest possible correctness with lucid com-
prehensibility, and to this end have redrawn them several times over the
course of seven years, I nevertheless still find that the two goals cannot
be perfectly allied. Where they collided, I have let clarity prevail."[57]

What did von Baer mean by letting "clarity prevail"? In the following remarks on the use of pictorial resources in his developmental
series, I show that he selected specific visual strategies to create an

impression of progressive development in the viewer's eye. At the forefront of this undertaking was the images' power to guide the gaze; their correctness took second place. Although von Baer had no doubt that "obvious inaccuracies" must be avoided,[58] it becomes clear that he constituted his model of development artistically, exploiting a whole spectrum of visual resources to construct the individual figures in his series and the relationships between them. Letters, colors, lines, sections — all were consciously deployed, separately and in combination, to weave the images into a series while highlighting whatever was conducive to understanding development and masking whatever was not.

The first of von Baer's two plates shows a total of ten figures against a neutral background, six of them arranged in the lower half of the plate into lines of three pairs each (figure 10.1). The upper half is dominated by a large image on the left, flanked on the right by a small figure and two more stacked one above the other. Apart from the one on the top left, the figures are more or less the same size. They describe segments of a circle's radius. The figures in the lower part of the plate are arranged horizontally in two rows, which are connected by pairs of images labeled with Arabic numerals (in the upper row) and Roman numerals (in the lower one).

Plate II (figure 10.2) continues the sequence, but implements the series more rigorously than plate I. On the same principle as in the first plate, we see two horizontal rows. The figures are still arranged vertically in pairs, but unlike the preceding plate, here, the figures increase gradually in size. In the horizontal arrangement of both plates, the upper row shows transverse sections, the lower row longitudinal sections. The sections in all the figures are made at the same locations in the embryo, this being the only way to keep the figures compatible with each other:

> So that the cross sections, when viewed in a row, will explain gradual formation, they are selected from the third and fourth day, but all from that area of the body in which the intestine has not yet closed. Cross sections from the same embryos in the foremost or hindmost part of the body would give a very different view. Fig. 8 is to be regarded as a section made just behind the vitelline duct.[59]

The transverse and longitudinal sections are aligned vertically to form pairs that each show the same state in a distinct spatial visualization.

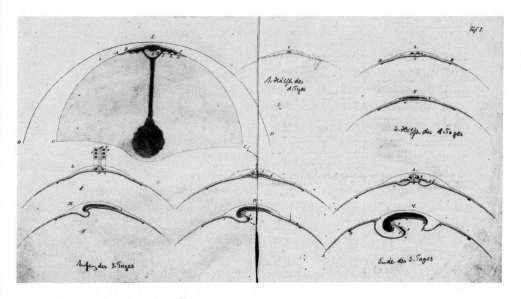

Figure 10.1 Plate I, von Baer, *Über Entwickelungsgeschichte der Thiere*, 1828.

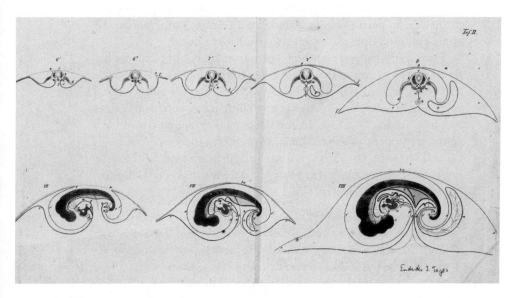

Figure 10.2 Plate II, von Baer, *Über Entwickelungsgeschichte der Thiere*, 1828.

The form of the structure in the sections is by no means always portrayed exactly as von Baer actually saw it:

> Figures V and VI, because they are longitudinal sections in the animal's median plane, could not portray the heart as being as long as it really is at this time, with its appendages, but only the length of its central portion. Likewise, in the cross sections of the last period, one will find the extent of the membranous parts of the abdominal wall less impressive than might have been expected upon reading the descriptive text.[60]

Instead, the drawing is adjusted to fit the representational conventions of the section. Where necessary, the actual length is reduced, the curvature straightened out, the appearances of longitudinal and cross section aligned. The figures are drawn at a magnification of approximately 6x — an enlargement that is crucial to understanding development, and not only because the observations were carried out with magnifying glasses. Von Baer explains this with an example. If adult hens are drawn life-size next to one other in a single plate, comparing the images will yield very few differences. However, if young embryos are magnified to the same size and placed next to one another, the impression of disparity is so strong that, "quite apart from the faster or slower progress of development as a whole, one would notice the greatest differences and would not believe that these embryos could develop into the same form."[61]

Development is therefore produced, first, by the arrangement of the embryos one next to the other and, second, by the strong enlargement of the embryonal structures. What von Baer describes here is an effect of perceptual physiology. High magnification results in a shift in the viewer's perspective and thus potentially a new perception of the object. Enlargement as an iconographic strategy allows von Baer to highlight certain structural elements in the embryo, guiding the viewer's gaze toward those structures that are most relevant to his construction of the developmental process.

The same purpose is served by another, immediately striking technique, the use of color. Throughout the series, red, yellow, and black lines are clearly distinguished and initially describe the shallow curve of a circle. From picture to picture, the lines become more strongly curved and begin to outline the contours of a body. The colors have two kinds of impact. On the one hand, they run through the series of figures like a ribbon, enabling the comparison of forms from image to image. On the other, they delimit one structure from the rest, emphasizing the individual element and its significance:

> In all the plates... the germ layers are indicated by three differently colored lines, the mucous layer being colored yellow, the vascular layer red, and the serous layer black. These same colors are retained in the parts into which the layers are transformed. But where the section encounters a real blood vessel, this is indicated by vermilion, whereas the vascular layer as such is designated by carmine. In order to distinguish the body's venous system, which appears later, from the vitelline veins (which belong to the portal system), the former is drawn in blue. The vitelline vessels, in contrast, are drawn in vermilion, whether they are arteries or veins.[62]

The use of colors, again, serves to orient the viewer's gaze. Color makes individual elements easily distinguishable even to the unpracticed eye, and it is equally effective in binding particular structures together across multiple images, suggesting a continuity in their change.

The line of sight is supported by the numbering of the figures, which bundles the figures vertically into pairs and horizontally into a sequence. The figures continue across both plates, running from 1 to 8. The first plate illustrates development in the first two days, the second the development of the next three days. The cross sections with their Arabic numerals and the longitudinal ones with their Roman numerals are connected by the fact that "numbers of the same value" always correspond.[63] The exceptions are figs. 6 and 7, each of which has two versions (6 and 6', 7 and 7').

Von Baer also distinguishes between solid and dotted lines. Whereas the colored lines are always solid, other lines within the body of the embryo and on the circle's outermost curve are broken. These dotted lines designate the vitelline membrane and the lateral vessels.[64] The choice of lines makes it possible to separate the different planes of the embryo. All the vessels shown with dotted lines are outside the center of the embryo, being included "so that the eye can quickly find orientation as to what lies in the median plane and what outside it."[65]

Within each figure, particular structures are designated by letters, which differ between the longitudinal and cross sections, but remain constant within each series of sections. Von Baer was concerned to ensure that within the rows, "the designation of one figure can easily be transferred to the next, especially since the figures are positioned such that they easily elucidate each other."[66] The sequence of letters underlines once again the sequentiality of development. However, von Baer has selected in advance which structures within the individual figures should be tracked. He does not label every structure in every picture, but only those he wishes the viewer to pursue, so as to avoid making the picture "obscure by overloading it."[67]

Colors, lines, dots, numbers, and alphabets in a continuous series, mutual delimitation, and selective reduction: all these techniques come together to order the embryo visually in its spatial environment. The purpose of the pictures is "to portray all the parts in their relative positions."[68] Yet a portrayal that truly did justice to all the spatial relationships within the embryo would exceed the capacities not only of verbal description — as becomes all too clear when reading von Baer's opus — but also of the image. Von Baer therefore made additional decisions of detail on how to represent the embryo's spatial order:

> For this reason, 1), in the longitudinal sections, no account has been taken of the sideways curvature of the head and tail end that appears from the third day on; instead, the mid area of the body is regarded as a flat surface. Equally, 2), in all these sections, the allantois is regarded as being situated in the middle area of the body, as is the heart with its various parts. Finally, 3), from the third day on, the later embryos have been a little straightened out, but in such a way that the embryo of the fifth day always appears more bent than that of the fourth day, and this more than the three-day-old embryo, thanks to which their mutual relationship is less disrupted.[69]

Von Baer here explains how he transferred the physical dimensions of the embryo onto the flat surface of the paper. In the transition from three dimensions to two, he simplified some of the spatial relationships, for example, by having the longitudinal sections represent the embryo as a simple plane. In fact, one of the most striking characteristics of embryogenesis is the way that the body of the embryo bends from the third day so that its head and tail move out of the straight line. Where von Baer does show this contraction, his portrayal follows not the degree of flexure he observed, but the *relative* degree. As a result, his pictures account only for the embryo's increasing curvature from one figure to the next, and not for the way it actually curves. When von Baer writes that he wants his images to give the viewer orientation, this means he does not reproduce only the spatial relationships as he found them in the embryo's body, but also the spatial relationships that constituted the body relative to its representation.

The inextricable entwinement of image and observation, seeing and representing, becomes even more obvious in the compilation of longitudinal and cross sections. By combining a horizontal with a vertical axis, the illustrations guide the viewer's eye in zigzags through the changes of the embryo in space and time. In a rhythmic alternation between the two aspects, the viewer winds the flat surface of the picture back into three-dimensional space in order to reconstruct the folding and flexure of the membranes. At the same time, the gaze, hurrying onward, extends this twisting into time and moves forward in step with the changes that are gradually coursing through the structures.

"The Rhythm of Their Organization"

What, then, is a developmental series? How does it evoke the impression of continuous transformation, and what makes it such a special iconographic form? We have seen how Pander's and von Baer's pictorial series constituted development as a set of visual relations. In the series, the constant change that traverses the embryo becomes a law-governed, ordered sequence of phenomena. The central principle underlying the series is the law of repetition and variation. Each picture is almost identical to the one that precedes it and the one that succeeds it; at the same time, it deviates from them in crucial points, usually in a single structural element. Compositional techniques are chosen to emphasize the transformation of that structure to the maximum: the background is neutral, and each image in the series essentially repeats the one before — the format, detail selected, perspective, size, and positioning of the object remain the same, and the figures are sharply divided into foreground and background by color and other graphic means. In Herold's pictures of the butterfly, for example, the emergence of the reproductive organs occurs as an explosion of color in the middle of the picture, against the foil of a body that has become pure copy. The individual image in the series is carefully chosen from a large number of representations, based on the criterion of how all the pictures in the series will interrelate.

In the relationship of forms within the series, the interval or gap between the pictures acquires constitutive force. It is the connection between the shown and the not shown, the bridge between fullness and emptiness, that generates change, and the spatial alternation of image and gap produces a new ordering of time. That order, the principle of the series, is its rhythm. Rhythm is the regularity, the intrinsic oscillation between repetition and variation, between past and prospective future. In this sense, the revolutionary aspect of the embryological developmental series does not lie in its using a linear arrangement of images to portray a linear, chronological forward movement of time. On the contrary, the form of temporality constituted by the developmental series is a rhythmical and periodic one: the organism is endowed with an autonomous, fundamental temporal order that is embedded into the chronological flux of time, but is not identical with it.

Writing many years after he had completed *Über Entwickelungsgeschichte der Thiere*, von Baer commented explicitly on the role of rhythm for biology.[70] He carried out a thought experiment in which man's perception of nature changes depending on the length of his life—whether a human life lasts just a few minutes or for eternity. In both cases, von Baer concluded, the only lasting things are "the forms of the life processes; *what* they form perishes continually."[71] Because the measure of time is always only relative, time is not a suitable magnitude for thinking about nature. Organic life can be grasped only as an order of time that is measured by its own intrinsic temporality. Von Baer's essay goes on to compare the forms of the life processes with music and more precisely with rhythm. The "organic life process," he writes, "develops by constantly continuing to build the body itself, to which end it absorbs simple substances from external nature. But it forms its body and builds it according to its own type and rhythm."[72]

It is not, therefore, their material substance that distinguishes fish, birds, and mammals, but the rhythm of their organization—"it is always earthly substance, re-formed according to different rhythms."[73] Living organisms differ primarily in the particular order, proper to each, by which they bind time and overcome it: "We regard the life process not as a result of organic construction, but as the rhythm, so to speak, the melody, according to which the organic body builds and rebuilds itself."[74] Just as a sequence of notes cannot become a melody without rhythm, explains von Baer, so life cannot exist without rhythm:

In organisms, the individual components are constructed according to the type and rhythm of the pertinent life process and by the manner of its effect, so that they cannot serve a different life process. For this reason, I believe it possible, comparing them with musical ideas or themes, to call the various life processes ideas of creation that construct their bodies themselves. What we name harmony and melody in music, is here type (conjunction of the parts) and rhythm (sequence of the formations).[75]

Seeking to counter a purely materialistic interpretation of life, in this 1860 text von Baer insists that the rhythm obeyed by the physiology of living organisms is not the *outcome* of the processes in the body, but their "guide and arbiter."[76]

Conclusion

The period from 1760 to 1830 saw the emergence of biology as the science of life. Historians regard this period—the *Sattelzeit*, to use Reinhart Koselleck's term—as an era of profound transformation encompassing many domains of life and knowledge, its most conspicuous feature being a new, distinctive kind of temporality. Koselleck's studies in the history of concepts show that notions of movement (*Bewegungsbegriffe*) such as progress, history, revolution, and acceleration formed the parameters of the new era. The opening of time into the future, the world's dynamization and acceleration, was complemented by the discovery that it had a history. The transformation at the threshold of the nineteenth century fundamentally changed experiences and attitudes: lived, experienced time no longer simply *was*; it became the means and result of human action and human technology.[1] This new perception of time, historians of science have argued, was the crucial step enabling the science of biology to take shape. They generally refer to the epoch as an era of the dynamization of nature.[2] Put simply, this means that around 1800, the living world began to be transposed into time. It was endowed with a past and a future, its changes channeled into a linear, chronological process that was always advancing, always unfinished.

The paradigm of temporalization has a long history. In the 1930s, Arthur O. Lovejoy traced the story of the "great chain of being" and showed that in the late eighteenth century, the multiplicity and plenitude of the world was reinterpreted, "converting the once immutable Chain of Being into the program of an endless Becoming."[3] In 1966, Michel Foucault's seminal *The Order of Things* examined the transition from natural history to the history of nature. Foucault argued that until the end of the eighteenth century, there was no such thing as "life," only living beings. Around 1800, the situation was reversed: now, only universal life existed, and individual beings were demoted to the status of ephemeral phenomena, to "variables of the living world."[4] Taking his cue from Foucault, Wolf Lepenies identified the "end of natural history" at the end of the eighteenth century. Under the "pressure of experience and the compulsion to empiricize," wrote Lepenies, the horizontally expanding territory of natural history also had to move vertically, down into time.[5] Since the 1960s and 1970s, the complex of dynamization, temporalization, and historicization around 1800 has been addressed in detail, and in the history of science today, the notion of nature's temporalization or dynamization has become almost universally accepted.[6]

Certainly, there is little doubt that the period around 1800 comprised a striking historical caesura. This book, however, began from the premise that concepts such as temporalization and dynamization are inadequate to describe the historical changes that were necessary for the new biology, and particularly for the new developmental thinking, to become established. What exactly does it mean to say that the new biology was more dynamic than the old natural history? Although the hypothesis of temporalization evidently names a profound change in the history of culture and science around 1800, at best it lacks explanatory power, while at worst it is misleading—for it is untenable to claim that natural history before 1800 had no temporal dimension. Around 1700, scholars such as Swammerdam and Malpighi were already taking an interest in phenomena such as embryogenesis and insect metamorphosis, which were evidently processes over time.[7] As Arno Seifert argued in the early 1980s, it would be wrong to interpret the concept of temporalization as the basis of "a fundamental transformation of perception and consciousness at the threshold of modernity."[8] Seifert rightly notes that in the late eighteenth century, nature did not require temporalization, "because it by no means lacked a temporal dimension."[9] Accordingly, more recent historiography has spoken instead of a "pluralization of temporal discourses" in the eighteenth century.[10]

What was new around 1800 was not time as such, but the way in which time was perceived and inscribed into nature. In this sense, the term "dynamization" is probably even more misleading than "temporalization," since it implies that a previously rigid and well-structured nature, liberated at last from the corset of order, now broke free.

Here I have argued that on the contrary, one of the new biology's essential features was precisely *not* the liberation from temporal restrictions, but the search for ways to order and circumscribe time. Similarly, the novelty of the concept of development in this period was not so much the irruption of time into the world of organic life as a new vision of order in a constantly changing world—in other words, of order under the conditions of incessant change. Development thus became a *structure* of time, a structure that I have called rhythm.

The concept of rhythm emerging between 1760 and 1830 can be outlined as follows. Put in the most abstract terms, rhythm was a relationship in time that, however, was not measured in *units* of time. To be more specific, rhythm was, first, a temporal structure that did not ensue according to temporal units. This distinguished it from meter or beat, produced by dividing up a series according to fixed counts and their precise repetition.[11] The second fundamental principle of rhythm around 1800 was the tension between wholeness and interruption arising from the duality of rhythm as a unified movement always perceived in its immediate wholeness, yet also as composed of individual elements. This composite construction was the source of a central characteristic of rhythm: it was a forward-oriented motion. In rhythm, one tension resolves into the next, resulting in a movement that presses ceaselessly onward—a pivotal aspect of thinking on development. The temporalization argument has generally focused on the historicization of thinking around 1800, arguing that every existence was now furnished with its own history and that only knowledge of its history could enable it to be understood in its current suchness. Rhythm, in contrast, generated a movement into the future. At the same time, it had a repetitive dimension, so that the future did not take shape randomly, but only ensued on the basis of past repetition. Third, then, rhythm was conceptualized as a bidirectional structure of time.

Fourth, rhythm did not mean uniformity, but regularity. It was a completely autonomous movement, neither dependent on causal connections between its individual elements nor teleologically directed at a preestablished goal. Instead, rhythm was variable at every moment of its interrupted sequence. There could be variation at each of its pauses without forfeiting the structure of regularity. Fifth, and this element was key to the conceptualization of development as a rhythmic process, rhythm constituted its autonomy by always positing simultaneously the individual components and the movement as a whole. It was an ordering pattern that emphasized each element as an independent unit — yet these elements were not absolute entities, but acquired their meaning through a network that was structural through and through, made of the links that preceded and succeeded them.

As I have shown in my comments on the observational practices of Wolff, Pander, and von Baer, the history of rhythm was also associated with a new episteme of observation. Crucial to the new practices was the deployment of an "epigenetic iconography," seeking to visualize what had never been observed as such. Rhythm existed not solely in its pulses, but also in its pauses; it was always also rupture, interval, or even silence — in iconographical terms, the pictures' "in between." Constituting an in-between space of this kind had far-reaching consequences for ways of looking at the living world. It meant pointing to the hidden, to dimensions that both existed and did not exist, to the contents of the intervals. It was precisely at these tipping points that the critical variation took place and by means of them that rhythm took control over the flux of time.

Asking why developmental series were so successful as a form of representing embryological processes, one answer offered by this book is that they were founded on the episteme of motion. The iconography of motion and instructional graphics had worked on that principle long before 1800, representing complex processes of movement through a concise series of snapshots in time. The movement that the viewer was supposed to be able to perform by following the instructions was not fully depicted in the scanty images themselves; the series called on the viewer to think and execute the movements *between* the figures, as well. Organic development, too, was represented as a rhythmical movement by first dividing up its individual components, then reassembling them according to a rhythmical rule. Only in this way could a previously disparate collection of individual observations appear as a naturally flowing movement. Once again, what was constitutive was precisely what was not shown.

The conception of development as rhythmical motion thus added a new dimension to the biological world. That dimension made it possible to understand the peculiarity of organic life as something that was structurally bound, but that simultaneously brought forth its own uniqueness again and again — in the hidden interstices, the gaps of perception.

Notes

Translator's note: Here and throughout the book, all translations of sources are my own unless otherwise attributed.

INTRODUCTION

1. Ignaz Döllinger, *Was ist Absonderung und wie geschieht sie?: Eine akademische Abhandlung von Dr. Ignaz Döllinger* (Würzburg: Nitribitt, 1819), p. 40.

2. Friedrich Nietzsche, "Rhythmische Untersuchungen," in *Friedrich Nietzsche, Werke. Kritische Gesamtausgabe, II.3: Vorlesungsaufzeichnungen (SS 1870–SS 1871)*, ed. Fritz Bornmann and Mario Carpitella (Berlin: de Gruyter, 1993), p. 322.

3. *Ibid.*, p. 325; see also p. 322.

4. *Ibid.*, p. 338.

5. *Ibid.*, pp. 322 and 338.

6. See Friederike Felicitas Günther, *Rhythmus beim frühen Nietzsche* (Berlin: de Gruyter, 2008), and the discussion later in this chapter.

7. Janina Wellmann, "Wie das Formlose Formen schafft: Bilder in der Haller-Wolff-Debatte und die Anfänge der Embryologie um 1800," *Bildwelten des Wissens: Kunsthistorisches Jahrbuch für Bildkritik* 1.2 (2003), pp. 105–15; see also Chapter 8.

8. I return to this historiographical paradigm at the end of the book.

9. The term *Biologie* is first found in Theodor Gustav August Roose, *Grundzüge der Lehre von der Lebenskraft* (Braunschweig: Thomas, 1797); Karl Friedrich Burdach, *Propädeutik zum Studium der gesammten Heilkunst* (Leipzig: Breitkopf und Härtel, 1800); Gottfried Reinhold Treviranus, *Biologie, oder Philosophie der lebenden Natur für Naturforscher und Ärzte*, vol. 1. (Göttingen: J. F. Röwer, 1802–1803), which carries the word in its title; and Jean Baptiste P. A. Lamarck, *Recherches sur l'organisation des corps vivants* (Paris: Maillard, 1802). See Peter McLaughlin, "Naming Biology," *Journal of the History of Biology* 35.1 (2002), pp. 1–4; Ilse Jahn, *Grundzüge der Biologiegeschichte* (Jena: Fischer, 1990), p. 298; Ilse Jahn and Erika Krauße (eds.), *Geschichte der Biologie: Theorien, Methoden, Institutionen, Kurzbiographien*, 3rd, rev. ed. (Heidelberg: Spektrum, 2000); Walter Baron, "Die Entwicklung der Biologie im 19.

Jahrhundert und ihre geistesgeschichtlichen Voraussetzungen," *Technikgeschichte* 33 (1966), pp. 307-28.

10. See, for example, Immanuel Kant, *Critique of the Power of Judgment*, trans. Paul Guyer and Eric Matthews (1790; Cambridge: Cambridge University Press, 2000); Timothy Lenoir, *The Strategy of Life: Teleology and Mechanics in Nineteenth-Century German Biology* (Chicago: University of Chicago Press, 1989); Lenoir, "Kant, von Baer und das kausal-historische Denken in der Biologie," *Berichte zur Wissenschaftsgeschichte* 8 (1985), pp. 99-114. See also Chapters 4 and 6.

11. Janice Joan Schall, "Rhythm and Art in Germany 1900-1930," PhD thesis, University of Texas at Austin, 1989, p. 14. Schall's text unfortunately remains unpublished. There is an enormous amount of research on rhythm for the period around 1900. On art, see, for example, Karin von Maur, *Vom Klang der Bilder* (Munich: Prestel, 1999) and Christiane Dessauer-Reiners, *Das Rhythmische bei Paul Klee: Eine Studie zum genetischen Bildverfahren* (Worms: Wernersche Verlagsgesellschaft, 1996).

12. Émile Jaques-Dalcroze, *Der Rhythmus als Erziehungsmittel für das Leben und die Kunst: Sechs Vorträge* (Basel: Helbing & Lichtenhahn, 1907). On Jaques-Dalcroze, see Marie-Laure Bachmann, *La rythmique Jaques-Dalcroze: Une éducation par la musique et pour la musique* (Neuchâtel: La Baconnière, 1984); Irwin Spector, *Rhythm and Life: The Work of Emile Jaques-Dalcroze* (Stuyvesant, NY: Pendragon, 1990); Elsa Findlay, *Rhythm and Movement: Applications of Dalcroze Eurythmics* (Evanston, IL: Summy-Birchard, 1971).

13. Karl Bücher, *Arbeit und Rhythmus* (Leipzig: B. G. Teubner, 1896). Other works from this period are Charles Féré, *Travail et plaisir: Nouvelles études expérimentales de psycho-mécanique* (Paris: Alcan, 1904); Adolf Carpe, *Der Rhythmus: Sein Wesen in der Kunst und seine Bedeutung im musikalischen Vortrage* (Leipzig: Gebrüder Reinecke, 1903); O.-L. Forel, *Le rythme: Etude psychologique* (Leipzig: J. A. Barth, 1920). See also Günter Schmölders, "Vom Rhythmus der wirtschaftlichen Aktivität," *Studium generale* 2 (1949), pp. 104-12; Karl Joel, "Der säkulare Rhythmus der Geschichte: Aus der Einleitung einer Geschichtsphilosophie," *Jahrbuch für Soziologie. Eine internationale Sammlung* 1 (1925), pp. 137-65. On the wider field of ergonomics and new notions of society and the body, see Anson Rabinbach, *The Human Motor: Energy, Fatigue, and the Origins of Modernity* (Berkeley: University of California Press, 1992), esp. chapter 6.

14. Schall, "Rhythm and Art," pp. 374-76; see also Maur, *Vom Klang der Bilder*, pp. 8-9; Dessauer-Reiners, *Das Rhythmische*, pp. 17-18.

15. On eurhythmics, see Gertrud Bünner and Peter Röthig (eds.), *Grundlagen und Methoden rhythmischer Erziehung*, 2nd, rev. ed. (Stuttgart: Klett, 1975); Röthig, "Rhythmus und Bewegung: Eine Analyse aus Sicht der Leibeserziehung," PhD thesis, University of Tübingen, 1966. On developments up to and during Nazism, see Helmut Günther, "Historische Grundlinien der deutschen Rhythmusbewegung," in Bünner and Röthig, *Grundlagen und Methoden*; Michael Golston, "'Im Anfang war der Rhythmus': Rhythmic

Incubations in Discourses of Mind, Body, and Race from 1850–1944," *Stanford Electronic Humanities Review* (1996), http://web.stanford.edu/group/SHR/5-supp/text/golston. html; Golston, *Rhythm and Race in Modernist Poetry and Science* (New York: Columbia University Press, 2008).

16. Exceptions are Paula Viterbo's study of rhythm as a method of birth control in the 1930s: Viterbo, "I Got Rhythm: Gershwin and Birth Control in the 1930s," *Endeavour* 28.1 (2004), pp. 30–35; Kenton Kroker's history of sleep research: Kroker, *The Sleep of Others and the Transformations of Sleep Research* (Toronto: University of Toronto Press, 2007); indirectly also Cornelius Borck, *Hirnströme: Eine Kulturgeschichte der Elektroenzephalographie* (Göttingen: Wallstein, 2005).

17. Wilhelm Wundt, *Völkerpsychologie: Eine Untersuchung der Entwicklungsgesetze von Sprache, Mythus und Sitte*, 2nd, rev. ed., 10 vols. (Leipzig: Engelmann, 1900–1920), vol. 1, part 2, 399–400. On the links between psychological experimentation, physiology, and new versification in the English language, see Jason R. Rudy, *Electric Meters: Victorian Physiological Poetics* (Athens: Ohio University Press, 2009); Jason David Hall, "Mechanized Metrics: From Verse Science to Laboratory Prosody, 1880–1918," *Configurations* 17.3 (2009), pp. 285–308. On rhythm research in psychology, see the historical outline by Albert Spitznagel, "Zur Geschichte der psychologischen Rhythmusforschung," in Gisa Aschersleben and Katharina Müller (eds.), *Rhythmus: Ein interdisziplinäres Handbuch* (Bern: Hans Huber, 2000). For a brief outline from 1900 to the present day, see Alf Gabrielsson, "Experimental Research on Rhythm," *The Humanities Association Review* 30.1–2 (1979), pp. 69–92; Gabrielsson, "Rhythm in Music," in Manfred Clynes and James R. Evans (eds.), *Rhythm in Psychological, Linguistic, and Musical Processes* (Springfield, IL: C. C. Thomas, 1986). An overview of the early phase of experimental research is offered by a bibliography compiled between 1910 and 1920 by Christian A. Ruckmich, which includes no fewer than 714 titles: Ruckmich, "A Bibliography of Rhythm" and three supplements, *The American Journal of Psychology* 24 (1913)–35 (1918). For the debate on this research, see Ruckmich, "The Role of Kinaesthesis in the Perception of Rhythm," *The American Journal of Psychology* 24 (1913), pp. 305–59; P. F. Swindle, "On the Inheritance of Rhythm," *The American Journal of Psychology* 24 (1913), pp. 180–203. On modern music psychology, see Helga de la Motte-Haber, *Ein Beitrag zur Klassifikation musikalischer Rhythmen: Experimentalpsychologische Untersuchungen* (Cologne: Volk, 1968); Motte-Haber, *Handbuch der Musikpsychologie* (Laaber: Laaber-Verlag, 1985).

18. Paul Kammerer, *Das Gesetz der Serie: Eine Lehre von den Wiederholungen im Lebens- und im Weltgeschehen* (Stuttgart: Deutsche Verlags-Anstalt, 1919); Ernst Mach, *Untersuchungen über den Zeitsinn des Ohres* (Vienna: n.p., 1865).

19. One exception is a short essay by Hans-Jörg Rheinberger in which the notion of rhythm resonates, even if there is no discussion of the rhythm of the organism. Rheinberger distinguishes three temporal structures of the organism: "period, span, and field." Rheinberger,

"Zeit und Biologie," in Georg Christoph Tholen and Michael O. Scholl (eds.), *Zeit-Zeichen: Aufschübe und Interferenzen zwischen Endzeit und Echtzeit* (Weinheim: VCH, 1990), p. 134.

20. See Gerold Baier, *Rhythmus: Tanz in Körper und Gehirn* (Reinbek bei Hamburg: Rowohlt, 2001); Russell Foster and Leon Kreitzman, *Rhythms of Life: The Biological Clocks That Control the Daily Lives of Every Living Thing* (New Haven, CT: Yale University Press, 2005); or the numerous journals devoted to this field, such as *Biological Rhythm Research*, *Chronobiology International*, *Journal of Biological Rhythms*, or *Journal of Circadian Rhythms*.

21. The most recent of these modifications is his 2003 revision of the entry on rhythm in the German historical dictionary of aesthetic terms: Wilhelm Seidel, s.v. "Rhythmus," in Karlheinz Barck (ed.), *Ästhetische Grundbegriffe: Historisches Wörterbuch in sieben Bänden* (Stuttgart: Metzler, 2003); also Seidel, *Über Rhythmustheorien der Neuzeit* (Bern: Francke, 1975); Seidel, *Rhythmus: Eine Begriffsbestimmung* (Darmstadt: Wissenschaftliche Buchgesellschaft, 1976); Seidel, "Rhythmus, Metrum, Takt," in Ludwig Finscher (ed.), *Die Musik in Geschichte und Gegenwart: Allgemeine Enzyklopädie der Musik begründet von Friedrich Blume*, 2nd, rev. ed., Sachteil 8 (Kassel: Bärenreiter/Stuttgart: Metzler, 1998); Seidel, "Zu Theorie und Ästhetik der Taktarten," in Günter Fleischauer, et al. (eds.), *Tempo, Rhythmik, Metrik, Artikulation in der Musik des 18. Jahrhunderts* (Blankenburg: Stiftung Kloster Michaelstein, 1998).

22. For more detail, see Seidel, *Über Rhythmustheorien der Neuzeit*, pp. 13, 85–134; Seidel, *Rhythmus: Eine Begriffsbestimmung*, p. 86.

23. Rudolf Steglich, "Über Wesen und Geschichte des Rhythmus," *Studium generale* 2.3 (1949), p. 145.

24. *Ibid.*

25. *Ibid.*, p. 46. The twentieth century was a second turning point, though here, Steglich mentions only music and the establishment of a "thrusting rhythm" (*Stoßrhythmus*), a "thrust accent" (*Stoßakzent*) or "forward-thrusting impetus." *Ibid.*, p. 149.

26. On developments in linguistics, which I cannot address here, see the overview in Isabel Zollna, "Der Rhythmus in der geisteswissenschaftlichen Forschung," *Zeitschrift für Literaturwissenschaft und Linguistik* 24.96 (1994), pp. 12–52; also Christoph Küper, *Sprache und Metrum: Semiotik und Linguistik des Verses* (Tübingen: Niemeyer, 1988); Küper, *Meter, Rhythm, and Performance. Metrum, Rhythmus, Performanz* (Frankfurt am Main: Lang, 2002). On orality and performance research within literary studies, opening up an important dimension of poetry and rhythm through attention to the voice and the situational and physical aspects of poetic language, see Henri Meschonnic, *Critique de rythme: Anthropologie historique du langage* (Paris: Verdier, 1982); Paul Zumthor, "Le rythme dans la poésie orale," *Langue française* 56 (1982), pp. 114–27; Zumthor, *La poésie et la voix dans la civilisation médiévale* (Paris: Presses universitaires de France, 1984); Jost Trier, "Rhythmus," *Studium generale* 2.3 (1949), pp. 135–41; Reinhart Meyer-Kalkus, *Stimme und Sprechkünste im 20. Jahrhundert* (Berlin: Akademie, 2001); Hans-Ulrich Gumbrecht, "Rhythm and Meaning," trans. William Whobrey, in Hans Ulrich Gumbrecht and K. Ludwig Pfeiffer (eds.), *Materialities*

of Communication (Stanford, CA: Stanford University Press, 1994). For a survey, see Kai Christian Ghattas, *Rhythmus der Bilder: Narrative Strategien in Text- und Bildzeugnissen des 11. bis 13. Jahrhunderts* (Cologne: Böhlau, 2009). See also Chapter 5.

27. Franz Norbert Mennemeier, "Rhythmus: Ein paar Daten und Überlegungen grundsätzlicher Art zu einem in gegenwärtiger Literaturwissenschaft vernachlässigten Thema," *Literatur für Leser* 4 (1990), p. 229.

28. Wolfgang Kayser, *Geschichte des deutschen Verses*, 2nd ed. (Munich: Francke, 1971), p. 113.

29. Zollna, "Der Rhythmus," p. 17.

30. For example, Steven Paul Scher (ed.), *Literatur und Musik: Ein Handbuch zur Theorie und Praxis eines komparatistischen Grenzgebietes* (Berlin: Schmidt, 1984); John Neubauer, *The Emancipation of Music from Language: Departure from Mimesis in Eighteenth-Century Aesthetics* (New Haven, CT: Yale University Press, 1986); Christine Lubkoll, *Mythos Musik: Poetische Entwürfe des Musikalischen in der Literatur um 1800* (Freiburg: Rombach, 1995); Caroline Welsh, *Hirnhöhlenpoetiken: Theorien zur Wahrnehmung in Wissenschaft, Ästhetik und Literatur um 1800* (Freiburg: Rombach, 2003); Alexander von Bormann, "Der Töne Licht: Zum frühromantischen Programm der Wortmusik," in Ernst Behler and Jochen Hörisch (eds.), *Die Aktualität der Frühromantik* (Paderborn: Schöningh, 1987); Helmut Müller-Sievers, "'... wie es keine Trennung gibt': Zur Vorgeschichte der romantischen Musikauffassung," *Athenäum: Jahrbuch für Romantik* 2 (1992), pp. 33–54. Rhythm does not feature in Manfred Frank's seminal study *Das Problem "Zeit" in der deutschen Romantik: Zeitbewußtsein und Bewußtsein von Zeitlichkeit in der frühromantischen Philosophie und in Tiecks Dichtung* (Munich: Winkler, 1972). See also Frank, *Unendliche Annäherung: Die Anfänge der philosophischen Frühromantik* (Frankfurt am Main: Suhrkamp, 1997). Neither does it play an important role in the work of Peter Szondi and Ernst Behler or in Bruno Markwardt's standard work on the history of German poetics: Peter Szondi, *Poetik und Geschichtsphilosophie*, 2 vols. (Frankfurt am Main: Suhrkamp, 1974); Ernst Behler, *Frühromantik* (Berlin: de Gruyter, 1992); Behler, "Natur und Kunst in der frühromantischen Theorie des Schönen," *Athenäum: Jahrbuch für Romantik* 2 (1992), pp. 7–32; Bruno Markwardt, *Geschichte der deutschen Poetik*, 5 vols. (Berlin: de Gruyter, 1937–1967); also Walter Jaeschke and Helmut Holzhey (eds.), *Früher Idealismus und Frühromantik: Der Streit um die Grundlagen der Ästhetik (1795-1805)* (Hamburg: Meiner, 1990).

31. Barbara Naumann, *Musikalisches Ideen-Instrument: Das Musikalische in Poetik und Sprachtheorie der Frühromantik* (Stuttgart: Metzler, 1990), pp. 143, 149–50, 209–15; Naumann, *Die Sehnsucht der Sprache nach der Musik: Texte zur musikalischen Poetik um 1800* (Stuttgart: Metzler, 1994), pp. 256–57; Naumann, "Kopflastige Rhythmen: Tanz ums Subjekt bei Schelling und Cunningham," in Barbara Naumann (ed.), *Rhythmus: Spuren eines Wechselspiels in Künsten und Wissenschaften* (Würzburg: Königshausen & Neumann, 2005). See also Bettine Menke, "Rhythmus und Gegenwart: Fragmente der Poetik um 1800," in Patrick

Primavesi and Simone Mahrenholz (eds.), *Geteilte Zeit: Zur Kritik des Rhythmus in den Kün-sten* (Schliengen: Argus, 2005).

32. Clémence Couturier-Heinrich, *Aux origines de la poésie allemande: Les théories du rythme des Lumières au Romantisme* (Paris: CNRS Éditions, 2004), p. 9.

33. On this, see Seidel, "Rhythmus," p. 292. On the etymology, see also Trier, "Rhyth-mus," and Seidel, *Rhythmus: Eine Begriffsbestimmung*, p. 15.

34. Émile Benveniste, *Problems in General Linguistics*, trans. Mary Elizabeth Meek (1951; Coral Gables, FL: University of Miami Press, 1971), pp. 286–87.

35. Plato, "Laws," 664e, trans. A. E. Taylor, in *The Collected Dialogues*, ed. Edith Hamil-ton and Huntingdon Cairns (New York: Pantheon, 1961), p. 1261.

36. Especially for music, the theory of feet was further developed on the basis of Aris-toxenus's propositions. In the third century BCE, Aristides Quintilianus distinguished musical feet according to their character (as calm, pleasing, enthusiastic, etc.). St. Augus-tine was the last important theorist in antiquity to discuss rhythm in music: in the fourth century CE, he proposed an ontology of rhythm that ascended from the lowest (bodily) to the highest (spiritual). See Seidel, *Rhythmus: Eine Begriffsbestimmung*.

37. Ludwig Klages, *Vom Wesen des Rhythmus*, 4th ed. (1933; Bonn: Bouvier, 2000), p. 23.

38. *Ibid.*, pp. 47, 57, 53.

39. *Ibid.*, pp. 87 and 71.

40. Alfred North Whitehead, *An Enquiry Concerning the Principles of Natural Knowledge*, 2nd ed. (Cambridge: Cambridge University Press, 1925), pp. 195–96.

41. *Ibid.*, pp. 196 and 197.

42. *Ibid.*, p. 198.

43. John Dewey, *Art as Experience* (1934; New York: Penguin, 1985), p. 14.

44. *Ibid.*, p. 156.

45. *Ibid.*, p. 172.

46. *Ibid.*, pp. 176 and 177.

47. Susanne K. Langer, *Feeling and Form: A Theory of Art Developed from Philosophy in a New Key* (London: Routledge & Kegan Paul, 1953), p. 128.

48. See *ibid.*, p. 112.

49. *Ibid.*, p. 120; see also p. 115.

50. *Ibid.*, pp. 109 and 126.

51. *Ibid.*, pp. 127–29.

52. Gaston Bachelard, *Dialectic of Duration*, trans. Mary McAllester Jones (1950; Man-chester: Clinamen, 2000), p. 17.

53. *Ibid.*, p. 20.

54. *Ibid.*, p. 21.

55. *Ibid.*, p. 39.

56. *Ibid.*, p. 89: "duration is not a datum but something that is made."

57. Émile Durkheim, *The Elementary Forms of the Religious Life*, trans. Joseph Ward Swain (1912; Mineola, NY: Dover, 2008), p. 216; see also p. 349.

58. Marcel Mauss, *The Manual of Ethnography*, trans. Dominique Lussier (1926; New York: Berghahn, 2007), p. 84.

59. Marcel Mauss, *Sociology and Psychology: Essays*, trans. Ben Brewster (London: Routledge, 1979), p. 22.

60. André Leroi-Gourhan, *Gesture and Speech*, trans. Anne Bostock Berger (1964; Cambridge, MA: MIT Press, 1993), p. 283.

61. Jacques Derrida, *Margins of Philosophy*, trans. Alan Bass (1972; Chicago: University of Chicago Press, 1982), p. 18.

62. *Ibid.*, p. 17.

63. *Ibid.*, p. 13.

64. This interest crosses disciplinary boundaries, finding expression in several interdisciplinary special issues. See Julian Henriques, Milla Tiainen, and Paso Väliaho (eds.), "Rhythm, Movement, Embodiment," special issue, *Body & Society* 20.3-4 (2014); Régis Debray (ed.), "Rythmes," special issue, *Médium* 41.4 (2014); Michael Cowan and Laurent Guido (eds.), "Rythmer/Rhythmize," special issue, *Intermédialités: Histoire et théorie des arts, des lettres et des techniques* 16 (2010); "Rhythmanalyses," special issue, *Multitudes* 46 (2011); Yasuhiro Sakamoto and Reinhart Meyer-Kalkus (eds.), "Bild, Ton, Rhythmus," special issue, *Bildwelten des Wissens* 10.2 (2014). See also see the work of the French philosopher and historian Pascal Michon, especially his project of a modern encyclopedia of rhythm research at www.rhuthmos.eu, and that of the medievalist Jean-Claude Schmitt, for example, Marie Formarier and Jean-Claude Schmitt (eds.), *Rythmes et croyances au Moyen Âge* (Bordeaux: Ausonius, 2014).

65. Naumann, *Rhythmus*, pp. 8–9.

66. Patrick Primavesi and Simone Mahrenholz, "Einleitung," in Simone Mahrenholz and Patrick Primavesi (eds.), *Zeiterfahrung und ästhetische Wahrnehmung* (Schliengen: Argus, 2005), p. 24.

67. See Julian Henriques, Milla Tiainen, and Pasi Väliaho, "Rhythm Returns: Movement and Cultural Theory," *Body & Society* 20.3-4 (2014), pp. 3–29; the edited collections Christa Brüstle et al. (eds.), *Aus dem Takt: Rhythmus in Kunst, Kultur und Natur* (Bielefeld: transcript Verlag, 2005); Primavesi and Mahrenholz, *Geteilte Zeit*; Naumann, *Rhythmus*; also Ghattas, *Rhythmus der Bilder*.

68. See, for example, Ernst Müller and Falko Schmieder (eds.), *Begriffsgeschichte der Naturwissenschaften: Zur historischen und kulturellen Dimension naturwissenschaftlicher Konzepte* (Berlin: de Gruyter, 2008). Similarly, Staffan Müller-Wille and Hans-Jörg Rheinberger (eds.), *Heredity Produced: At the Crossroads of Biology, Politics, and Culture, 1500–1870* (Cambridge, MA: MIT Press, 2007); Ohad Parnes, Ulrike Vedder, and Stefan Willer, *Das Konzept der Generation: Eine Kultur- und Wissenschaftsgeschichte* (Frankfurt am Main: Suhrkamp, 2008).

69. Recent years have seen renewed interest in Romanticism, including among Anglophone scholars. For the German context, see Dalia Nassar (ed.), *The Relevance of Romanticism: Essays on German Romantic Philosophy* (Oxford: Oxford University Press, 2014); Nassar, *The Romantic Absolute: Being and Knowing in Early Romantic Philosophy, 1795-1804* (Chicago: University of Chicago Press, 2014); Laurie Johnson (ed.), "The New German Romanticism," special issue, *Seminar* 50.3 (2014); Mattias Pirholt (ed.), *Constructions of German Romanticism: Six Studies* (Uppsala: Uppsala Universitet, 2011); Nicholas Saul (ed.), *The Cambridge Companion to German Romanticism* (Cambridge: Cambridge University Press, 2009). On science and literature, see Robert Mitchell, *Experimental Life: Vitalism in Romantic Science and Literature* (Baltimore, MD: Johns Hopkins University Press, 2013), and Theresa M. Kelley, *Clandestine Marriage: Botany and Romantic Culture* (Baltimore, MD: Johns Hopkins University Press, 2012). For the French context, see John Tresch, *The Romantic Machine: Utopian Science and Technology after Napoleon* (Chicago: University of Chicago Press, 2012).

70. Johannes Hegener, *Die Poetisierung der Wissenschaften bei Novalis dargestellt am Prozeß der Entwicklung von Welt und Menschheit* (Bonn: Bouvier, 1975); Dennis F. Mahoney, *Die Poetisierung der Natur bei Novalis: Beweggründe, Gestaltung, Folgen* (Bonn: Bouvier, 1980); Jocelyn Holland, *German Romanticism and Science: The Procreative Poetics of Goethe, Novalis, and Ritter* (New York: Routledge, 2009).

71. Interest in the phenomenon and concept of rhythm permeates and unifies an epoch that historiography otherwise scatters into a kaleidoscope of aesthetic, literary, and musical labels: late Enlightenment, *Sturm und Drang*, early Romanticism, Weimar classicism, Jena (or Heidelberg, high, late, etc.) Romanticism, or, for music, Baroque, Rococo, Viennese classicism, and Romanticism. See, for example, Gerhard Schulz, *Die deutsche Literatur zwischen Französischer Revolution und Restauration*, vol. 7.1 of *Geschichte der deutschen Literatur*, ed. Helmut de Boor and Richard Newald, 2nd, rev. ed. (Munich: Beck, 2000); for music, Carl Dahlhaus (ed.), *Die Musik des 18. Jahrhunderts* (Laaber: Laaber-Verlag, 1985).

72. Stefan Washausen, Bastian Obermayer, Guido Brunnett, Hans-Jürg Kuhn, and Wolfgang Knabe, "Apoptosis and Proliferation in Developing, Mature, and Regressing Epibranchial Placodes," *Developmental Biology* 278.1 (2005), pp. 86–102.

CHAPTER ONE: LITERARY FORM

1. On the following points, see the seminal works of Winfried Menninghaus and Hans-Heinrich Hellmuth: Menninghaus, "Dichtung als Tanz," *Comparatio* 2.3 (1991), pp. 129–50; Menninghaus, "Klopstocks Poetik der schnellen 'Bewegung,'" in Friedrich Gottlieb Klopstock, *Gedanken über die Natur der Poesie: Dichtungstheoretische Schriften*, ed. Winfried Menninghaus (Frankfurt am Main: Insel, 1989); Hellmuth, *Metrische Erfindung und metrische Theorie bei Klopstock* (Munich: Fink, 1973); also Johann-Nikolaus Schneider, *Ins Ohr geschrieben: Lyrik als akustische Kunst zwischen 1750 und 1800* (Göttingen: Wallstein, 2004).

NOTES TO PAGES 37-39

On Klopstock in the context of eighteenth-century poetics, see Wilhelm Große, *Studien zu Klopstocks Poetik* (Munich: Fink, 1977).

2. Friedrich Gottlieb Klopstock, "Gedanken über die Natur der Poesie" (1759), in *Gedanken über die Natur der Poesie*, p. 181.

3. *Ibid.*, p. 180. In this quotation and throughout the book, emphasis is original unless otherwise noted.

4. Gerhard Kaiser, "Denken und Empfinden: Ein Beitrag zur Sprache und Poetik Klopstocks," in Heinz Ludwig Arnold (ed.), *Friedrich Gottlieb Klopstock* (Munich: Edition Text+Kritik, 1981), p. 23, in which Kaiser situates Klopstock within the rationalist tradition.

5. On the genesis and inconsistencies of Klopstock's theory, see Hellmuth, *Metrische Erfindung*.

6. Friedrich Gottlieb Klopstock, "Vom deutschen Hexameter" (1779), in *Gedanken über die Natur der Poesie*, p. 129.

7. Klopstock as a maker of verse, especially in the period 1764–1767, was far ahead of Klopstock as a theorist (in the later years, 1764–1779); see Hellmuth, *Metrische Erfindung*, p. 240.

8. The terms were developed in 1779 in the fragment "Grundsätze der Verskunst," located in the second part of the treatise "Vom deutschen Hexameter."

9. Klopstock, "Vom deutschen Hexameter," p. 126.

10. *Ibid.*, p. 131.

11. *Ibid.*, p. 127.

12. Thus Hellmuth, *Metrische Erfindung*, p. 232. The foundations of rhythm—or in the case of Klopstock, *Zeitausdruck* and *Tonverhalt*—remain the subdivision of the verse line based on ancient Greek, assuming fixed syllable durations (short or long), as opposed to definition by accents.

13. Friedrich Gottlieb Klopstock, "Vom Sylbenmaße" (1771), in *Klopstocks sämmtliche Werke*, ed. Anton Leberecht Back and Albert Richard Constantin Spindler, vol. 15 (Leipzig: Fleischer, 1830), p. 228.

14. See Klopstock, "Vom deutschen Hexameter," p. 100; also Hellmuth, *Metrische Erfindung*, p. 213.

15. Friedrich Gottlieb Klopstock, "Grammatische Gespräche" (1794), in *Klopstocks sämmtliche Werke*, vol. 13, p. 127.

16. On this and Klopstock's dissociation from the rhetorical tradition of poetry, see Menninghaus, "Dichtung als Tanz," p. 132.

17. Klopstock hoped that this "German" (i.e., Klopstockian) hexameter would outshine its Greek model. See Hellmuth, *Metrische Erfindung*, pp. 219–22.

18. Klopstock, "Vom deutschen Hexameter," p. 128.

19. For Klopstock, the feeling that was to be moved by poetry was not the individual

sensation of one person; sensation formed part of a general substance of truth. On this, and on the relationship of poetry to religion, see Große, *Studien zu Klopstocks Poetik*, esp. pp. 80–129, 136.

20. Menninghaus, "Dichtung als Tanz," pp. 130–37.

21. The position of the words in poetry, for example, can make a thought more immediately graspable: "A good position, or one making it possible to follow what belongs together in the thought, not only allows one to think the periods more clearly than a less good position does, but also to think them more quickly.... Being *quicker* is by no means a small matter, and in representation it is a very great one." Klopstock, "Von der Wortfolge" (1779), in *Gedanken über die Natur der Poesie*, p. 174. See also Menninghaus, "Klopstocks Poetik," p. 174.

22. Menninghaus, "Dichtung als Tanz," p. 143.

23. Friedrich Gottlieb Klopstock, "Skating," in *Odes of Klopstock from 1747 to 1780*, trans. William Nind (1771; London: William Pickering, 1848), p. 192.

24. See Mark Emanuel Amtstätter, *Beseelte Töne: Die Sprache des Körpers und der Dichtung in Klopstocks Eislaufoden* (Tübingen: Niemeyer, 2005). The great importance that Klopstock attributed to dance can be seen in his allegory "Von dem Range der schönen Künste und der schönen Wissenschaften" [Of the rank of the arts and the sciences], in Arno Sachse (ed.), *Klopstock: Eine Auswahl aus Werken, Briefen und Berichten* (1758; Berlin: Verlag der Nation, 1956), in which dance appears as an unexpected third party in the dispute between the fine arts (which appeal to the senses) and the sciences (which appeal directly to the heart and reason), leaving open the judgment in favor of one side or the other.

25. Rhythm has frequently been discussed as a structuring aesthetic principle in literature: Ulrich Gaier, *Der gesetzliche Kalkül: Hölderlins Dichtungslehre* (Tübingen: Niemeyer, 1962); Annette Hornbacher, *Blume des Mundes: Zu Hölderlins poetologisch-poetischem Sprachdenken* (Würzburg: Königshausen & Neumann, 1995); Jean-Luc Nancy, *Des lieux divins*; suivi de Calcul du poète (Mauvezin: Trans-Europ-Repress, 1997); Patrick Primavesi, "Das Reißen der Zeit: Rhythmus und Zäsur in Hölderlins 'Anmerkungen,'" in Simone Mahrenholz and Patrick Primavesi (eds.), *Zeiterfahrung und ästhetische Wahrnehmung* (Schliegen: Argus, 2005); Winfried Menninghaus, *Hälfte des Lebens: Versuch über Hölderlins Poetik* (Frankfurt am Main: Suhrkamp, 2005); Boris Previšić, *Hölderlins Rhythmus: Ein Handbuch* (Frankfurt am Main: Stroemfeld, 2008); Kathrin H. Rosenfield, *Antigone: Sophocles' Art, Hölderlin's Insight*, trans. Charles B. Duff (Aurora, CO: Davies Group, 2009); Anita-Mathilde Schrumpf, *Sprechzeiten: Rhythmus und Takt in Hölderlins Elegien* (Göttingen: Wallstein, 2011). However, scholarship has not so far examined poetic calculation as a physiological category and developmental law that also embraces organic life.

26. On the life of the poet, ending tragically in madness, see Uwe Beyer, *Hölderlin: Lesarten seines Lebens, Dichtens und Denkens* (Würzburg: Königshausen & Neumann, 1997); Günter Mieth, *Friedrich Hölderlin: Dichter der bürgerlich-demokratischen Revolution*, 2nd ed. (Würzburg: Königshausen & Neumann, 2001); Gerhard Kurz, *Mittelbarkeit*

und Vereinigung: Zum Verhältnis von Poesie, Reflexion und Revolution bei Hölderlin (Stuttgart: Metzler, 1975); Friedrich Beissner, *Hölderlin: Reden und Aufsätze* (Cologne: Böhlau, 1969); Alexander Honold, *Nach Olympia: Hölderlin und die Erfindung der Antike* (Berlin: Vorwerk 8, 2002).

27. In 1934–1935, Martin Heidegger published one of the first philosophical interpretations of Hölderlin's poetry: see Heidegger, *Hölderlin's Hymns "Germania" and "The Rhine,"* trans. William McNeill and Julia Ireland (Bloomington: Indiana University Press, 2014). Hölderlin's poetry is also treated as philosophy by Michael Konrad, *Hölderlins Philosophie im Grundriß* (Bonn: Bouvier, 1967); Uwe Beyer, *Mythologie und Vernunft: Vier philosophische Studien zu Friedrich Hölderlin* (Tübingen: Niemeyer, 1993); Dieter Henrich, *Der Grund im Bewußtsein: Untersuchungen zu Hölderlins Denken (1794–1795)* (Stuttgart: Klett-Cotta, 1991). From the perspective of philosophy around 1800, see especially Véronique M. Fóti, *Epochal Discordance: Hölderlin's Philosophy of Tragedy* (Albany: State University of New York Press, 2006); Sieglinde Grimm, *"Vollendung im Wechsel": Hölderlins Verfahrungsweise des poetischen Geistes als poetologische Antwort auf Fichtes Subjektphilosophie* (Tübingen: Francke, 1997); Violetta L. Waibel, *Hölderlin und Fichte 1794–1800* (Paderborn: Schöningh, 2000).

28. Friedrich Hölderlin to Immanuel Niethammer, February 24, 1796, in Hölderlin, *Essays and Letters: Friedrich Hölderlin*, trans. Thomas Pfau (Albany: State University of New York Press, 1988), pp. 131–32.

29. *Ibid.*, p. 131.

30. On Hölderlin's poetic theory, see especially research from the 1950s and 1960s by Meta Corssen, "Der Wechsel der Töne in Hölderlins Lyrik," *Hölderlin-Jahrbuch* (1951), pp. 19–49; Lawrence J. Ryan, *Hölderlins Lehre vom Wechsel der Töne* (Stuttgart: Kohlhammer, 1960); Gaier, *Der gesetzliche Kalkül*; Konrad, *Hölderlins Philosophie.* Among more recent works, see Fred Lönker, *Welt in der Welt: Eine Untersuchung zu Hölderlins "Verfahrungsweise des poetischen Geistes"* (Göttingen: Vandenhoeck & Ruprecht, 1989); Dietrich Mathy, "'Harmonisch entgegengesetzt eines': Zur Wiederholungsfigur in Hölderlins Dialektik des Kalkulablen," in Dietrich Mathy and Carola Hilmes (eds.), *Dasselbe noch einmal: Die Ästhetik der Wiederholung* (Opladen: Westdeutscher Verlag, 1998); Elena Polledri, *". . . immer bestehet ein Maas": Der Begriff des Maßes in Hölderlins Werk* (Würzburg: Königshausen & Neumann, 2002); Uta Degner, *Bilder im Wechsel der Töne: Hölderlins Elegien und "Nachtgesänge"* (Heidelberg: Winter, 2008), esp. pp. 47–56. On the musical conception of poetic tone or key, see James H. Donelan, *Poetry and the Romantic Musical Aesthetic* (New York: Cambridge University Press, 2008).

31. Friedrich Hölderlin, "Remarks on 'Oedipus'" (1804), in *Essays and Letters*, p. 101.

32. *Ibid.* (Translation emended.)

33. Friedrich Hölderlin, "Remarks on 'Antigone,'" (1804), in *Essays and Letters*, p. 109.

34. See Nancy, *Des lieux divins*, p. 70: "But the poetics of contact demands distance, which is the essence of touching. Toucher and touched are distinct; touching is *discreet/*

discrete, or it does not exist. Feeling is not possible except through the distance of a *propriety*—consent and reserve, each the measure of the other."

35. Friedrich Hölderlin, "On Religion" (c. 1797), in *Essays and Letters*, p. 92.

36. This is explained in Friedrich Hölderlin, "On the Difference of Poetic Modes" (c. 1800), in *Essays and Letters*.

37. *Ibid.*, p. 83.

38. Hölderlin, "Remarks on 'Oedipus,'" pp. 101–102. (Translation emended.)

39. *Ibid.* (Translation emended.)

40. Hölderlin analyzes the caesura through the works of Sophocles. See also Hornbacher, *Die Blume*, pp. 234–43. Hölderlin's own poetry, especially his late hymnic style, implements these principles; *ibid.*, p. 284.

41. This is exemplified by Hölderlin's encounter with Homeric epic; see *ibid.*, pp. 62–76. The basic tone or key of a poem cannot be expressed in a "pure" form, but is inevitably fixed in language. It exists only once poured into the mold of poetry and can be found only in the interplay of art-characters. See Friedrich Hölderlin, "On the Operations of the Poetic Spirit," in *Essays and Letters*, p. 84. On the mind's resolution of the tension between basic tone and art-character, see Corssen, "Der Wechsel der Töne."

42. Hölderlin, "On the Operations of the Poetic Spirit," pp. 66–67. (Translation emended.)

43. See Hornbacher, *Die Blume*, pp. 76–83. On Hölderlin's poetry in the context of Klopstock and Heinse, see Polledri, "... *immer bestehet ein Maas*," pp. 137–43.

44. Hölderlin, "On the Operations of the Poetic Spirit," p. 71.

45. *Ibid.*, p. 67.

46. See Gerhard H. Müller, "Wechselwirkung in the Life and Other Sciences: A Word, New Claims and a Concept around 1800... and Much Later," in Stefano Poggi and Maurizio Bossi (eds.), *Romanticism in Science: Science in Europe 1790–1840* (Dordrecht: Kluwer, 1994); Peter Kapitza, *Die frühromantische Theorie der Mischung: Über den Zusammenhang von romantischer Dichtungstheorie und zeitgenössischer Chemie* (Munich: Hueber, 1968); see also Chapter 6.

47. Johann Servatius Doutrepont, "Ueber den Wechsel der thierischen Materie," *Archiv für die Physiologie* 4.3 (1800), p. 490.

48. Johann Heinrich Ferdinand von Autenrieth, "Bemerkungen über die Verschiedenheit beyder Geschlechter und ihrer Zeugungsorgane, als Beytrag zu einer Theorie der Anatomie," *Archiv für die Physiologie* 7.1 (1807), p. 129.

49. Johann Christian Reil, "Veränderte Mischung und Form der thierischen Materie, als Krankheit oder nächste Ursache der Krankheitszufälle betrachtet," *Archiv für die Physiologie* 3 (1799), p. 429.

50. "Ueber die verschiedenen Arten (modi) des Vegetationsprocesses in der animalischen Natur, und die Gesetze, durch welche sie bestimmt werden," *Archiv für die Physiologie* 6.1 (1805), p. 125.

51. Johann Friedrich Blumenbach, *Über den Bildungstrieb und das Zeugungsgeschäfte* (Göttingen: Dieterich, 1781), pp. 12–13; translation expanded from *An Essay on Generation*, trans. A. Crichton (London: T. Cadell, 1792), p. 20.

52. On the concept of drive in Hölderlin's work and its contemporary usages, see Ulrike Enke, "Der 'Trieb in uns, das Ungebildete zu bilden': Der Begriff 'Bildungstrieb' bei Blumenbach und Hölderlin," *Hölderlin-Jahrbuch* 30 (1996–1997), pp. 102–18; Jeffrey Barnouw, "'Der Trieb, bestimmt zu werden': Hölderlin, Schiller und Schelling als Antwort auf Fichte," *Deutsche Vierteljahrsschrift für Literaturwissenschaft und Geistesgeschichte* 46 (1972), pp. 248–93; Sigrid Oehler-Klein and Manfred Wenzel, "Reizbarkeit— Bildungstrieb—Seelenorgan: Aspekte der Medizingeschichte der Goethezeit," *Hölderlin-Jahrbuch* 30 (1996–1997), pp. 83–101; Stefan Büttner, "Natur—Ein Grundwort Hölderlins," *Hölderlin-Jahrbuch* 26 (1988–1989), pp. 224–47. On the relationship of nature and art as a harmonious opposition, see also Patrizia Hucke, *Entgegengesetzte Wechselwirkungen: Hölderlins "Grund zum Empedokles"* (Würzburg: Königshausen & Neumann, 2006).

53. Friedrich Hölderlin, "Brief an den Bruder, 4.6.1799 (Brief Nr. 179)," in *Friedrich Hölderlin: Briefe*, ed. Adolf Beck (Stuttgart: Cotta, 1954), p. 328.

54. Friedrich Hölderlin, "Handschriftlich überlieferte Bruchstücke," in *Friedrich Hölderlin: Sämtliche Werke, 3: Hyperion*, ed. Friedrich Beissner (Stuttgart: Kohlhammer, 1957), p. 194.

55. Hölderlin, "Brief an den Bruder, 4.6.1799," p. 329.

56. *Ibid.*

57. Friedrich Hölderlin, "The Perspective from Which We Have to Look at Antiquity," in *Essays and Letters*, pp. 39–40.

58. Hölderlin, "Brief an den Bruder, 4.6.1799," p. 328.

59. Friedrich Hölderlin, "The Ground for Empedocles," in *Essays and Letters*, p. 53.

60. Hölderlin, "The Perspective," p. 40.

61. *Ibid.*

62. *Ibid.*

63. Goethe to Charlotte von Stein, December 14, 1786, from Rome, in Johann Wolfgang von Goethe, "Goethes Werke: Briefe. Italienische Reise August 1786–Juni 1788," *Goethes Werke*, vol. 4.8 (Weimar: Böhlau, 1890), p. 94. Moritz's rather miserable life contrasts with his literary reputation in the German-speaking world today. For an introduction, see Hans Joachim Schrimpf, *Karl Philipp Moritz* (Stuttgart: Metzler, 1980); Mark Boulby, *Karl Philipp Moritz: At the Fringe of Genius* (Toronto: University of Toronto Press, 1979). On Moritz as the founder of the psychological novel, see Hugo Eybisch, *Anton Reiser: Untersuchungen zur Lebensgeschichte von Karl Philipp Moritz und zur Kritik seiner Autobiographie* (Leipzig: R. Voigtländer, 1909); Lothar Müller, *Die kranke Seele und das Licht der Erkenntnis: Karl Philipp Moritz' Anton Reiser* (Frankfurt am Main: Athenäum, 1987); Alo Allkemper, *Ästhetische Lösungen: Studien zu Karl Philipp Moritz* (Munich: Fink, 1990). From 1783 to 1793, Moritz

edited the first German journal of empirical and analytical psychology, *Magazin zur Erfahrungsseelenkunde*; see Raimund Bezold, *Popularphilosophie und Erfahrungsseelenkunde im Werk von Karl Philipp Moritz* (Würzburg: Königshausen & Neumann, 1984); Monika Class, "K. P. Moritz's Case Poetics: Aesthetic Autonomy Reconsidered," *Literature and Medicine* 32.1 (2014), pp. 46–73. On his work in the context of the Berlin Enlightenment and popular philosophy, see Alessandro Costazza, *Schönheit und Nützlichkeit: Karl Philipp Moritz und die Ästhetik des 18. Jahrhunderts* (Frankfurt am Main: Lang, 1996), pp. 165–98; Sean Franzel, "'Hear him! hört ihn!': Scholarly Lecturing in Berlin and the Popular Style of Karl Philipp Moritz," *Goethe Yearbook* 19 (2012), pp. 93–115; Boulby, *Karl Philipp Moritz*; Martin Fontius and Anneliese Klingenberg (eds.), *Karl Philipp Moritz und das 18. Jahrhundert: Bestandsaufnahmen, Korrekturen, Neuansätze* (Tübingen: Niemeyer, 1995). Moritz's uneasy relationship with Pietism is the theme of the classic study by Robert Minder, *Glaube, Skepsis, Rationalismus dargestellt aufgrund der autobiographischen Schriften von Karl Philipp Moritz* (Frankfurt am Main: Suhrkamp, 1974). On Moritz and Romanticism, see Ulrich Hubert, *Karl Philipp Moritz und die Anfänge der Romantik: Tieck, Wackenroder, Jean Paul, Friedrich und August Wilhelm Schlegel* (Frankfurt am Main: Athenäum, 1971).

64. On Moritz's aesthetics, see Thomas P. Saine's reading of it as "aesthetic theodicy," Saine, *Die ästhetische Theodizee: Karl Philipp Moritz und die Philosophie des 18. Jahrhunderts* (Munich: Fink, 1971), and the converse interpretation as "aesthetic anthropodicy" by Schrimpf, *Karl Philipp Moritz*. Moritz's aesthetics as a form of resistance to his painful experience of reality is studied by Peter Rau, *Identitätserinnerung und ästhetische Rekonstruktion: Studien zum Werk von Karl Philipp Moritz* (Frankfurt am Main: R. G. Fischer, 1983); Jörg Bong, *"Die Auflösung der Disharmonien": Zur Vermittlung von Gesellschaft, Natur und Ästhetik in den Schriften Karl Philipp Moritz'* (Frankfurt am Main: Lang, 1993); Allkemper, *Ästhetische Lösungen*; with reference to Kant: Peter Szondi, *Poetik und Geschichtsphilosophie* (Frankfurt am Main: Suhrkamp, 1974), vol. 1, esp. pp. 82–98; as a description of modernity: Elliott Schreiber, *The Topography of Modernity: Karl Philipp Moritz and the Space of Autonomy* (Ithaca, NY: Cornell University Press, 2012). In the context of the eighteenth century's philosophical and aesthetic currents, see Costazza, *Schönheit und Nützlichkeit*; Costazza, *Genie und tragische Kunst: Karl Philipp Moritz und die Ästhetik des 18. Jahrhunderts* (Frankfurt am Main: Lang, 1999).

65. Johann Wolfgang von Goethe, letter of May 30, 1791, in "Goethes Werke: Briefe. Weimar, Oberitalien, Schlesien, Weimar 18. Juni 1788–8. August 1792," *Goethes Werke*, vol. 4.9 (Weimar: Böhlau, 1890). The friendship and reciprocal influence between Moritz and Goethe has been evaluated differently. David E. Wellbery calls Moritz's treatise "one of the founding documents of classical aesthetic doctrine" and stresses Goethe's filiations with Moritz. Wellbery, "On the Logic of Change in Goethe's Work," *Goethe Yearbook* 21 (2014), pp. 1–21; see also Costazza, *Schönheit und Nützlichkeit*, pp. 16–26. For a nuanced picture of the two men's complex relationship, see Jutta Eckle, *"Er ist wie ein jüngerer Bruder von*

mir": Studien zu Johann Wolfgang von Goethes Wilhelm Meisters theatralische Sendung und Karl Philipp Moritz' Anton Reiser (Würzburg: Königshausen & Neumann, 2003).

66. Johann Wolfgang von Goethe, *Italian Journey*, trans. W. H. Auden and Elizabeth Mayer (1829; Harmondsworth: Penguin, 1970), p. 386.

67. Karl Philipp Moritz, "Versuch einer Vereinigung aller schönen Künste und Wissenschaften unter dem Begriff des in sich selbst Vollendeten" (1785), in *Schriften zur Ästhetik und Poetik: Kritische Ausgabe*, ed. Hans Joachim Schrimpf (Tübingen: Niemeyer, 1962), p. 4.

68. Karl Philipp Moritz, "Über die bildende Nachahmung des Schönen" (1788), in *Schriften zur Ästhetik und Poetik*, p. 79.

69. Moritz, "Versuch einer Vereinigung," p. 6.

70. Moritz, "Über die bildende Nachahmung," p. 78.

71. *Ibid.*, p. 77.

72. Moritz's aesthetics of autonomy has been interpreted as a "theory of the organic artwork," Johannes Nohl, "Karl Philipp Moritz und Goethe," in Noa Kiepenheuer (ed.), *Vierzig Jahre Kiepenheuer 1910–1950: Ein Almanach* (Weimar: Kiepenheuer, 1952), p. 206, and thus as a "radical" innovation in aesthetics, Tzvetan Todorov, *Theories of the Symbol*, trans. Catherine Porter (Oxford: Blackwell, 1982), p. 153. However, his image of nature also rests on a no less radical reevaluation of organic life. On Moritz's concept of nature as a polarity of construction and destruction, see Bong, *"Die Auflösung der Disharmonien"*; as a totality of nature, see Rau, *Identitätserinnerung*; as a backdrop to the formation of the individual and society, see again Bong, *"Die Auflösung der Disharmonien,"* esp. pp. 91–98; Wolfgang Grams, *Karl Philipp Moritz: Eine Untersuchung zum Naturbegriff zwischen Aufklärung und Romantik* (Opladen: Westdeutscher Verlag, 1992); Jürgen Fohrmann, "'Bildende Nachahmung': Über die Bedeutung von 'Bildung' und 'Ordnung' als Prinzipien der Moritzschen Ästhetik," in Fontius and Klingenberg, *Karl Philipp Moritz*.

73. On the concept of rhythm in Moritz's work, see Hans Joachim Schrimpf, "Vers ist tanzhafte Rede: Ein Beitrag zur deutschen Prosodie aus dem 18. Jahrhundert," in William Foerste and Karl Heinz Borck (eds.), *Festschrift für Jost Trier zum 70. Geburtstag* (Cologne: Böhlau, 1964). Brief comments can be found in Saine's introduction to Moritz, *Versuch einer deutschen Prosodie* (1786; Darmstadt: Wissenschaftliche Buchgesellschaft, 1975); Clémence Couturier-Heinrich, *Aux origines de la poésie allemande: Les théories du rythme des Lumières au Romantisme* (Paris: CNRS Éditions, 2004), pp. 174–77; on Moritz's language in general, Adrian Aebi Farahmand, *Die Sprache und das Schöne: Karl Philipp Moritz' Sprachreflexionen in Verbindung mit seiner Ästhetik* (Berlin: de Gruyter, 2012).

74. On *Versuch einer deutschen Prosodie*, see Schrimpf, "Vers ist tanzhafte Rede." Boulby locates this text "at the very centre of Moritz's life-work"; Boulby, *Karl Philipp Moritz*, p. 150.

75. On the enhancement of the appeal of verse by coordinating or contrasting the rhythmic and the semantic order, see Moritz, *Versuch einer deutschen Prosodie*, p. 44.

76. *Ibid.*, p. 43.

77. *Ibid.*, pp. 27-28.

78. *Ibid.*, p. 25.

79. *Ibid.*

80. *Ibid.*

81. *Ibid.*

82. *Ibid.*, pp. 43-44; see also p. 83.

83. A different interpretation is offered by Barbara Thums, who approaches the relationship of aesthetics and biology in Moritz's work through the concepts of surface and ornament. Thums, "Das feine Gewebe der Organisation: Zum Verhältnis von Biologie und Ästhetik in Karl Philipp Moritz' Kunstautonomie- und Ornamenttheorie," *Zeitschrift für Ästhetik und Allgemeine Kunstwissenschaft* 49.2 (2004), pp. 237-60.

84. See Karl Philipp Moritz, "Die Signatur des Schönen: In wie fern Kunstwerke beschrieben werden können?" (1788), in *Schriften zur Ästhetik und Poetik.*

85. Karl Philipp Moritz, "Zufälligkeit und Bildung: Vom Isoliren, in Rücksicht auf die schönen Künste überhaupt" (1789), in *Schriften zur Ästhetik und Poetik*, p. 116. In 1790, Moritz's *Götterlehre oder mythologische Dichtungen der Alten* appeared, interpreting the history of Greek mythology as the product of the poetic imagination and as a continuing process of the emergence of form out of formlessness, in which the contradictory is unified in the higher language of poetry.

86. Moritz, "Zufälligkeit und Bildung," p. 116.

87. Moritz, "Die Signatur des Schönen," p. 97.

88. *Ibid.*, p. 98.

89. *Ibid.*

90. For an introduction to Novalis's life and work, see Friedrich Hiebel, *Novalis: German Poet—European Thinker—Christian Mystic*, 2nd, rev. ed. (Chapel Hill: University of North Carolina Press, 1954); John Neubauer, *Novalis* (Boston: Twayne, 1980), For a bibliography and literature review, see Herbert Uerlings, *Friedrich von Hardenberg, genannt Novalis: Werk und Forschung* (Stuttgart: J. B. Metzler, 1991). On reception, see Uerlings's edited collection *"Blüthenstaub": Rezeption und Wirkung des Werkes von Novalis* (Tübingen: Niemeyer, 2000).

91. Novalis, *Fichte Studies*, trans. Jane Kneller (1795-1796; Cambridge: Cambridge University Press, 2003), no. 456, p. 145.

92. *Ibid.*, no. 455, p. 145.

93. *Ibid.*, no. 224, p. 64, and no. 284, p. 101.

94. *Ibid.*, no. 456, p. 145.

95. Novalis, *Notes for a Romantic Encyclopaedia (Das Allgemeine Brouillon)*, ed. and trans. David W. Wood (1772-1801; Albany: State University of New York Press, 2007), no. 448, p. 73. (Translation emended.)

96. On Novalis's reception of contemporary writing on music, such as the works of C. G. Schocher, see *ibid.*, no. 382, p. 57, also no. 245, p. 36–37, no. 367, p. 55. On Röschlaub, *ibid.*, no. 446, p. 71. See also Hans-Joachim Mähl's introduction to the German edition: "Das Allgemeine Brouillon (Materialien zur Enzyklopädistik 1798/99)," in *Novalis: Schriften, 3. Das philosophische Werk II*, ed. Richard Samuel with Hans-Joachim Mähl and Gerhard Schulz, 3rd, rev. ed. (Stuttgart: Kohlhammer, 1983), p. 233, and the explanatory notes on pp. 917–18, 930–31, 936. On Novalis's interest in illness, see Theodor Haering, *Novalis als Philosoph* (Stuttgart: Kohlhammer, 1954), pp. 508–16; Johannes Hegener, *Die Poetisierung der Wissenschaften bei Novalis dargestellt am Prozeß der Entwicklung von Welt und Menschheit* (Bonn: Bouvier, 1975), pp. 475–81, 496–500.

97. Novalis, *Philosophical Writings*, ed. and trans. Margaret Mahoney Stoljar (Albany: State University of New York Press, 1997), p. 55.

98. Essential strands of this project are developed in Novalis's theoretical text "Das Allgemeine Brouillon," a collection of notes of his own thoughts and quotations from his reading. It has been shown that Novalis followed contemporary scientific developments with interest and deployed new findings in his texts. See Mähl's introduction in Novalis, "Das Allgemeine Brouillon," pp. 207–41. Jonas Maatsch emphasizes the influence of scientific taxonomy on Novalis's encylopedic project, which he calls an "attempted morphology of knowledge." Maatsch, *"Naturgeschichte der Philosopheme": Frühromantische Wissensordnungen im Kontext* (Heidelberg: Winter, 2008), p. 13. Novalis's knowledge of contemporary science was wide ranging; see Herbert Uerlings (ed.), *Novalis und die Wissenschaften* (Tübingen: Niemeyer, 1997), including Uerlings's own essay in that volume, "Novalis und die Wissenschaften: Forschungsstand und Perspektiven"; Erk F. Hansen, *Wissenschaftswahrnehmung und -umsetzung im Kontext der deutschen Frühromantik* (Frankfurt am Main: Lang, 1992); Dalia Nassar, "'Idealism Is Nothing but Genuine Empiricism': Novalis, Goethe, and the Ideal of Romantic Science," *Goethe Yearbook* 18 (2011), pp. 67–96. For individual sciences, see Irene Bark, *"Steine in Potenzen": Konstruktive Rezeption der Mineralogie bei Novalis* (Tübingen: Niemeyer, 1999), on mineralogy; Jürgen Daiber, *Experimentalphysik des Geistes: Novalis und das romantische Experiment* (Göttingen: Vandenhoeck & Ruprecht, 2001), on experimentation in literature and science; Ralf Liedtke, *Das romantische Paradigma der Chemie: Friedrich von Hardenbergs Naturphilosophie zwischen Empirie und alchemischer Spekulation* (Paderborn: Mentis, 2003), on chemistry; Joyce S. Walker, "Romantic Chaos: The Dynamic Paradigm in Novalis's 'Heinrich von Ofterdingen,'" *German Quarterly* 66.1 (1993), pp. 43–59, on physics and mathematics.

99. Again, scholars have frequently mentioned the significance of rhythm in Novalis's work, but without recognizing that his notion of rhythm was rooted in contemporary physiology. See, for example, Barbara Naumann, *Musikalisches Ideen-Instrument: Das Musikalische in Poetik und Sprachtheorie der Frühromantik* (Stuttgart: Metzler, 1990), esp. pp. 208–15. Conversely, Novalis's "poetization of the sciences" or "poetization of nature" has been discussed

without reference to the role of rhythm: Hegener, *Die Poetisierung der Wissenschaften*; Dennis F. Mahoney, *Die Poetisierung der Natur bei Novalis: Beweggründe, Gestaltung, Folgen* (Bonn: Bouvier, 1980); Ulrich Gaier, *Krumme Regel: Novalis' "Konstruktionslehre des schaffenden Geistes" und ihre Tradition* (Tübingen: Niemeyer, 1970); Hans-Joachim Mähl, *Die Idee des goldenen Zeitalters im Werk des Novalis* (Heidelberg: Winter, 1965); Haering, *Novalis als Philosoph*; also John Neubauer, "Nature as Construct," in Frederick Amrine (ed.), *Literature and Science as Modes of Expression* (Dordrecht: Kluwer, 1989); Nikolaus Lohse, *Dichtung und Theorie: Der Entwurf einer dichterischen Transzendentalpoetik in den Fragmenten des Novalis* (Heidelberg: Winter, 1988). On Novalis's "body poetics," see Nicholas Saul, "'Poëtisierung des Körpers': Der Poesiebegriff Friedrich von Hardenbergs (Novalis) und die anthropologische Tradition," in Herbert Uerlings (ed.), *Novalis: Poesie und Poetik* (Tübingen: Niemeyer, 2004). Although Maatsch views Novalis's encylopedics as an attempt to understand the regularities of knowledge's growth, he does not identify rhythm as its constitutive law of relationships, but shifts the role of assembling knowledge to the observer, primarily the genius and his analogous, "physiognomical" way of knowing. Maatsch, *"Naturgeschichte der Philosopheme,"* esp. pp. 219-45.

100. The classic study of the philosophical dimension in Novalis's oeuvre is Haering, *Novalis als Philosoph*.

101. Novalis, *Fichte Studies*, p. 135-36.

102. Novalis, *Philosophical Writings*, p. 55.

103. See also Chapter 4. Novalis's concept of generation is discussed from the perspective of the instrument or "self-instrument" in Jocelyn Holland, *German Romanticism and Science: The Procreative Poetics of Goethe, Novalis, and Ritter* (New York: Routledge, 2009), pp. 56-112.

104. Novalis, "Briefe von Novalis," in *Novalis: Schriften, 4. Tagebücher, Briefwechsel, Zeitgenössische Zeugnisse*, ed. Richard Samuel (Darmstadt: Wissenschaftliche Buchgesellschaft, 1975), p. 246.

105. Novalis, *Notes for a Romantic Encyclopaedia*, no. 382, p. 57.

106. Literary criticism has discussed Schlegel's approach as a "natural history of art" or "epigenetic literary history" without considering the parallel developments in natural history itself. See Claudia Becker, *"Naturgeschichte der Kunst": August Wilhelm Schlegels ästhetischer Ansatz im Schnittpunkt zwischen Aufklärung, Klassik und Frühromantik* (Munich: Fink, 1998); John Neubauer, "Epigenetische Literaturgeschichten bei August Wilhelm und Friedrich Schlegel," in Reinhard Wegner (ed.), *Kunst—die andere Natur* (Göttingen: Vandenhoeck & Ruprecht, 2004); Couturier-Heinrich, *Aux origines*, pp. 55-72; also Ernst Behler, "Lyric Poetry in the Early Romantic Theory of the Schlegel Brothers," in Angela Esterhammer (ed.), *Romantic Poetry* (Amsterdam: John Benjamins, 2002); Georg K. Braungart, "Die Lyriktheorie August Wilhelm Schlegels," in Peter Wiesinger (ed.), *Akten des X. Internationalen Germanistikkongresses Wien 2000: "Zeitenwende"—Die Germanistik auf dem Weg vom 20. ins 21. Jahrhundert* (Bern: Lang, 2002).

14. *Ibid.*, pp. 35-36.

15. *Ibid.*, p. 45.

16. See the later chapters of this book, especially Chapters 4, 5, 6, 9, and 10.

17. Johann Nikolaus Forkel, *Allgemeine Geschichte der Musik*, ed. Othmar Wessely (1788–1801; Graz: Akademische Druck- und Verlagsanstalt, 1967), vol. 2, pp. 387-88.

18. Around 1800, the word "rhythm" was not used in the broad sense that is usual today. It was primarily found in the domain of rhetoric and poetics, whereas in music (especially in Germany), *Takt*—"beat," "meter," or "measure"—often described a nonmetrical ordering of time. The modern use of the term "rhythm" in music can be found only later in the nineteenth century, probably influenced by the writings of Moritz Hauptmann (see Seidel, "Rhythmus, Metrum, Takt," and Seidel, "Rhythmus"), and the term *Takt* itself has a long and complicated history in music theory. It was also around 1800 that the system of musical notation took on more or less its present-day form. Crucial to this process was the standardization of measures and time signatures. I will not discuss this matter further here; it has been addressed by numerous music theorists, such as Claudia Maurer Zenck, *Vom Takt: Untersuchungen zur Theorie und kompositorischen Praxis im ausgehenden 18. und beginnenden 19. Jahrhundert* (Vienna: Böhlau, 2001); Ernst Apfel and Carl Dahlhaus, *Studien zur Theorie und Geschichte der musikalischen Rhythmik und Metrik* (Munich: Musikverlag Katzbichler, 1974). On *melos* and rhythm, see also Dieter Mersch, "Maß und Differenz: Zum Verhältnis von Mélos und Rhythmós im europäischen Musikdenken," in Simone Mahrenholz and Patrick Primavesi (eds.), *Zeiterfahrung und ästhetische Wahrnehmung* (Schliengen: Argus, 2005).

19. Daniel Gottlob Türk, *Klavierschule*, ed. Erwin R. Jacobi (1789; Kassel: Bärenreiter, 1962), p. 89.

20. *Ibid.*

21. Forkel, *Allgemeine Geschichte der Musik*, vol. 2, p. 386.

22. *Ibid.*

23. *Ibid.*, p. 388.

24. Gottfried Hermann, *Handbuch der Metrik* (Leipzig: Fleischer, 1799), p. 20.

25. *Ibid.*

26. Forkel, *Allgemeine Geschichte der Musik*, vol. 2, p. 387.

27. *Ibid.*

28. *Ibid.*

29. Heinrich Christoph Koch, *Versuch einer Anleitung zur Composition* (1782-1793; Hildesheim: Olms, 1969), p. 270. On Koch, see Nancy Kovaleff Baker and Thomas Christensen (eds.), *Aesthetics and the Art of Musical Composition in the German Enlightenment: Selected Writings of Johann Georg Sulzer and Heinrich Christoph Koch* (Cambridge: Cambridge University Press, 1995), which includes a translation of the essay introducing vol. 2, part 1, of the *Versuch* (1787); Stephan Maulbetsch, "Die Kunst, Töne zu verbinden: Heinrich Christoph Koch als Komponist und Theoretiker," *Mozart Studien* 12 (2003), pp. 217-55.

30. Heinrich Christoph Koch, *Musikalisches Lexikon*, ed. Nicole Schwindt (1802; Kassel: Bärenreiter, 2001), p. 1257.

31. Koch, *Versuch*, pp. 275–76.

32. *Ibid.*, p. 277.

33. *Ibid.*, pp. 277–78.

34. See, for example, Nancy G. Siraisi, "The Music of Pulse in the Writings of Italian Academic Physicians (Fourteenth and Fifteenth Centuries)," *Speculum* 50.4 (1975), pp. 689–710; Leofranc Holford-Strevens, "The Harmonious Pulse," *The Classical Quarterly* n.s. 43.2 (1993), pp. 475–79. My account draws on Werner Friedrich Kümmel, "Puls und Musik (16.–18. Jahrhundert)," *Medizinhistorisches Journal* 4.4 (1968), pp. 269–93, and Kümmel, *Musik und Medizin: Ihre Wechselbeziehungen in Theorie und Praxis von 800 bis 1800* (Freiburg: Alber, 1977).

35. Kümmel counts Avicenna's *Canon of Medicine* and Pietro d'Abano's *Conciliator* among the most influential medieval texts. Kümmel, *Musik und Medizin*, pp. 30–31.

36. On the transition from the music of the spheres to the notion of resonance and the figure of mood in the musical aesthetics of the second half of the eighteenth century, see Caroline Welsh, "Die 'Stimmung' im Spannungsfeld zwischen Natur- und Geisteswissenschaften: Ein Blick auf deren Trennungsgeschichte aus der Perspektive einer Denkfigur," *NTM: Zeitschrift für Geschichte der Wissenschaft, Technik und Medizin* 17 (2009), pp. 135–69; Welsh, *Hirnhöhlenpoetiken: Theorien zur Wahrnehmung in Wissenschaft, Ästhetik und Literatur um 1800* (Freiburg: Rombach, 2003); Welsh, "Nerven-Saiten-Stimmung: Zum Wandel einer Denkfigur zwischen Musik und Wissenschaft 1750–1800," *Berichte zur Wissenschaftsgeschichte* 31 (2008), pp. 113–29.

37. See Kümmel, *Musik und Medizin*, pp. 54–60.

38. On the linking of body and rhythm as reflected in the centrality of dance to eighteenth-century composition, see Leonard G. Ratner, "Eighteenth-Century Theories of Musical Period Structure," *Musical Quarterly* 42.4 (1956), pp. 439–54.

39. Johann Philipp Kirnberger, *The Art of Strict Musical Composition*, trans. David Beach and Jurgen Thym (1776–1779; New Haven, CT: Yale University Press, 1982), p. 375.

40. *Ibid.*

41. *Ibid.*, p. 381–82.

42. *Ibid.*, p. 375.

43. *Ibid.*, p. 382.

44. *Ibid.*, p. 384.

45. *Ibid.*, pp. 391 and 397. For more detail, see pp. 384–403. On Kirnberger's concept of measure in its historical context, see Maurer Zenck, *Vom Takt*, pp. 11–27 and 156–82.

46. Kirnberger, *Art of Strict Musical Composition*, p. 403.

47. *Ibid.*, p. 404.

48. *Ibid.*, p. 416.

49. *Ibid.*, p. 404. Rest points and caesuras are essential to the perception of rhythm, allowing the listener to perceive rhythmical units within the musical flow. Like Hölderlin's caesura or "counter-rhythmic rupture" in literature, in music it is the pause "that provides the ear with a small rest point and concludes the meaning of the phrase" so that the ear can gather the sounds into a comprehensible unit. *Ibid.* See also pp. 405-406. On the close relationship between music's rhythmical constitution and the order of versification in poetry, see *ibid.*, p. 404.

50. *Ibid.*, p. 404.

51. *Ibid.*

52. Johann Georg Sulzer, *Allgemeine Theorie der schönen Künste in einzelnen, nach alphabetischer Ordnung der Kunstwörter auf einander folgenden Artikeln abgehandelt*, 5 vols. (1792; Hildesheim: Olms, 1994).

53. Sulzer's oeuvre includes scientific, educational, psychological, and aesthetic writings. See Johann Georg Sulzer, *Dialogues on the Beauty of Nature and Moral Reflections on Certain Topics of Natural History*, trans. Eric Miller (1750 and 1745, respectively; Lanham, MD: University Press of America, 2005); Sulzer, *Kurzer Begriff aller Wissenschaften und andern Theile der Gelehrsamkeit, worin jeder nach seinem Innhalt, Nuzen und Vollkommenheit kürzlich beschrieben wird*, 2nd, rev. ed. (Leipzig: Langenheim, 1759). There are collected editions of Sulzer's writing on educational theory and his most important essays: Sulzer, *Pädagogische Schriften*, ed. Willibald Klinte (Langensalza: Beyer, 1922); Sulzer, *Vermischte philosophische Schriften aus den Jahrbüchern der Akademie der Wissenschaften zu Berlin gesammelt* (1773-1781; Hildesheim: Olms, 1974). Selections from his *Allgemeine Theorie der schönen Künste* are translated in Baker and Christensen, *Aesthetics*. On Sulzer's life, see his autobiography: Sulzer, *Johann Georg Sulzers Lebensbeschreibung von ihm selbst aufgesetzt* (Berlin: n.p., 1809) and works by his contemporaries: J. H. S. Formey, "Éloge de M. Sulzer," *Nouveaux Mémoires de l'Académie Royale des Sciences et Belles Lettres*, 1779 (1781), pp. 45-60; Johann Caspar Hirzel, *Hirzel an Gleim über Sulzer den Weltweisen* (Zurich: J. C. Füssli, 1779); Friedrich von Blankenburg, "Einige Nachrichten von dem Leben und den Schriften des Herrn Johann Georg Sulzer," in Johann Georg Sulzer, *Eine Fortsetzung der vermischten philosophischen Schriften desselben: Nebst einigen Nachrichten von seinem Leben und seinen sämtlichen Werken* (1773; Hildesheim: Olms, 1974); also Hans Wili, *Johann Georg Sulzer: Persönlichkeit und Kunstphilosophie* (St. Gallen: Ostschweiz, 1954), See also the introduction in Baker and Christensen, *Aesthetics*, pp. 3-24.

54. On Sulzer's concept of rhythm, see Clémence Couturier-Heinrich, *Aux origines de la poésie allemande: Les théories du rythme des Lumières au Romantisme* (Paris: CNRS Éditions, 2004), pp. 39-48; on his theory of art, see Sulzer, *Allgemeine Theorie der schönen Künste*, especially the entry on "Künste, schöne Künste," *ibid.*, vol. 3, pp. 72-98; also Matthew Riley, "Civilizing the Savage: Johann Georg Sulzer and the 'Aesthetic Force' of Music," *Journal of the Royal Musical Association* 127.1 (2002), pp. 1-22; Baker and Christensen, *Aesthetics*; Wili,

Johann Georg Sulzer; Johannes Dobai, *Die bildenden Künste in Johann Georg Sulzers Ästhetik: Seine "Allgemeine Theorie der Schönen Künste"* (Winterthur: Stadtbibliothek, 1978), esp. pp. 17–65. From the perspective of an aesthetics of emotion, see Caroline Torra-Mattenklott, *Metaphorologie der Rührung: Ästhetische Theorie und Mechanik im 18. Jahrhundert* (Munich: Fink, 2002), esp. pp. 118–23, 227–93. Sulzer's pioneering role in the psychological and anthropological aspect of the German Late Enlightenment is discussed by Wolfgang Riedel, "Erkennen und Empfinden: Anthropologische Achsendrehung und Wende zur Ästhetik bei Johann Georg Sulzer," in Hans-Jürgen Schings (ed.), *Der ganze Mensch: Anthropologie und Literatur im 18. Jahrhundert. DFG-Symposion 1992* (Stuttgart: Metzler, 1994). See also Anna Tumarkin, *Der Ästhetiker Johann Georg Sulzer* (Frauenfeld: Huber, 1933); Johan van der Zande, "Orpheus in Berlin: A Reappraisal of Johann Georg Sulzer's Theory of the Polite Arts," *Central European History* 28.2 (1995), pp. 175–208. An early bibliography of works by and about Sulzer can be found in Sulzer, *Allgemeine Theorie der schönen Künste*, vol. 1, pp. vii–xix.

55. *Ibid.*, vol. 4, p. 96.

56. *Ibid.*, vol. 3, p. 423; translation by Baker and Christensen, *Aesthetics*, p. 82.

57. *Ibid.*, vol. 4, p. 97.

58. *Ibid.*, p. 98.

59. *Ibid.*, pp. 98–99.

60. *Ibid.*, p. 100.

61. *Ibid.*

62. *Ibid.*, p. 96.

63. Johann Georg Sulzer, "Untersuchung über den Ursprung der angenehmen und unangenehmen Empfindungen," in *Vermischte philosophische Schriften*, p. 32.

64. *Ibid.*, p. 27.

65. Sulzer, *Allgemeine Theorie der schönen Künste*, vol. 4, p. 91.

66. *Ibid.*

67. *Ibid.*, p. 92.

68. *Ibid.*, p. 93.

69. *Ibid.*, p. 100.

CHAPTER THREE: RHYTHMICAL PRODUCTIVITY IN SCHELLING'S PHILOSOPHY OF NATURE AND ART

1. F. W. J. Schelling, *The Philosophy of Art*, ed. and trans. Douglas W. Stott (1859; Minneapolis: University of Minnesota Press, 2009), p. 13.

2. *Ibid.*, p. 16.

3. *Ibid.*, p. 202.

4. *Ibid.*, p. 17.

5. *Ibid.*, p. 116.

6. F. W. J. Schelling, *First Outline of the System of the Philosophy of Nature*, ed. and trans. Keith R. Peterson (1799; Albany: State University of New York, 2004), p. 15.

7. Of the abundant literature, the following should be mentioned: the lucid introduction by Andrew Bowie, *Schelling and Modern European Philosophy* (London: Routledge, 1993); Frederick C. Beiser, *German Idealism: The Struggle Against Subjectivism, 1781–1801* (Cambridge, MA: Harvard University Press, 2002); Dale E. Snow, *Schelling and the End of Idealism* (Albany: State University of New York Press, 1996); Lara Ostaric (ed.), *Interpreting Schelling: Critical Essays* (Cambridge: Cambridge University Press, 2014); Dalia Nassar, *The Romantic Absolute: Being and Knowing in Early Romantic Philosophy, 1795–1804* (Chicago: University of Chicago Press, 2014); Hans Jörg Sandkühler, *F. W. J. Schelling* (Stuttgart: Metzler, 1998); Xavier Tilliette, *Schelling: Une philosophie en devenir: Le système vivant 1794–1821* (Paris: Vrin, 1970), For a survey, see Hans Michael Baumgartner and Wilhelm G. Jacobs (eds.), *Philosophie der Subjektivität?: Zur Bestimmung des neuzeitlichen Philosophierens* (Stuttgart–Bad Cannstatt: Frommann-Holzboog, 1993). On *Naturphilosophie*, see Michael Rudolphi, *Produktion und Konstruktion: Zur Genese der Naturphilosophie in Schellings Frühwerk* (Stuttgart–Bad Cannstatt: Frommann-Holzboog, 2001); Reinhard Heckmann, Hermann Krings, and Rudolf W. Meyer (eds.), *Natur und Subjektivität: Zur Auseinandersetzung mit der Naturphilosophie des jungen Schelling* (Stuttgart–Bad Cannstatt: Frommann-Holzboog, 1985); Joseph L. Esposito, *Schelling's Idealism and Philosophy of Nature* (Lewisburg, PA: Bucknell University Press, 1977). In the context of Kant's and Hegel's philosophy, see Wolfgang Bonsiepen, *Die Begründung einer Naturphilosophie bei Kant, Schelling, Fries und Hegel: Mathematische versus spekulative Naturphilosophie* (Frankfurt am Main: Klostermann, 1997); Thomas Bach, *Biologie und Philosophie bei C. F. Kielmeyer und F. W. J. Schelling* (Stuttgart–Bad Cannstatt: Frommann-Holzboog, 2001). On the presumed present-day relevance of Schelling's idea of nature and the organism, see Marie-Luise Heuser-Keßler, *Die Produktivität der Natur: Schellings Naturphilosophie und das neue Paradigma der Selbstorganisation in den Naturwissenschaften* (Berlin: Duncker & Humblot, 1986). This position has been criticized by, among many others, Bernd-Olaf Küppers, *Natur als Organismus: Schellings frühe Naturphilosophie und ihre Bedeutung für die moderne Biologie* (Frankfurt am Main: Klostermann, 1992), p. 15. On Schelling's encounter with the sciences of his day, see Manfred Durner, Francesco Moiso, and Jörg Jantzen, *Wissenschaftshistorischer Bericht zu Schellings naturphilosophischen Schriften 1797–1800* (Stuttgart: Frommann-Holzboog, 1994).

8. F. W. J. Schelling, *System of Transcendental Idealism*, trans. Peter Heath (1800; Charlottesville: University Press of Virginia, 1978), p. 12.

9. In the 1960s, Dieter Jähnig published the standard study on Schelling's philosophy of art in which he blames Hegel's "virtually normative" status in the area of aesthetics for the neglect of Schelling. Jähnig, *Schelling. Die Kunst in der Philosophie*, 2 vols. (Pfullingen: Neske, 1966–1969), vol. 2, p. 323n3. On the philosophy of art, see also Devin Zane

Shaw, *Freedom and Nature in Schelling's Philosophy of Art* (New York: Continuum, 2010); Paul Gordon, *Art as the Absolute: Art's Relation to Metaphysics in Kant, Fichte, Schelling, Hegel, and Schopenhauer* (London: Bloomsbury, 2015); Bernhard Barth, *Schellings Philosophie der Kunst: Göttliche Imagination und ästhetische Einbildungskraft* (Freiburg: Alber, 1991); Lothar Knatz, *Geschichte, Kunst, Mythos: Schellings Philosophie und die Perspektive einer philosophischen Mythologie* (Würzburg: Königshausen & Neumann, 1999); Emil L. Fackenheim, "Schelling's Philosophy of the Literary Arts," *Philosophical Quarterly* 4.17 (1954), pp. 310–26; Manfred Frank, *Einführung in die frühromantische Ästhetik: Vorlesungen* (Frankfurt am Main: Suhrkamp, 1989), pp. 137–230; Berbeli Wanning, *Konstruktion und Geschichte: Das Identitätssystem als Grundlage der Kunstphilosophie bei F. W. J. Schelling* (Frankfurt am Main: Haag u. Herchen, 1988); Peter Szondi, *Poetik und Geschichtsphilosophie II* (Frankfurt am Main: Suhrkamp, 1974). On the concept of art from the perspective of the late philosophy of mythology, see Jochen Hennigfeld, *Mythos und Poesie: Interpretationen zu Schellings "Philosophie der Kunst" und "Philosophie der Mythologie"* (Meisenhein am Glan: Hain, 1973).

10. The concept of rhythm has hitherto been studied only in the context of individual art forms, especially music and poetry. The importance of rhythm in Schelling's work is noted by Barbara Naumann, "Kopflastige Rhythmen: Tanz ums Subjekt bei Schelling und Cunningham," in Naumann (ed.), *Rhythmus: Spuren eines Wechselspiels in Künsten und Wissenschaften* (Würzburg: Königshausen & Neumann, 2005). On Schelling and music, see Enrico Fubini, *The History of Music Aesthetics*, trans. Michael Hatwell (Basingstoke: Macmillan, 1990), pp. 272–74; Wanning, *Konstruktion und Geschichte*, pp. 113–21; Frank, *Einführung*, pp. 212–18; Ian Biddle, "F. W. J. Schelling's *Philosophie der Kunst*: An Emergent Semiology of Music," in Ian Bent (ed.), *Music Theory in the Age of Romanticism* (Cambridge: Cambridge University Press, 1996); Herbert M. Schueller, "Schelling's Theory of the Metaphysics of Music," *Journal of Aesthetics and Art Criticism* 15.4 (1957), pp. 461–76. On poetry, see Wanning, *Konstruktion und Geschichte*, pp. 168–79; Szondi, *Poetik*, pp. 262–71; on poetic rhythm, Fackenheim, "Schelling's Philosophy," pp. 318–19.

11. Schelling, *First Outline*, p. 202.

12. See Küppers, *Natur als Organismus*, pp. 48–56. For a purely dialectical interpretation of Schelling's notion of the organism, see Werner Hartkopf, *Studien zu Schellings Dialektik* (Meisenheim am Glan: Hain, 1986), pp. 85–100.

13. Schelling, *First Outline*, pp. 47–48.

14. *Ibid.*, p. 37.

15. *Ibid.*

16. *Ibid.*, p. 38.

17. F. W. J. Schelling, "Von der Weltseele, eine Hypothese der höheren Physik zur Erklärung des allgemeinen Organismus" (1798), in *F. W. J. v. Schelling. Werke: Auswahl*

in drei Bänden, 1: Schriften zur Naturphilosophie, ed. Otto Weiß (Leipzig: Eckardt, 1907), p. 445.

18. *Ibid.*, pp. 610–11.

19. *Ibid.*, p. 635.

20. *Ibid.*

21. *Ibid.*, p. 636.

22. *Ibid.*, p. 635.

23. *Ibid.*, p. 623.

24. *Ibid.*

25. *Ibid.*, p. 645.

26. *Ibid.*, p. 662.

27. F. W. J. Schelling, "Treatise Explicatory of the Idealism in the *Science of Knowledge*" (1797–1798), in *Idealism and the Endgame of Theory: Three Essays by F. W. J. Schelling*, trans. Thomas Pfau (Albany: State University of New York Press, 1994), p. 92.

28. *Ibid.*, p. 93.

29. *Ibid.*, p. 92.

30. *Ibid.*, p. 90.

31. *Ibid.*, p. 91.

32. *Ibid.*, p. 92.

33. *Ibid.*, p. 93.

34. *Ibid.*, p. 91.

35. Schelling, *Philosophy of Art*, p. 111.

36. On the various genres, see Wanning, *Konstruktion und Geschichte*.

37. Schelling, *Philosophy of Art*, p. 111.

38. *Ibid.*, p. 110.

39. *Ibid.*, p. 111.

40. *Ibid.*, p. 116.

41. *Ibid.*, p. 17.

42. *Ibid.*, p. 116.

43. *Ibid.*, p. 202.

44. *Ibid.*

45. *Ibid.*, p. 205.

46. *Ibid.*

47. *Ibid.*

48. *Ibid.*, p. 208.

49. *Ibid.*, p. 112.

50. *Ibid.*, p. 205.

51. *Ibid.*, p. 206.

52. Ibid., p. 110.

53. Ibid.

54. Ibid., p. 111.

55. Ibid.

CHAPTER FOUR: FORMS OUT OF FORMLESSNESS

1. Little is known about Wolff's life. See Julius Schuster, "Caspar Friedrich Wolff: Leben und Gestalt eines deutschen Biologen," *Sitzungsberichte der Gesellschaft der naturforschenden Freunde zu Berlin* (1936), pp. 175–95; Georg Uschmann, *Caspar Friedrich Wolff: Ein Pionier der modernen Embryologie* (Leipzig: Urania, 1955); Robert Herrlinger, "C. F. Wolffs 'Theoria generationis' (1759): Die Geschichte einer epochemachenden Dissertation," *Zeitschrift für Anatomie und Entwicklungsgeschichte* 121 (1959), pp. 245–70; Ilse Jahn, "Caspar Friedrich Wolff (1743–1794)," in Ilse Jahn and Michael Schmitt (eds.), *Darwin & Co.: Eine Geschichte der Biologie in Portraits* (Munich: Beck, 2001). On the dissertation, see Herrlinger, "C. F. Wolffs 'Theoria generationis'"; Abba E. Gaissinovitch, "Notizen von C. F. Wolff über die Bemerkungen der Opponenten zu seiner Dissertation," *Wissenschaftliche Zeitschrift der Friedrich-Schiller-Universität Jena, Mathematisch-naturwissenschaftliche Reihe* 3–4 (1956–1957), pp. 121–24; Jahn, "Wer regte Caspar Friedrich Wolff (1734–1794) zu seiner Dissertation Theoria generationis (1759) an?" *Philosophia Scientiae*, cahier spécial 2 (1998–1999), pp. 35–54; William Morton Wheeler, "Caspar Friedrich Wolff" (1898), in Jane Maienschein (ed.), *Defining Biology: Lectures from the 1890s* (Cambridge, MA: Harvard University Press, 1986); also Olaf Breidbach, "Einleitung: Zur Mechanik der Ontogenese," in Caspar Friedrich Wolff, *Theoria generationis: Ueber die Entwicklung der Pflanzen und Thiere. I., II. und III. Theil*, ed. and trans. Paul Samassa (1759; Thun: Deutsch, 1999); Reinhard Mocek, "Caspar Friedrich Wolffs Epigenesis-Konzept—ein Problem im Wandel der Zeit," *Biologisches Zentralblatt* 114 (1995), pp. 179–90. More attention has been paid to Wolff's St. Petersburg research on teratology, which is still held, unpublished, in Russian archives. Russian scholars of the 1960s worked intensively on the published and manuscript sources and proposed the hypothesis of Wolff's transformism, largely untenable for historians of science today. On the St. Petersburg papers, see Karl Ernst Baer, "Ueber den litterärischen Nachlass von Caspar Friedrich Wolff, ehemaligem Mitgliede der Akademie der Wissenschaften zu St. Petersburg," *Bulletin de la classe physico-mathématique de l'Académie Impériale des Sciences de Saint-Pétersbourg* 5.105–106 (1847), pp. 130–60; Modzalevsky, "Liste der Manuskripte von C. F. Wolff in der Akademie der Wissenschaften in Petersburg," *Vestnik Akademii Nauk SSSR* 3 (1933), pp. 59–66; V. Schütz, "Kaspar Friedrich Wolff in Russland," *Experientia* 4 (1947), pp. 465–67. On the Russian research, see also Gaissinovitch, *K. F. Vol'f i ucenie o razvitii organizmov* (Moscow: Izdatel'stvo akademii nauk SSSR, 1961); Gaissinovitch, "C. F. Wolff on Variability and Heredity," *History and Philosophy of the Life Sciences* 12 (1990), pp. 179–201; Boris Evgen'evic Raikov, "Caspar Friedrich Wolff," *Zoologische Jahrbücher / Abteilung für Systematik, Ökologie und Geographie der Tiere* 91.4 (1964), pp. 555–626; Tatjana A. Lukina,

"Caspar Friedrich Wolff und die Petersburger Akademie der Wissenschaften," in Kurt Mothes and Joachim-Hermann Scharf (eds.), *Beiträge zur Geschichte der Naturwissenschaften und der Medizin: Festschrift für Georg Uschmann* (Leipzig: Barth, 1975).

2. See, for example, Elke Witt, "Form—A Matter of Generation: The Relation of Generation, Form, and Function in the Epigenetic Theory of Caspar F. Wolff," *Science in Context* 21.4 (2008), pp. 649–64.

3. See Shirley A. Roe, *Matter, Life and Generation: Eighteenth-Century Embryology and the Haller-Wolff Debate* (Cambridge: Cambridge University Press, 1981), p. 110; Jane Marion Oppenheimer, *Essays in the History of Embryology and Biology* (Cambridge, MA: MIT Press, 1967), p. 141; Elizabeth B. Gasking, *Investigations into Generation 1651–1828* (Baltimore, MD: Johns Hopkins University Press, 1967), pp. 97–106; Emanuel Rádl, *Geschichte der biologischen Theorien in der Neuzeit*, 2nd, rev. ed. (1913; Hildesheim: Olms, 1970), vol. 1, p. 246; Olivier Rieppel, *Fundamentals of Comparative Biology* (Basel: Birkhäuser, 1988), pp. 32–34; also William Coleman, *Biology in the Nineteenth Century: Problems of Form, Function, and Transformation* (New York: Wiley, 1971), pp. 41–47; Karen Detlefsen, "Explanation and Demonstration in the Haller-Wolff Debate," in Justin E. H. Smith (ed.), *The Problem of Animal Generation in Early Modern Philosophy* (Cambridge: Cambridge University Press, 2006); Ina Goy, "Epigenetic Theories: Caspar Friedrich Wolff and Immanuel Kant," in Ina Goy and Eric Watkins (eds.), *Kant's Theory of Biology* (Berlin: de Gruyter, 2014).

4. Helmut Müller-Sievers, *Epigenesis: Naturphilosophie im Sprachdenken Wilhelm von Humboldts* (Paderborn: Schöningh, 1993), p. 45.

5. Thomas Haffner, "Die Epigenesisanalogie in Kants Kritik der reinen Vernunft," PhD thesis, University of the Saarland, 1997, p. 95.

6. Aristotle, *Generation of Animals*, trans. A. L. Peck (1942; Cambridge, MA: Harvard University Press, 2000).

7. See Henry George Liddell and Robert Scott, *A Greek-English Lexicon* (Oxford: Clarendon Press of Oxford University Press, 1996), vol. 1 pp. 343, 621–23, 627. On the term "epigenesis" as distinct from very recent research in "epigenetics," see Linda van Speybroeck, Dani de Waele, and Gertrudis van de Vijver, "Theories in Early Embryology: Close Connections between Epigenesis, Preformationism, and Self-Organization," *Annals of the New York Academy of Sciences* 981 (2002), pp. 7–49.

8. William Harvey, "Anatomical Exercises on the Generation of Animals" (1651), in *The Works of William Harvey*, trans. Robert Willis (1847; New York: Johnson, 1965), p. 334. On Harvey, see Kenneth D. Keele, *William Harvey: The Man, the Physician, and the Scientist* (London: Nelson, 1965); Walter Pagel, *William Harvey's Biological Ideas: Selected Aspects and Historical Background* (Basel: Hafner, 1967).

9. For example, Howard B. Adelmann, *Marcello Malpighi and the Evolution of Embryology*, (Ithaca, NY: Cornell University Press, 1966), vol. 2, p. 764; Jacques Roger, *The Life Sciences in Eighteenth-Century French Thought*, ed. Keith R. Benson, trans. Robert Ellrich (1963;

Stanford, CA: Stanford University Press, 1997), pp. 89–96; in detail, Gasking, *Investigations*; Francis Joseph Cole, *Early Theories of Sexual Generation* (Oxford: The Clarendon Press of Oxford University Press, 1930), p. 132; Olivier Rieppel, "Atomism, Epigenesis, Preformation and Preexistence: A Clarification of Terms and Consequences," *Biological Journal of the Linnean Society* 28 (1986), pp. 331–41; François Duchesneau, *La physiologie des lumières: Empirisme, modèles et théories* (The Hague: Nijhoff, 1982); Pagel, *William Harvey's Biological Ideas*; Oppenheimer, *Essays*; Jane Maienschein, *Embryos Under the Microscope: The Diverging Meanings of Life* (Cambridge, MA: Harvard University Press, 2014).

10. See Roger, *The Life Sciences*; Charles W. Bodemer, "Regeneration and the Decline of Preformationism in Eighteenth Century Embryology," *Bulletin of the History of Medicine* 38 (1964), pp. 20–31; Peter J. Bowler, "Preformation and Pre-Existence in the Seventeenth Century: A Brief Analysis," *Journal of the History of Biology* 4.2 (1971), pp. 221–44; Rieppel, "Atomism"; Clara Pinto-Correia, *The Ovary of Eve: Egg and Sperm and Preformation* (Chicago: University of Chicago Press, 1997).

11. Pierre Louis Moreau de Maupertuis, *The Earthly Venus*, trans. S. B. Boas (1745; New York: Johnson Reprint Corporation, 1966); John Turberville Needham, *Nouvelles observations microscopiques, avec des découvertes intéressantes sur la composition & la décomposition des corps organisés* (Paris: Ganeau, 1750); Georges-Louis Leclerc Buffon, *Natural History: General and Particular*, trans. William Smellie (1749; Edinburgh: William Creech, 1780), vol. 2.

12. See Justin E. H. Smith (ed.), *The Problem of Animal Generation in Early Modern Philosophy* (Cambridge: Cambridge University Press, 2006); Bodemer, "Regeneration"; Charles E. Dinsmore (ed.), *A History of Regeneration Research: Milestones in the Evolution of a Science* (Cambridge: Cambridge University Press, 1991); Virginia P. Dawson, *Nature's Enigma: The Problem of the Polyp in the Letters of Bonnet, Trembley and Reaumur* (Philadelphia: American Philosophical Society, 1987); Aram Vartanian, "Trembley's Polyp, La Mettrie, and Eighteenth-Century Materialism," *Journal of the History of Ideas* 11 (1950), pp. 259–86; John Farley, *The Spontaneous Generation Controversy from Descartes to Oparin* (Baltimore, MD: Johns Hopkins University Press, 1979); Shirley A. Roe, "John Turberville Needham and the Generation of Living Organisms," *Isis* 74 (1983), pp. 159–84; Roe, "Needham's Controversy with Spallanzani: Can Animals Be Reproduced from Plants?," in Giuseppe Montalenti (ed.), *Lazzaro Spallanzani e la biologia del settecento: Teorie, esperimenti, istituzioni scientifiche* (Florence: L. S. Olschki, 1982); Paula Gottdenker, "Three Clerics in Pursuit of 'Little Animals,'" *Clio medica* 14.3-4 (1980), pp. 213–24; Marc J. Ratcliff, "Clandestinité, autorité et expérimentalisme: Styles et querelles de la génération spontanée de Trevoux (1735) à Réaumur (1757)," *Medicina nei Secoli arte e scienza* 15.2 (2003), pp. 319–48. On monstrosities, see Patrick Tort, *L'ordre et les monstres: Le débat sur l'origine des déviations anatomiques au XVIIIe siècle* (Paris: Syllepse, 1998); Michael Hagner (ed.), *Der falsche Körper: Beiträge zu einer Geschichte der Monstrositäten* (Göttingen: Wallstein, 1995); Hagner, "Enlightened Monsters," in William

Clark, Jan Golinski, and Simon Schaffer (eds.), *The Sciences in Enlightened Europe* (Chicago: University of Chicago Press, 1999).

13. Albrecht von Haller, *Sur la formation du cœur dans le poulet sur l'œil, sur la structure du jaune &c.* (Lausanne: Marc-Mich. Bousquet & Comp., 1758). Haller changed his position on epigenesis several times; see Chapter 8.

14. Roe, *Matter*, p. 89. A similar line can be found in Gasking, *Investigations*, pp. 102–104; Oppenheimer, *Essays*, p. 174; Roger, *The Life Sciences*, p. 498. In his study of epigenesis in literature and philosophy, Helmut Müller-Sievers calls the "obliteration of preformation by epigenesis . . . a purely textual event," Müller-Sievers, *Self-Generation: Biology, Philosophy, and Literature around 1800* (Stanford: Stanford University Press, 1997), p. 5.

15. Roe, *Matter*, p. 156; see also pp. 148 and 150.

16. *Ibid.*, p. 156.

17. Gasking, *Investigations*, p. 151.

18. *Ibid.*, p. 164; see also Roe, *Matter*, p. 151.

19. Roe, *Matter*, p. 152.

20. *Ibid.*, p. 151.

21. *Ibid.*, pp. 152 and 155; also Gasking, *Investigations*, pp. 151, 160, 161.

22. See Timothy Lenoir, "The Göttingen School and the Development of Transcendental *Naturphilosophie* in the Romantic Era," *Studies in the History of Biology* 5 (1981), pp. 111–205; Lenoir, "Kant, Blumenbach, and Vital Materialism in German Biology," *Isis* 71 (1980), pp. 77–108; Lenoir, "Kant, von Baer und das kausal-historische Denken in der Biologie," *Berichte zur Wissenschaftsgeschichte* 8 (1985), pp. 99–114; Lenoir, *The Strategy of Life: Teleology and Mechanics in Nineteenth-Century German Biology* (Chicago: University of Chicago Press, 1989). Lenoir's approach has attracted some criticism; see K. L. Caneva, "Teleology with Regrets," *Annals of Science* 47 (1990), pp. 291–300; Robert J. Richards, "Kant and Blumenbach on the Bildungstrieb: A Historical Misunderstanding," *Studies in the History and Philosophy of Biology and the Biomedical Sciences* 31.1 (2000), pp. 11–32; John H. Zammito, "The Lenoir Thesis Revisited: Blumenbach and Kant," *Studies in History and Philosophy of Biological and Biomedical Sciences* 43.1 (2012), pp. 120–32.

23. Lenoir, *Strategy of Life*, pp. 28–29.

24. Recent scholarship has framed the period from the eighteenth to the twentieth century and contemporary science more widely, as a transition from generation to reproduction. See Susanne Lettow (ed.), *Reproduction, Race, and Gender in Philosophy and the Early Life Sciences* (Albany: State University of New York Press, 2014); Bettina Bock von Wülfingen et al. (eds.), "Temporalities of Reproduction," special issue, *History and Philosophy of the Life Sciences* 37.1 (2015).

25. Around 1900, the term "rhythm" surfaced in several areas of biomedical research at once. The same applies for experimental psychology, for example, in the work of Wilhelm Wundt: see Wundt, *Grundriß der Psychologie* (Leipzig: Engelmann, 1896); Wundt,

Völkerpsychologie: Eine Untersuchung der Entwicklungsgesetze von Sprache, Mythus und Sitte, 10 vols. (Leipzig: Engelmann, 1900-1920), vols. 1 and 3, and for the arts, vol. 10; Wundt, *Grundzüge der physiologischen Psychologie* (Leipzig: Engelmann, 1874). In 1903, the Dutch physician Karel Frederik Wenckebach published the treatise *Die Arrythmie als Ausdruck bestimmter Funktionsstörungen Eine physiologisch-klinische Studie* (Leipzig: W. Engelmann, 1903). Wenckebach appears to have been among the first to introduce the term "arrhythmia" and established the semantic field of rhythm/arrhythmia as key terms in cardiac research. See Berndt Lüderitz, "History of Cardiac Rhythm Disorders," *Zeitschrift für Kardiologie* 91, suppl. 4 (2002), pp. 4/50–4/55.

26. Caspar Friedrich Wolff, *Über die Bildung des Darmkanals im bebrüteten Hühnchen*, trans. Johann Friedrich Meckel (1768; Halle: Rengersche Buchhandlung, 1812).

27. Karl Ernst von Baer, *Über Entwickelungsgeschichte der Thiere: Beobachtung und Reflexion*. 2 vols. (1828 and 1837; Brussels: Culture et civilisation, 1967, vol. 2, p. 121. One of the few studies addressing Wolff's treatise is Jean-Claude Dupont, "Pre-Kantian Revival of Epigenesis: Caspar Friedrich Wolff's *De formatione intestinorum* (1768–69)," in Philippe Hunemann (ed.), *Understanding Purpose: Kant and the Philosophy of Biology* (Rochester, NY: University of Rochester Press, 2007).

28. Wolff, *Theoria generationis*, part 1, § 32, p. 9.

29. *Ibid.*, part 1, p. 3.

30. *Ibid.*, part 1, § 11, p. 5. Wolff was particularly influenced by the rationalist philosophy of Christian Wolff (1679-1754), a teacher at the University of Halle, where Wolff himself studied from 1755 to 1759. See Gasking, *Investigations*, pp. 97–98; Haffner, "Die Epigenesis-analogie," pp. 102–11.

31. Wolff, *Theoria Generationis*, part 1, § 8, pp. 4–5.

32. In the first two sections of *Theorie von der Generation*, Wolff summarizes earlier positions on generation and their shortcomings. Caspar Friedrich Wolff, *Theorie von der Generation: In zwei Abhandlungen erklärt und bewiesen. Theoria generationis* (1764 and 1759; Hildesheim: Olms, 1966), pp. 2–34; esp. pp. 7–10, 13–14, 19–22, 26, 34; for his position on preformation, see pp. 43–44.

33. Wolff, *Theoria Generationis*, part 1, § 5, p. 4. Opinions on Wolff's methodology have diverged. Jane Oppenheimer locates Wolff's contribution in his empirical observations, which he carried out despite (and not because of) his "abstruse reasoning," while William M. Wheeler takes the converse view: "Wolff's method . . . was, if anything, more admirable than his observations." Oppenheimer, *Essays*, p. 133; Wheeler, "Caspar Friedrich Wolff," p. 204. Olaf Breidbach refers to an "analytical gaze" that broke open the preformationist schema. Breidbach, "Die Geburt des Lebendigen—Embryogenese der Formen oder Embryogenese der Natur?: Anmerkungen zum Bezug von Embryologie und Organismustheorien vor 1800," *Biologisches Zentralblatt* 114 (1995), p. 194.

34. Caspar Friedrich Wolff, "Von der eigenthümlichen und wesentlichen Kraft der

vegetabilischen sowohl als auch der animalischen Substanz," in *Zwo Abhandlungen über die Nutritionskraft welche von der Kayserlichen Academie der Wissenschaften in St. Petersburg den Preis getheilt erhalten haben* (St. Petersburg: Kayserliche Akademie der Wissenschaften, 1789), p. 42, see also p. 49.

35. *Ibid.*, p. 49.

36. Wolff, *Theorie von der Generation*, p. 191–92; see also Wolff, "Von der eigenthümlichen," p. 48.

37. Wolff refers to Stephen Hales, who, he says, demonstrated this at the beginning of the century, Wolff, *Theoria Generationis*, part 1, § 1, p. 11; see Hales, *La statique des végétaux et l'analyse de l'air: Expériences nouvelles liés à la Societé Royale de Londres* (Paris: Debure l'aîné, 1735).

38. Wolff, *Theoria Generationis*, part 2, § 241, p. 59.

39. Wolff, *Theoria Generationis*, part 1, § 13, p. 5. See also Wolff, "Von der eigenthümlichen," p. 61: "Nutrition is thus increase or replacement of the substance out of which the parts of a plant or an animal are formed, without anything being thereby changed in the organization, structure, or construction of the parts."

40. Wolff, *Theoria Generationis*, part 1, § 54, pp. 34–35, and § 58, pp. 37–38.

41. *Ibid.*, part 1, § 15, p. 5.

42. *Ibid.*, part 1, § 16, p. 6.

43. *Ibid.*, part 1, § 95, p. 57; § 106, p. 62; § 115, p. 66; § 165, pp. 88–89.

44. Wolff, "Von der eigenthümlichen," p. 75. This was one of Wolff's last works. It includes contributions by Johannn Friedrich Blumenbach and Ignaz Born along with Wolff's commentary on the prize essay question of the St. Petersburg Academy of Sciences, initiated by Wolff several times between 1782 and 1789, which called for an explanation of the nature of organic forces.

45. Wolff, *Über die Bildung*, p. 149.

46. Wolff, *Theoria Generationis*, part 1, § 6, p. 13.

47. *Ibid.*, Part 1, § 5, p. 12.

48. See the Latin original of the *Theoria* reproduced in Wolff, *Theorie von der Generation*, part 1, § 5, p. 13 (*guttulas sphaericas, cylindricas humorum*), also §§ 6, 8, 9, 13, 14, pp. 13–14 (*vesiculae*), § 20, p. 16 (*foraminula, cellulas*), §§ 21, 22, 29, p. 16–17 (*bulla, poris*), § 30, p. 19 (*globuli*), § 36, p. 21 (*in majori vesicula minores cellulae*).

49. Wolff, *Theoria Generationis*, part 1, § 6, p. 13; see also § 21–22, pp. 17–18, and § 34, p. 22.

50. *Ibid.*, part 1, § 10, p. 14.

51. *Ibid.*, part 1, § 20, p. 16.

52. *Ibid.*, part 1, § 21, p. 17.

53. *Ibid.*, part 1, § 23, p. 18.

54. *Ibid.*; also part 1, § 34, p. 22, and Wolff, *Theorie von der Generation*, pp. 147 and 156–58.

55. Wolff, *Theoria Generationis*, part 1, § 27, p. 19.

56. *Ibid.*, part 1, § 28, p. 19.

57. *Ibid.*, part 1, § 28, p. 20.

58. Wolff, *Theorie von der Generation*, p. 131.

59. Wolff, "Von der eigenthümlichen," p. 27.

60. Wolff, *Theoria Generationis*, part 1, § 11, p. 14.

61. *Ibid.*, part 1, § 29, p. 20.

62. *Ibid.*, part I, § 22, p. 18.

63. *Ibid.*, part 1, § 34, p. 22.

64. *Ibid.*, part 1, § 35, p. 23.

65. *Ibid.*

66. On many occasions, historians have interpreted Wolff's vesicles as cells. Older literature in the history of science, especially, sees Wolff as foreshadowing Schwann's cell theory. See Raikov, "Caspar Friedrich Wolff, pp. 564, 577-78; Schuster, "Caspar Friedrich Wolff," p. 184; Uschmann, *Caspar Friedrich Wolff*; Wheeler, "Caspar Friedrich Wolff," p. 205. More recent research has moved away from this interpretation. See Gasking, *Investigations*, pp. 100-101; Breidbach "Einleitung," p. xv; and especially studies on the concept of protoplasm, such as Miklos Lambrecht, "Die Schwierigkeiten der biologischen Erkenntnis: Protoplasma-Konzeptionen von Christian Friedrich Wolff bis Jan Evangelista Purkinje," in Burchard Thaler and Wolfram Kaiser (eds.), *Johann Andreas Segner (1704-1777) und seine Zeit* (Halle: Abt. Wissenschaftspublizistik d. Martin-Luther-Univ., 1977).

67. Wolff, *Theorie von der Generation*, p. 133.

68. *Ibid.*, p. 132.

69. *Ibid.*, p. 133.

70. Wolff, *Über die Bildung*, p. 58.

71. *Ibid.*, p. 68.

72. See *ibid.*, pp. 70-80.

73. The following account is set out in *ibid.*, pp. 101-32.

74. *Ibid.*, pp. 133-40.

75. *Ibid.*, pp. 132-33.

76. *Ibid.*

77. *Ibid.*, p. 145.

78. *Ibid.*, p. 146.

79. *Ibid.*, p. 147-48.

80. *Ibid.*, p. 149.

81. *Ibid.*, p. 148.

82. *Ibid.*

83. *Ibid.*, p. 150.

84. *Ibid.*

85. Karl Philipp Moritz, "Versuch einer Vereinigung aller schönen Künste und Wissenschaften unter dem Begriff des in sich selbst Vollendeten" (1785), in *Schriften zur Ästhetik und Poetik: Kritische Ausgabe*, ed. Hans Joachim Schrimpf (Tübingen: Niemeyer, 1962), p. 3.

86. See Wolff, *Über die Bildung*, p. 124.

87. *Ibid.*, pp. 183–84.

88. *Ibid.*, p. 187.

89. *Ibid.*, pp. 188–89.

90. *Ibid.*, p. 152.

CHAPTER FIVE: SENSE AND VERSE

1. See Ernst Cassirer, *Freiheit und Form* (1916), vol. 7 of Cassirer, *Gesammelte Werke*, ed. Birgit Recki (Hamburg: Meiner, 2003), pp. 194–95.

2. On Goethe's concept of metamorphosis, the history of the text, and Goethe's later interest in plant growth, see Wilhelm Troll, "Goethe in seinem Verhältnis zur Natur: Eine Einführung des Herausgebers," in *Goethes Morphologische Schriften*, ed. Wilhelm Troll (Jena: Diederichs, 1926); Alfred Kirchhoff, "Die Idee der Pflanzen-Metamorphose bei Wolff und Göthe," in *Zweiter Jahresbericht über die Luisenstädtische Gewerbeschule in Berlin* (Berlin: Schade, 1867); Adolph Hansen, *Goethes Metamorphose der Pflanzen: Geschichte einer botanischen Idee* (Giessen: A. Töpelmann, 1907); Adolf Portmann, "Goethe und der Begriff der Metamorphose," *Goethe-Jahrbuch* 90 (1973), pp. 11–12; Timothy Lenoir, "The Eternal Laws of Form: Morphotypes and the Conditions of Existence in Goethe's Biological Thought," in Frederick Amrine, Francis J. Zucker, and Harvey Wheeler (eds.), *Goethe and the Sciences: A Reappraisal* (Dordrecht: D. Reidel, 1987); Dorothea Kuhn, *Typus und Metamorphose: Goethe-Studien*, ed. Renate Grumach (Marbach: Deutsche Schillergesellschaft, 1988); Marie-Luise Kahler and Gisela Maul, *Alle Gestalten sind ähnlich: Goethes Metamorphose der Pflanzen* (Weimar: Klassikerstätten zu Weimar, 1991); Karl J. Fink, *Goethe's History of Science* (Cambridge: Cambridge University Press, 1991); Robert J. Richards, *The Romantic Conception of Life: Science and Philosophy in the Age of Goethe* (Chicago: University of Chicago Press, 2002); Astrida Orle Tantillo, *The Will to Create: Goethe's Philosophy of Nature* (Pittsburgh, PA: University of Pittsburgh Press, 2002), pp. 64–74; William Gray, "Goethe's *The Metamorphosis of Plants*: The Issue of Science and Poetry," *Archives of Natural History* 21.3 (1994), pp. 379–91; Dorothea von Mücke, "Goethe's Metamorphosis: Changing Forms in Nature, the Life Sciences, and Authorship," *Representations* 95 (2006), pp. 27–53; Thomas Pfau, "'All Is Leaf': Difference, Metamorphosis, and Goethe's Phenomenology of Knowledge," *Studies in Romanticism* 49.1 (2010), pp. 3–41. On art history, see Christa Lichtenstern, *Die Wirkungsgeschichte der Metamorphosenlehre Goethes: Von Philipp Otto Runge bis Joseph Beuys* (Weinheim: VCH Acta Humaniora, 1990).

3. Frequent mention has been made of rhythm in Goethe's notion of metamorphosis, but without placing it in the wider context of concepts of development and nature around

1800. See Andreas B. Wachsmuth, *Geeinte Zwienatur: Aufsätze zu Goethes naturwissenschaft-lichem Denken. Beiträge zur deutschen Klassik* (Berlin: Aufbau, 1966), p. 22; Klaudia Hilgers, *Entelechie, Monade und Metamorphose: Formen der Vervollkommnung im Werk Goethes* (Munich: Fink, 2002), p. 172; Lichtenstern, *Wirkungsgeschichte*, p. 1; Jocelyn Holland, *German Romanticism and Science: The Procreative Poetics of Goethe, Novalis, and Ritter* (New York: Routledge, 2009), p. 46. A study without reference to metamorphosis or rhythm, but discussing Goethe's musical understanding of the organism, is Frederick Amrine, "The Music of the Organism: Uexküll, Merleau-Ponty, Zuckerkandl, and Deleuze as Goethean Ecologists in Search of a New Paradigm," *Goethe Yearbook* 22 (2015), pp. 45–72.

4. Cassirer, *Freiheit und Form*, pp. 233–34. Rather than unity, scholars have regarded polarity and *Wechselwirkung* as the crucial explanatory categories in Goethe's work, for example Tantillo, *The Will to Create*; Wachsmuth, *Geeinte Zwienatur*. Peter Hanns Reill, however, does stress mediation as Goethe's chief concern. Reill, "Bildung, Urtyp and Polarity: Goethe and Eighteenth-Century Physiology," *Goethe Yearbook* 3 (1986), pp. 139–48, esp. p. 143.

5. Cassirer, *Freiheit und Form*, p. 234.

6. *Ibid.*, p. 238. *Gestalt* becomes a "basic biological concept" because it is capable of encompassing just that "peculiar interweaving of being and becoming, of permanence and change." *Ibid.*, p. 161. The biological *Gestalt* thus no longer belongs only to space, but also to time, and it is here that Cassirer finds Goethe's new "knowledge ideal" and his contribution to biology. *Ibid.*, p. 162.

7. *Ibid.*, p. 256.

8. Ernst Cassirer, *Das Erkenntnisproblem in der Philosophie und Wissenschaft der neueren Zeit, 4: Von Hegels Tod bis zur Gegenwart (1832–1932)* (1957), vol. 4 of Cassirer, *Gesammelte Werke*, ed. Birgit Recki (Hamburg: Meiner, 2000), p. 172.

9. Cassirer, *Freiheit und Form*, p. 278.

10. *Ibid.*, p. 169.

11. Goethe's papers include an outline of the arts that subsumes poetry into the categories of meter and rhythm. See Dorothea Kuhn, "Zu Goethes Theorie der Künste: Mit einem unveröffentlichten Schema Goethes," in Kuhn, *Typus und Metamorphose*.

12. As early as the mid-seventeenth century, William Harvey used the term "metamorphosis" in natural history to describe the generation of insects: "Some [animals], out of material previously concocted, and that has already attained its bulk, receive their forms and transfigurations; and all their parts are fashioned simultaneously, each with its distinctive characteristic, by the process called metamorphosis, and in this way a perfect animal is at once born." Harvey, "Anatomical Exercises on the Generation of Animals," in *The Works of William Harvey*, trans. Robert Willis (1651; New York: Johnson, 1965), p. 334. For the plant kingdom, in 1676 the Italian Giacomo Sinibaldi used the term for the first time in his *Plantarum metamorphosis*, and in Carl Linnaeus's work, it denotes the similarity of

forms such as leaves and blossoms. Later, Linnaeus compared the metamorphosis of plants to that of insects, in the sense of ecdysis or "unveiling." See Johann Wolfgang von Goethe, *Versuch die Metamorphose der Pflanzen zu erklären*, ed. Dorothea Kuhn (1790; Weinheim: Acta humaniora, 1984), p. 91; also Hansen, *Goethes Metamorphose*, pp. 181–219. The story of metamorphosis in natural history has not been studied, and the most detailed such research of which I am aware is Hansen, *Goethes Metamorphose*, p. 173–219; a brief historical note may be found in William Kirby and William Spence (eds.), *An Introduction to Entomology, or Elements of the Natural History of Insects*, 7th ed. (1815; London: Longman, 1858), pp. 31–41. Clemens Heselhaus, "Metamorphose-Dichtungen und Metamorphose-Anschauungen," *Euphorion* 47.2 (1953), pp. 121–46, remains only an eclectic collection.

13. Quoted in Johann Wolfgang von Goethe, *The Metamorphosis of Plants*, trans. Douglas Miller, (1790; Cambridge, MA: MIT Press, 2009), p. xxvi.

14. *Ibid.*, p. 11.

15. *Ibid.*, p. 19.

16. *Ibid.*, p. 14.

17. *Ibid.*, p. 6.

18. *Ibid.*, p. 16.

19. *Ibid.*

20. *Ibid.*, p. 22.

21. *Ibid.*, p. 24.

22. *Ibid.*, p. 28.

23. *Ibid.*, p. 36.

24. *Ibid.*, p. 52.

25. *Ibid.*, p. 60.

26. *Ibid.*, pp. 99–100.

27. *Ibid.*, p. 6.

28. *Ibid.*, p. 67.

29. *Ibid.*, p. 23.

30. *Ibid.*, p. 30.

31. Reprinted in the Leopoldina edition of Goethe's writings on science, Johann Wolfgang von Goethe, "Morphologische Hefte" (1817), in *Goethe: Die Schriften zur Naturwissenschaft. Vollständige mit Erläuterungen versehene Ausgabe im Auftrage der Deutschen Akademie der Naturforscher Leopoldina*, ed. Dorothea Kuhn (Weimar: Metzler, 1994), vol. 1.9, pp. 1–83. Much of the additional material is translated in Goethe, *Goethe's Botanical Writings*, pp. 21–29. For the textual history, see also Mücke, "Goethe's Metamorphosis."

32. Johann Wolfgang von Goethe, "Genesis of the Essay on the Metamorphosis of Plants," in *Goethe's Botanical Writings*, p. 166.

33. Johann Wolfgang von Goethe, "Our Objective Is Stated," in *Goethe's Botanical Writings*, p. 23.

34. Johann Wolfgang von Goethe, "The Content Is Given a Foreword," in *Goethe's Botanical Writings*, p. 27.

35. *Ibid.*, p. 28.

36. Goethe, *Metamorphosis*, p. 5–6.

37. *Ibid.*, p. 5.

38. *Ibid.*, p. 16.

39. *Ibid.*, p. 23.

40. *Ibid.*, p. 48.

41. *Ibid.*, p. 74.

42. *Ibid.*, p. 75.

43. *Ibid.*, p. 31.

44. *Ibid.*, p. 44.

45. *Ibid.* See also p. 60.

46. *Ibid.*, pp. 60–65.

47. *Ibid.*, p. 92.

48. See *ibid.*, p. 100.

49. *Ibid.*

50. Johann Wolfgang von Goethe, "My Discovery of a Worthy Forerunner," in *Goethe's Botanical Writings*, pp. 176–81.

51. *Ibid.*, p. 180.

52. Goethe, *Metamorphosis*, p. 23.

53. *Ibid.*, p. 30.

54. Johann Wolfgang von Goethe, "On Morphology," in *Goethe's Botanical Writings*, p. 102.

55. Goethe, *Metamorphosis*, p. 102.

56. *Ibid.*, p. 35.

57. *Ibid.*, p. 60.

58. *Ibid.*, p. 5.

59. *Ibid.*, p. 6.

60. *Ibid.*, p. 7.

61. Goethe, "Morphologische Hefte." According to his original plans, Goethe wanted the *Metamorphosis of Plants* to include more concrete detail on his general discussion of metamorphosis. Goethe, "Morphologische Hefte," p. 96.

62. *Ibid.*, p. 98.

63. *Ibid.*, p. 100.

64. *Ibid.*, p. 101.

65. *Ibid.*, p. 99; emphasis added.

66. Lichtenstern, *Wirkungsgeschichte*, p. 1.

67. *Ibid.*, p. 5. See also Wachsmuth, *Geeinte Zwienatur*, p. 283. Goethe's dual persona as a

NOTES TO PAGE 121

poet and a naturalist has fascinated Goethe scholars, with generations of essays dedicated to his views on art and nature. On Goethe's thinking on aesthetics, artists, and the work of art see, for example, Matthew Bell, *Goethe's Naturalistic Anthropology: Man and Other Plants* (Oxford: Clarendon Press, 1994); Sabine Schulze (ed.), *Goethe und die Kunst* (Ostfildern: Hatje, 1994); Kuhn, "Zu Goethes Theorie der Künste"; Günter Peters, *Der zerrissene Engel: Genieästhetik und literarische Selbstdarstellung im achtzehnten Jahrhundert* (Stuttgart: Metzler, 1982). On his notion of the picture, see Frank Fehrenbach, "'Das lebendige Ganze, das zu allen unsern geistigen und sinnlichen Kräften spricht': Goethe und das Zeichnen," in Matussek, *Goethe und die Verzeitlichung der Natur*. Goethe's role in the study of nature in general and biology in particular was vigorously debated by his contemporaries, and still is today. Judgments differ considerably according to the scientific paradigms dominant at the time concerned. In recent years, Goethe's concept of nature has been placed in the context of the temporalization and dynamization of nature around 1800. The consensus is that Goethe's thinking played a leading part in the "change of perspective from natural history to the history of nature that characterized the period around 1800," Lichtenstern, *Wirkungsgeschichte*, p. 1. The details of such appraisals vary, however, with Goethe interpreted as everything from a forerunner of Darwinism, to a transitionary figure in history, to the singular proponent of an approach not found in the later categories of evolutionary biology. See Wyder, *Goethes Naturmodell*; Matussek, *Goethe und die Verzeitlichung der Natur*; Richards, *The Romantic Conception*. Transformist interpretations of Goethe are criticized by Dorothea Kuhn, "Goethe's Relationship to the Theories of Development of His Time," in Amrine, Zucker, and Wheeler, *Goethe and the Sciences*; Lenoir, "The Eternal Laws"; Hansen, *Goethes Metamorphose*. Contradictory positions in Goethe's work are discussed by Wolfgang Schad, "Zeitgestalten der Natur: Goethe und die Evolutionsbiologie," in Matussek, *Goethe und die Verzeitlichung der Natur*. Uwe Pörksen encapsulates Goethe's position in the concept of time-space: Pörksen, "Raumzeit: Goethes Zeitbegriff aufgrund seiner sprachlichen Darstellung geologischer Ideen und ihrer Visualisierung," in Matussek, *Goethe und die Verzeitlichung der Natur*; in more detail, Pörksen, *Raumzeit: Goethes Zeitbegriff, abgelesen an seinen sprachlichen und zeichnerischen Naturstudien* (Stuttgart: F. Steiner, 1999). For Astrida Orle Tantillo, Goethe was "certainly not a precursor to Darwin," since although he thought in evolutionary categories, he did so "from a different perspective," Tantillo, *The Will to Create*, pp. 128–29. In this context, interpretations of Goethe's concept of the type (as a static category or as a merely abstract principle) are also relevant. See Ronald H. Brady, "Form and Cause in Goethe's Morphology," in Amrine, Zucker, and Wheeler, *Goethe and the Sciences*; Cassirer, *Das Erkenntnisproblem*. On Goethe's geological studies, see Wolf von Engelhardt, *Goethe im Gespräch mit der Erde: Landschaft, Gesteine, Mineralien und Erdgeschichte in seinem Leben und Werk* (Weimar: Böhlau, 2003). From an environmentalist perspective, see the recent special section edited by Dalia Nassar and Luke Fischer on Goethe and environmentalism, *Goethe Yearbook* 22 (2015).

68. Cassirer, *Freiheit und Form*, p. 216. Cassirer also observed that "Goethe's criticism of eighteenth-century natural history travels precisely the same path that had been taken by his critique of eighteenth-century poetology." Ernst Cassirer, *Goethe und die geschichtliche Welt* (1932; Hamburg: Meiner, 1995), p. 72.

69. Lichtenstern, *Wirkungsgeschichte*, pp. 4 and 1.

70. It is surprising how little research has been devoted to the concept of metamorphosis; in fact, the history of metamorphosis is as good as unwritten. What work there is can be found in literary studies, where a new interest is currently emerging. See Sabine Coelsch-Foisner and Michaela Schwarzbauer (eds.), *Metamorphosen: Akten der Tagung der Interdisziplinären Forschungsgruppe Metamorphosen an der Universität Salzburg* (Heidelberg: Winter, 2005); Herwig Gottwald and Holger Klein (eds.), *Konzepte der Metamorphose in den Geisteswissenschaften* (Heidelberg: Winter, 2005). Whereas the number of individual studies on particular authors, from antiquity to the present, is enormous, more broad-based approaches are rare. A taxonomy of the concept's use in the human sciences is attempted by Peter Kuon, "Metamorphose als geisteswissenschaftlicher Begriff," in Klein and Herwig, *Konzepte der Metamorphose*; metamorphosis as myth is studied by Pierre Brunel, *Le mythe de la métamorphose* (Paris: A. Colin, 1974); as linguistic critique by Irving Massey, *The Gaping Pig: Literature and Metamorphosis* (Berkeley: University of California Press, 1976); as a mental concept of identity by Harold Skulsky, *Metamorphosis: The Mind in Exile* (Cambridge, MA: Harvard University Press, 1981) and by Marina Warner, *Fantastic Metamorphoses, Other Worlds: Ways of Telling the Self* (Oxford: Oxford University Press, 2002); as an allegory by Bruce Clarke, *Allegories of Writing: The Subject of Metamorphosis* (Albany: State University of New York Press, 1995); as a literary motif by Pascal Nicklas, *Die Beständigkeit des Wandels: Metamorphosen in Literatur und Wissenschaft* (Hildesheim: Olms, 2002); and as a pagan phenomenon by Leonard Barkan, *The Gods Made Flesh: Metamorphosis and the Pursuit of Paganism* (New Haven, CT: Yale University Press, 1986). On metamorphosis in music, see Oswald Panagl, "Metamorphose: Bedeutungsprofil, Begriffsgeschichte, musikalisches Spektrum," in Gottwald and Klein, *Konzepte der Metamorphose*; in nineteenth- and twentieth-century art, Lichtenstern, *Wirkungsgeschichte*; Lichtenstern, *Vom Mythos zum Prozeßdenken: Ovid-Rezeption, Surrealistische Ästhetik, Verwandlungsthematik der Nachkriegskunst* (Weinheim: VCH Acta Humaniora, 1992), which includes a discussion of the concept in the context of the humanities. Nicklas looks at metamorphosis as a theme of science and literature across the period. Nicklas, *Die Best ändigkeit*.

71. More recent research stresses that Ovid's treatment of metamorphosis already represented "a particular stage in the process of differentiation" of myth, so that there is limited value in using Ovid as an absolute starting point. Herwig Gottwald, "Die Metamorphose aus mythostheoretischer und literaturwissenschaftlicher Sicht," in Gottwald and Klein, *Konzepte der Metamorphose*, p. 84. On metamorphosis in mythic thinking, see

also Cassirer, *The Philosophy of Symbolic Forms*, vol. 2, *Mythical Thought*, trans. Ralph Manheim (1925; New Haven, CT: Yale University Press, 1965).

72. See Nicklas, *Die Beständigkeit*, p. 126; also Barkan, *The Gods*; Heselhaus, "Metamorphose-Dichtungen"; Fink, *Goethe's History of Science*.

73. According to Nicklas, *Beständigkeit*, and Lichtenstern, *Vom Mythos*, this pattern did not change fundamentally in the twentieth century, which saw only a shift in art, literature, and psychology to privilege not the given, but the other, not creating identity, but creating uncertainty.

74. Goethe to Frau von Stein, December 20, 1786, in *Letters from Goethe*, trans. M. von Herzfeld and C. Melvil Sym (Edinburgh: Edinburgh University Press, 1957), p. 186.

75. Johann Wolfgang von Goethe, "History of the Manuscript," in *Goethe's Botanical Writings*, p. 168.

76. Wulf Segebrecht, "Sinnliche Wahrnehmung Roms: Zu Goethes 'Römischen Elegien,' unter besonderer Berücksichtigung der Fünften Elegie," in Segebrecht (ed.), *Gedichte und Interpretationen*, vol. 3, *Klassik und Romantik* (Stuttgart: Reclam, 1984), p. 52.

77. On the *Roman Elegies*, see, for example, Bernd Witte, Theo Buck, and Hans-Dietrich Dahnke (eds.), *Goethe Handbuch* (Stuttgart: J. B. Metzler, 1996-1999), vol. 1, pp. 225-32; Benedikt Jeßing, "Sinnlichkeit und klassische Ästhetik: Zur Konstituierung eines poetischen Programms im Gedicht," in Bernd Witte (ed.), *Gedichte von Johann Wolfgang Goethe* (Stuttgart: Reclam, 1998); Bell, *Goethe's Naturalistic Anthropology*; Reiner Wild, *Goethes klassische Lyrik* (Stuttgart: Metzler, 1999).

78. The poem first appeared in Schiller's *Musen-Almanach für das Jahr 1799*. It was reprinted in 1800 by Unger along with six other elegies under the title *Elegien II*, in *Goethes Neue Schriften*, and then again in *Zur Morphologie* in 1817. There, for the first time, the poem was embedded in Goethe's scientific writings. See Goethe, "History of the Printed Brochure," in *Goethe's Botanical Writings*, pp. 170-76; on the textual history, see also Mücke, "Goethe's Metamorphosis."

79. Goethe, "History of the Printed Brochure," pp. 171-72; see also Johann Wolfgang von Goethe, "Other Friendly Overtures," in *Goethe's Botanical Writings*, pp. 182-86, where Goethe argues that the conjunction of science and poetry had been "entirely outside the intellectual horizon of the time," which made them "the greatest adversaries" (p. 185).

80. See Maike Arz, "Die Metamorphose der Pflanzen," in Witte, Buck, and Dahnke, *Goethe Handbuch*, vol. 1.

81. *The Poems of Goethe, Translated in the Original Metres by Edgar Alfred Bowring* (London: John W. Parker, 1853), pp. 332-36. The original German can be consulted at http://gutenberg.spiegel.de/buch/johann-wolfgang-goethe-gedichte-3670/203.

82. See Günter Müller, "Goethes Elegie 'Die Metamorphose der Pflanzen': Versuch einer morphologischen Interpretation," *Deutsche Vierteljahrsschrift für Literaturwissenschaft und Geistesgeschichte* 21 (1943), pp. 90-91.

83. See Reiner Wild, "Die Poetik der Natur," in Witte, *Gedichte von Johann Wolfgang Goethe*; Wild, *Goethes klassische Lyrik*, pp. 151–71; Karl Richter, "Wissenschaft und Poesie 'auf höherer Stelle' vereit: Goethes Elegie 'Die Metamorphose der Pflanzen,'" in Segebrecht, *Gedichte und Interpretationen*, vol. 3.

84. Wild, "Die Poetik der Natur," p. 165.

85. *Ibid.*, p. 166. See also Richter, "Wissenschaft und Poesie"; Tantillo, *The Will to Create*, pp. 187–88; Holland, *German Romanticism*, pp. 42–55. On poetry as a way of popularizing knowledge, see Michael Bies, "Staging the Knowledge of Plants: Goethe's Elegy 'The Metamorphosis of Plants,'" in Mary Helen Dupree and Sean B. Franzel (eds.), *Performing Knowledge, 1750–1850* (Berlin: de Gruyter, 2015). As a form of biosemiotics, see Kate Rigby, "Art, Nature, and the Poesy of Plants in the *Goethezeit*: A Biosemiotic Perspective," *Goethe Yearbook* 22 (2015), pp. 23–44.

86. Wild, "Die Poetik der Natur," p. 169.

87. *Ibid.*, p. 160; see also Arz, "Die Metamorphose."

88. See, for example, Bell, *Goethe's Naturalistic Anthropology*, esp. pp. 221–27. Interpretations informed by gender studies are offered by Holland, *German Romanticism*, and Lisbet Koerner, "Goethe's Botany: Lessons of a Feminine Science," *Isis* 84 (1993), pp. 470–95.

89. Günter Peters, "Das Schauspiel der Natur: Goethes Elegien 'Die Metamorphose der Pflanzen' und 'Euphrosyne' im Kontext einer Naturästhetik der szenischen Anschauung," *Poetica* 22 (1990), p. 67. On the symbolic aspect, see also Richter, "Wissenschaft und Poesie"; Wild, "Die Poetik der Natur."

90. See Peters, "Das Schauspiel der Natur," and Günter Peters, *Die Kunst der Natur: Ästhetische Reflexion in Blumengedichten von Brockes, Goethe und Gautier* (Munich: Fink, 1993).

91. Peters, "Das Schauspiel der Natur," p. 59.

92. *Ibid.*, p. 67.

93. Müller, "Goethes Elegie"; Gertrud Overbeck, "Goethes Lehre von der Metamorphose der Pflanzen und ihre Widerspiegelung in seiner Dichtung," *Publications of the English Goethe Society* n.s. 31 (1961), pp. 38–59.

94. Overbeck, "Goethes Lehre," p. 57; see also p. 46.

95. Müller, "Goethes Elegie," pp. 83 and 82.

96. Overbeck, "Goethes Lehre," p. 58.

97. Nicklas, *Die Beständigkeit*, p. 126.

98. *Ibid.*, p. 128.

99. *Ibid.*, p. 126.

100. See Hans Lösener, *Der Rhythmus der Rede: Linguistische und literaturwissenschaftliche Aspekte des Sprachrhythmus* (Tübingen: Niemeyer, 1999); Hans Ulrich Gumbrecht, "Rhythm and Meaning," trans. William Whobrey, in Hans Ulrich Gumbrecht and K. Ludwig Pfeiffer (eds.), *Materialities of Communication* (Stanford, CA: Stanford University Press, 1994); Gumbrecht, *Production of Presence: What Meaning Cannot Convey* (Stanford,

CA: Stanford University Press, 2004); Kai Christian Ghattas, *Rhythmus der Bilder: Narrative Strategien in Text- und Bildzeugnissen des 11. bis 13. Jahrhunderts* (Cologne: Böhlau, 2009), especially the discussion of rhythm in literary and linguistic studies on pp. 21–34; Henri Meschonnic, *Critique de rythme: Anthropologie historique du langage* (Paris: Verdier, 1982).

101. Hans Ulrich Gumbrecht, "Charms of the Distich: About Functions of Poetic Form in Goethe's *Römische Elegien*," in Gerhard Neumann and David E. Wellbery (eds.), *Die Gabe des Gedichts: Goethes Lyrik im Wechsel der Töne* (Freiburg: Rombach, 2008).

102. Goethe, "History of the Manuscript," p. 168.

103. Goethe discussed versification not only with Moritz and Schlegel, but also with Johann Heinrich Voß and Carl Ludwig Knebel. Wild, *Goethes klassische Lyrik*, pp. 187–89.

104. *Ibid.*, p. 190.

105. Overbeck, "Goethes Lehre," pp. 56–57.

106. Gumbrecht, "Charms of the Distich," p. 280.

107. See Leif Ludwig Albertsen, "Rom 1789, auch eine Revolution: Unmoralisches oder vielmehr Moralisches in den 'Römischen Elegien,'" *Goethe-Jahrbuch* 99 (1982), pp. 183–94.

108. On Goethe and the Linnaean system, see Chad Wellmon, "Goethe's Morphology of Knowledge, or the Overgrowth of Nomenclature," *Goethe Yearbook* 17 (2010), pp. 153–77.

109. Wild, *Goethes klassische Lyrik*, p. 191.

110. Olaf Breidbach has attempted this. He regards metamorphosis as an aesthetic concept and, accordingly, Goethe's text not as a "specialist botanical treatise," but as "his idea of a natural morphology portrayed through the example of botany," Breidbach, *Goethes Metamorphosenlehre* (Munich: Fink, 2006), pp. 307 and 130. Rather than solely a theory of nature, Goethe's metamorphosis is thus a "basic concept of his aesthetics" by means of which he was able to present his epistemology. Ibid., p. 14.

CHAPTER SIX: THE RHYTHM OF THE LIVING WORLD

1. See, for example, Friedrich Casimir Medicus, *Von der Lebenskraft, eine Vorlesung bei Gelegenheit des höchsten Namensfestes Sr. Kuhrfürstlichen Durchleucht von der Pfalz in der Kuhrpfälzisch-Theodorischen Akademie der Wissenschaften den 5. November 1774* (Mannheim: Hof- und akademische Buchdruckerei, 1774); Paul Joseph Barthez, *Nouveaux éléments de la science de l'homme* (Montpellier: Jean Martel, ainé, 1778); Johann Friedrich Blumenbach, *Über den Bildungstrieb* (Göttingen: Dieterich, 1791); Blumenbach, *Über den Bildungstrieb und das Zeugungsgeschäfte* (Göttingen: Dieterich, 1781); Carl Friedrich Kielmeyer, *Ueber die Verhältnisse der organischen Kräfte unter einander in der Reihe der verschiedenen Organisationen, die Geseze und Folgen dieser Verhältnisse* (Stuttgart: n.p., 1793); Heinrich Friedrich Link, *Ueber die Lebenskräfte in naturhistorischer Rücksicht und die Classification der Säugethiere* (Rostock: Stiller, 1795); Joachim Dietrich Brandis, *Versuch über die Lebenskraft* (Hannover: Hahn, 1795).

2. Peter McLaughlin, "Blumenbach und der Bildungstrieb: Zum Verhältnis von

epigenetischer Embryologie und typologischem Artbegriff," *Medizinhistorisches Journal* 17 (1982), p. 364.

3. Of the historical studies, see Jörg Jantzen, "Physiologische Theorien," in Manfred Durner, Francesco Moiso, and Jörg Jantzen (eds.), *Wissenschaftshistorischer Bericht zu Schellings naturphilosophischen Schriften 1797–1800* (Stuttgart: Frommann-Holzboog, 1994); François Duchesneau, *La physiologie des Lumières: Empirisme, modèles et théories* (The Hague: Nijhoff, 1982); Duchesneau, "Vitalism in Late Eighteenth-Century Physiology: The Cases of Barthez, Blumenbach and John Hunter," in W. F. Bynum and Roy Porter (eds.), *William Hunter and the Eighteenth-Century Medical World* (Cambridge: Cambridge University Press, 1985); Guido Cimino and François Duchesneau (eds.), *Vitalisms from Haller to the Cell Theory* (Florence: L. S. Olschki, 1997); Peter Hanns Reill, *Vitalizing Nature in the Enlightenment* (Berkeley: University of California Press, 2005); Timothy Lenoir, *The Strategy of Life: Teleology and Mechanics in Nineteenth-Century German Biology* (Chicago: University of Chicago Press, 1989); Lenoir, "The Göttingen School and the Development of Transcendental *Naturphilosophie* in the Romantic Era," *Studies in the History of Biology* 5 (1981), pp. 111–205; opposing Lenoir: John H. Zammito, "The Lenoir Thesis Revisited: Blumenbach and Kant," *Studies in History and Philosophy of Biological and Biomedical Sciences* 43.1 (2012), pp. 120–32; Andrea Gambarotto, "Vital Forces and Organization: Philosophy of Nature and Biology in Karl Friedrich Kielmeyer," *Studies in History and Philosophy of Biological and Biomedical Sciences* 48.A (2014), pp. 12–20; Joan Steigerwald, "Treviranus' Biology: Generation, Degeneration, and the Boundaries of Life," in Susanne Lettow (ed.), *Reproduction, Race, and Gender in Philosophy and the Early Life Sciences* (Albany: State University of New York Press, 2014); Kai Torsten Kanz (ed.), *Philosophie des Organischen in der Goethezeit: Studien zu Werk und Wirkung des Naturforschers Carl Friedrich Kielmeyer (1765–1844)* (Stuttgart: F. Steiner, 1994).

4. Reil studied medicine at Göttingen and Halle and in 1810 was offered a professorship at the newly founded University of Berlin. He edited *Archiv für die Physiologie* from 1796 to 1812, at first alone and later with Johann Heinrich Ferdinand Autenrieth. On Reil, see Reinhard Mocek, *Johann Christian Reil (1759–1813): Das Problem des Übergangs von der Spätaufklärung zur Romantik in Biologie und Medizin in Deutschland* (Frankfurt am Main: Lang, 1995); LeeAnn Hansen Le Roy, "Johann Christian Reil and Naturphilosophie in Physiology," PhD thesis, University of California, 1985; Wolfram Kaiser and Arina Völker (eds.), *Johann Christian Reil (1759–1813) und seine Zeit* (Halle: Abt. Wiss.-Publizistik d. Martin-Luther-Univ. Halle-Wittenberg, 1989); Lenoir, *Strategy of Life*, pp. 35–37.

5. See Johann Christian Reil, *Von der Lebenskraft* (1795; Leipzig: Barth, 1910), p. 25 n.2. Mikuláš Teich includes English translations of some excerpts: Mikuláš Teich and Dorothy M. Needham, *A Documentary History of Biochemistry, 1770–1940* (Cranbury, NJ: Associated University Presses, 1992), pp. 439–44.

6. Reil, *Von der Lebenskraft*, p. 26.

7. *Ibid.*, p. 23.

8. *Ibid.*, p. 24.

9. *Ibid.*, p. 26.

10. *Ibid.*, p. 68.

11. *Ibid.*, p. 66.

12. *Ibid.*, p. 30.

13. *Ibid.*

14. *Ibid.*, pp. 30–31.

15. *Ibid.*, p. 58.

16. *Ibid.*, p. 59.

17. See also *ibid.*, p. 31: "organic beings constantly change themselves through external stimuli and their own operations, and thus continually express different phenomena."

18. *Ibid.*

19. *Ibid.*, p. 35.

20. *Ibid.*, p. 44.

21. *Ibid.*, p. 11.

22. *Ibid.* Reil's term *Stimmung* alludes both to temper as a feeling or mood and to the tuning of an instrument. On *Stimmung* as a figure of thought straddling science and music, see Caroline Welsh, "Nerven-Saiten-Stimmung: Zum Wandel einer Denkfigur zwischen Musik und Wissenschaft 1750–1800," *Berichte zur Wissenschaftsgeschichte* 31 (2008), pp. 113–29.

23. Reil, *Von der Lebenskraft,* p. 72.

24. *Ibid.*, p. 93.

25. *Ibid.*, p. 31.

26. *Ibid.* pp. 72–73.

27. *Ibid.*, p. 69.

28. *Ibid.*, p. 74.

29. *Ibid.*, p. 72.

30. *Ibid.*

31. *Ibid.*, p. 74.

32. *Ibid.*, pp. 74–75.

33. *Ibid.*, pp. 76–77.

34. *Ibid.*, p. 79.

35. *Ibid.*

36. *Ibid.*, p. 76.

37. *Ibid.*, p. 90.

38. *Ibid.*, p. 80.

39. *Ibid.*, p. 92.

40. On physiology, see Karl Eduard Rothschuh, *Geschichte der Physiologie* (Berlin: Springer, 1953); Rothschuh, "Ursprünge und Wandlungen der physiologischen Denk-weisen im 19. Jahrhundert," in Wilhelm Treue and Kurt Mauel (eds.), *Naturwissenschaft,*

Technik und Wirtschaft im 19. Jahrhundert (Göttingen: Vandenhoeck & Ruprecht, 1976); Thomas Steele Hall, *Ideas of Life and Matter: Studies in the History of General Physiology, 600 B.C.–1900 A.D.*, 2 vols. (Chicago: University of Chicago Press, 1969); Duchesneau, *La physiologie*; Brigitte Lohff, *Die Suche nach der Wissenschaftlichkeit der Physiologie in der Zeit der Romantik* (Stuttgart: Fischer, 1990); Jantzen, "Physiologische Theorien"; Manfred Horstmanshoff, Helen King, and Claus Zittel (eds.), *Blood, Sweat and Tears: The Changing Concepts of Physiology from Antiquity into Early Modern Europe* (Leiden: Brill, 2012). On the relationship of anatomy and physiology, experiment and theory, see Andrew Cunningham, "The Pen and the Sword: Recovering the Disciplinary Identity of Physiology and Anatomy Before 1800. I: Old Physiology—the Pen," *Studies in History and Philosophy of Biological and Biomedical Sciences* 33.4 (2002), pp. 632–65; Cunningham, "The Pen and the Sword: Recovering the Disciplinary Identity of Physiology and Anatomy Before 1800. II: Old Anatomy—the Sword," *Studies in History and Philosophy of Biological and Biomedical Sciences* 34.1 (2003), pp. 51–76.

41. Caspar Friedrich Wolff, *Theoria generationis: Ueber die Entwicklung der Pflanzen und Thiere. I., II. und III. Theil*, ed. and trans. Paul Samassa (1759; Leipzig: Engelmann, 1896), part 2, p. 3.

42. Carl Friedrich Kielmeyer, "Ideen zu einer allgemeinen Geschichte und Theorie der Entwicklungserscheinungen der Organisationen" (1793-1794), in *Gesammelte Schriften*, ed. Fritz-Heinz Holler (Berlin: Keiper, 1938), p. 142.

43. See John R. Baker, "The Cell-Theory: A Restatement, History, and Critique," *Quarterly Journal of Microscopical Science* 89.1 (1949), pp. 103-25, and 90.1 (1949), pp. 87-108.

44. Kielmeyer, "Ideen," p. 147.

45. Johann Heinrich Ferdinand von Autenrieth, "Bemerkungen über die Verschiedenheit beyder Geschlechter und ihrer Zeugungsorgane, als Beytrag zu einer Theorie der Anatomie," *Archiv für die Physiologie* 7.1 (1807), p. 130.

46. Xavier Bichat, *General Anatomy, Applied to Physiology and the Practice of Medicine*, trans. Constant Coffyn, rev. ed. (1801; London: S. Highley, 1824), pp. xcv and xcviii.

47. D. Friedrich Tiedemann, *Zoologie: Zu seinen Vorlesungen entworfen* (Landshut: Weber, 1808), p. 56.

48. Karl Friedrich Burdach, *Der Mensch nach den verschiedenen Seiten seiner Natur* (Stuttgart: Balz, 1837), p. 492.

49. Karl Asmund Rudolphi, *Elements of Physiology*, trans. William Dunbar How (1821; London: Longman, 1825), vol. 1, p. 64.

50. *Ibid.*, p. 66.

51. "Ueber die verschiedenen Arten (modi) des Vegetationsprocesses in der animalischen Natur, und die Gesetze, durch welche sie bestimmt werden," *Archiv für die Physiologie* 6.1 (1805), pp. 121-22.

52. Autenrieth, "Bemerkungen," p. 129.

53. *Ibid.*, p. 137.

54. Const. Anast. Philites, "Von dem Alter des Menschen überhaupt und dem Marasmus senilis insbesondere," *Archiv für die Physiologie* 9.1 (1809), p. 27.

55. Andreas Sniadezki, *Theorie der organischen Wesen*, trans. Andreas Neubig (1804; Nürnberg: C. H. Zeh, 1821), p. 105.

56. *Ibid.*, p. 18.

57. Gottfried Reinhold Treviranus, *Biologie, oder Philosophie der lebenden Natur für Naturforscher und Ärzte* (Göttingen: J. F. Röwer, 1802–1803), vol. 1, p. 50.

58. Carl Friedrich Burdach, *Die Physiologie als Erfahrungswissenschaft*, 2nd, rev. ed. (1826; Leipzig: Voss, 1835), vol. 1, p. 6.

59. Treviranus, *Biologie*, p. 50.

60. Kielmeyer, "Ideen," p. 109.

61. *Ibid.*, p. 108.

62. *Ibid.*, p. 118.

63. *Ibid.*, p. 117.

64. *Ibid.*, pp. 117–18.

65. Samuel Christian Luca, *Grundriß der Entwickelungsgeschichte des menschlichen Körpers* (Marburg: Krieger, 1819), pp. 8–9.

66. *Ibid.*, p. 34.

67. Rudolphi, *Elements of Physiology*, vol. 1, p. 202.

68. Moritz Ernst Adolph Naumann, *Ueber die Grenzen zwischen Philosophie und Naturwissenschaften* (Leipzig: n.p., 1823), p. 163. On *Wechselwirkung* as a topos in Romantic science, see Gerhard H. Müller, "Wechselwirkung in the Life and Other Sciences: A Word, New Claims and a Concept around 1800 . . . And Much Later," in Stefano Poggi and Maurizio Bossi (eds.), *Romanticism in Science: Science in Europe 1790–1840* (Dordrecht: Kluwer, 1994).

69. "Ueber den Geschlechtsunterschied und dessen Einfluß auf die organische Form," *Die Horen: Eine Monatsschrift, herausgegeben von Schiller* 1.1 (1795), pp. 115–16.

70. Goethe to Frau von Stein, July 10, 1786, in *Goethes Werke: Briefe. 1. Januar 1785–24. Juli 1786*, *Goethes Werke*, vol. 4.7 (Weimar: Böhlau, 1891), p. 242.

71. Ignaz Döllinger, "Kreislaufe des Blutes," *Denkschriften der königlichen Academie der Wissenschaften zu München für die Jahre 1818, 1819 und 1821: Classe der Mathematik und Naturwissenschaften* 7 (1821), p. 178.

72. *Ibid.*

73. See also Döllinger, *Was ist Absonderung und wie geschieht sie?: Eine akademische Abhandlung von Dr. Ignaz Döllinger* (Würzburg: Nitribitt, 1819), p. 20.

74. Döllinger, "Kreislaufe des Blutes," p. 198.

75. *Ibid.*

76. *Ibid.*, p. 194.

77. Döllinger, *Was ist Absonderung*, p. 81; see also p. 194.

NOTES TO PAGE 146

78. *Ibid.*, p. 25.

79. Little is known of Döllinger's biography. He studied in Vienna with Georg Pro-
chaska. Appointed professor of medicine at the University of Bamberg in 1794, from
1803 to 1823 he taught physiology and pathology at the University of Würzburg, which
is where he became most influential. In 1823, he was elected to the Bavarian Academy of
Sciences, succeeding Thomas Samuel Sömmerring, and became rector of the university
when it moved from Landshut to Munich. The most important sources on Döllinger's
life are Karl Ernst von Baer, *Autobiography of Dr. Karl Ernst von Baer*, ed. Jane M. Oppenheimer, trans. H. Schneider (1886; Canton, MA: Science History Publications, 1986);
Eckhard Struck, "Ignaz Döllinger (1770–1841): Ein Physiologe der Goethe-Zeit in seinem
Leben und Werk," PhD thesis, University of Munich, 1977; Georg Sticker, "Entwicklungsgeschichte der medizinischen Fakultät," in Max Buchner (ed.), *Aus der Vergangen-
heit der Universität Würzburg: Festschrift zum 350jährigen Bestehen der Universität* (Berlin:
Springer, 1932).

80. Döllinger was interested in improving the microscope: Döllinger, *Nachricht von
einem verbesserten aplanatischen Mikroskop aus dem optischen Institute von Utzschneider und
Frauenhofer* (Munich: n.p., 1829), and in the building of anatomical theaters in Würzburg
and Munich: Döllinger, *Bericht von dem neuerbauten anatomischen Theater der Königlichen
Academie der Wissenschaften zu München* (Munich: Lindauer, 1826). He assembled a signifi-
cant collection of scientific specimens, now lost. See Philipp Franz von Walther, *Rede zum
Andenken an Ignaz Döllinger Dr. in der zur Feier des Allerhöchsten Namens- und Geburtstages Sr.
Majestät des Königs am 25. August 1841 gehaltenen öffentlichen Sitzung der königl. Bayerischen
Akademie der Wissenschaften* (Munich: Wolff, 1841), pp. 65–67.

81. Despite his importance, Döllinger's name barely appears in the standard works on
the history of biology around 1800. Edward Stuart Russell mentions him only as an initia-
tor, Russell, *Form and Function: A Contribution to the History of Animal Morphology* (1916; Chicago: University of Chicago Press, 1982), p. 113; see also Frederick B. Churchill, "The Rise of
Classical Descriptive Embryology," in Scott F. Gilbert (ed.), *A Conceptual History of Modern
Embryology* (New York: Plenum, 1991), p. 2. He is absent from Lynn K. Nyhart's genealogy of
the most important actors, Nyhart, *Biology Takes Form: Animal Morphology and the German
Universities, 1800–1900* (Chicago: University of Chicago Press, 1995), esp. pp. 1–32 (his name
appears just once, on p. 47). Timothy Lenoir regards Döllinger as a vital materialist, this
being his connection to von Baer. Lenoir, *Strategy of Life*, pp. 65–71.

82. See von Baer, *Autobiography*, pp. 135–44. Von Baer speaks of Döllinger's "precision
and lucidity, always presenting the essential and never containing a superfluous word,"
and describes him as a philosopher capable of awakening his pupils' "joy of studying" by
themselves. *Ibid.*, pp. 130 and 121. Despite close collaboration, his students published under
their own names, making it virtually impossible to determine Döllinger's part in their
writings. Von Baer describes an ideal scientist who "cared more about accretion to his

knowledge than to his name." *Ibid.*, p. 132.

83. Walther, *Rede zum Andenken*, pp. 100–101; emphasis added.

84. These include the programmatic *Ueber den Werth und die Bedeutung der vergleichen-den Anatomie* (Würzburg: Nitribitt, 1814), the influential *Was ist Absonderung* (1819) and "Kreislaufe des Blutes" (1821), and the speech "Von den Fortschritten, welche die Physiologie seit Haller gemacht hat: Eine Rede gelesen zur Feier des allerhöchsten Namensfestes Sr. Majestät des Königs am 12ten October 1824," in Cajetan von Weiller (ed.), *Fünfter Bericht über die Arbeiten der königl. baier. Akademie der Wissenschaften in München vom October bis December 1924* (Munich: Akademie der Wissenschaften, 1824). Döllinger also published several textbooks, including Döllinger, *Grundriß der Naturlehre des menschlichen Organismus: Zum Gebrauche bey seinen Vorlesungen* (Bamberg: Goebhardt, 1805); Döllinger, *Grundzüge der Physiologie* (Regensburg: Manz, 1836); Döllinger, *Grundzüge der Physiologie der Entwicklung des Zell-, Knochen- und Blutsystems* (Regensburg: Manz, 1842).

85. Döllinger, "Von den Fortschritten," p. 11.

86. Döllinger, *Was ist Absonderung*, p. 45.

87. Döllinger, "Kreislaufe des Blutes," pp. 206–207.

88. On the issue of stasis and movement, see Paul Feyerabend, *Wissenschaft als Kunst* (Frankfurt am Main: Suhrkamp, 1984), pp. 123–32.

89. Döllinger, *Was ist Absonderung*, p. 21.

90. Döllinger, "Kreislaufe des Blutes," p. 179.

91. *Ibid.*, p. 181.

92. *Ibid.*, p. 182–83.

93. *Ibid.*, pp. 184–85.

94. *Ibid.*, pp. 188–89.

95. *Ibid.*, p. 189.

96. *Ibid.*, p. 183.

97. *Ibid.*, p. 192.

98. Döllinger, *Was ist Absonderung*, p. 21.

99. Döllinger, "Kreislaufe des Blutes," p. 208.

100. *Ibid.*, p. 207.

101. *Ibid.*, pp. 206–207.

102. *Ibid.*, p. 208.

103. Döllinger, *Was ist Absonderung*, p. 23.

104. Döllinger, "Kreislaufe des Blutes," p. 206.

105. *Ibid.*, p. 199.

106. *Ibid.*, p. 200.

107. Caspar Friedrich Wolff had already propounded this thesis. See *ibid.*, pp. 189 and 178.

108. *Ibid.*, p. 193.

109. *Ibid.*, p. 194.

110. Döllinger, *Was ist Absonderung*, p. 23.

111. *Ibid.*, p. 78.

112. *Ibid.*, pp. 76–77.

113. Döllinger, "Von den Fortschritten," p. 13.

114. Döllinger, *Was ist Absonderung*, p. 35.

115. *Ibid.*, pp. 16–17.

116. Döllinger, *Ueber den Werth*, p. 21.

117. Döllinger, *Was ist Absonderung*, p. 13.

118. Döllinger, *Ueber den Werth*, p. 18.

119. Döllinger, *Was ist Absonderung*, pp. 40–41.

120. *Ibid.*

121. *Ibid.*

122. *Ibid.*, p. 51.

123. Döllinger, *Ueber den Werth*, p. 17.

CHAPTER SEVEN: THE ICONOGRAPHY OF MOTION

1. The history of instructional graphics has not yet been examined in detail by either historians or art historians, so little is known today about the origins, context, and history of this visual genre and thus a whole tradition of graphic art, with all its historical and epistemic significance. Ernst Gombrich was one of the first to examine the iconographic potential of instructional graphics: Gombrich, "Pictorial Instructions," in *The Uses of Images: Studies in the Social Function of Art and Visual Communication* (London: Phaidon, 1999). On the design of instructions, see Paul Mijksenaar and Piet Westendorp, *Open Here: The Art of Instructional Design* (New York: Joost Elffers, 1990). On informational graphics and techniques, see Edward R. Tufte, *The Visual Display of Quantitative Information* (Cheshire, CT: Graphics Press, 1984); Tufte, *Envisioning Information* (Cheshire, CT: Graphics Press, 1990); Tufte, *Visual Explanations: Images and Quantities, Evidence and Narrative* (Cheshire, CT: Graphics Press, 1997); Dominic McIver Lopes, "Directive Pictures," *The Journal of Aesthetics and Art Criticism* 63 (2004), pp. 189–96.

2. See the seminal study by Jörg Jochen Berns, *Film vor dem Film: Bewegende und bewegliche Bilder als Mittel der Imaginationssteuerung in Mittelalter und Früher Neuzeit* (Marburg: Jonas, 2000). Berns studies "image-sequencing procedures" (p. 9) at the threshold of the medieval to the modern period, describing the meditational iconography of the *Arma Christi* and the iconography of military drill as "strategies to guide the imagination." According to Berns, the two pictorial worlds had in common the education of body and soul, which they achieved iconographically by using segmentation to reduce complexity. For Berns, pictorial sequencing is therefore a "strategy of self-influence that was originally intended theologically and mnemonically." *Ibid.*, p. 10. Its objective was to channel

and manipulate the imagination.

3. On portrayals of motion in the ancient world, see Henriette Antonia Groenewegen-Frankfort, *Arrest and Movement: An Essay on Space and Time in the Representational Art of the Ancient Near East* (1951; Cambridge, MA: Belknap Press of Harvard University Press, 1987); Luca Giuliani, *Image and Myth: A History of Pictorial Narration in Greek Art*, trans. Joseph O'Donnell (Chicago: University of Chicago Press, 2013); Marc Azéma, *La préhistoire du cinema: Origines paléolithiques de la narration graphique et du cinématographe* (Paris: Errance, 2011). On the Renaissance, especially Leonardo, see Alexander Perrig, "Leonardo: Die Rekonstruktion menschlicher Bewegung," *Freiburger Universitätsblätter* 138 (1997), pp. 67-100; Martin Kemp, "Die Zeichen lesen: Zur graphischen Darstellung von physischer und mentaler Bewegung in den Manuskripten Leonardos," in Frank Fehrenbach (ed.), *Leonardo da Vinci: Natur im Übergang* (Munich: Fink, 2002); Fehrenbach, *Licht und Wasser: Zur Dynamik naturphilosophischer Leitbilder im Werk Leonardo da Vincis* (Tübingen: Ernst Wasmuth, 1997); and more recently Fehrenbach, "Leonardo's Liquid Bodies," in Victor Stoichita (ed.), *Inner and Outer Body / Le corps transparent* (Rome: "L'Erma" di Bretschneider, 2013); Gottfried Schramm, "Leonardo: Bewegung und Ruhe," *Freiburger Universitätsblätter* 138 (1997), pp. 9-50; Franz-Joachim Verspohl, "Die Entdeckung der Schönheit des Körpers—Von seiner maßästhetischen Normierung zu seiner bewegten Darstellung," in Richard van Dülmen (ed.), *Die Erfindung des Menschen: Schöpfungsträume und Körperbilder 1500-2000* (Vienna: Böhlau, 1998).

4. Leonardo da Vinci, *The Notebooks of Leonardo da Vinci*, ed. Jean Paul Richter (1883; New York: Dover, 1970), vol. 2, p. 286.

5. Leonardo da Vinci, *Leonardo on Painting*, ed. Martin Kemp (New Haven, CT: Yale University Press, 2001), pp. 132-33.

6. Scholarship on the history of the military, martial practices, historical forms of movement, and sport, for example, has almost completely neglected manuals of movement, specifically, instructional graphics, as sources. An exception is Sydney Anglo's study, although he limits himself to the treatises of medieval and Renaissance masters of arms. The tradition of later military treatises is only touched upon, and there is no comment on graphic instruction in other areas such as dance or vaulting. Anglo, *The Martial Arts of Renaissance Europe* (New Haven, CT: Yale University Press, 2000), especially chapter 2.

7. The earliest surviving fencing manual is thought to be an anonymous German manuscript of the late thirteenth or early fourteenth century; Anglo, *The Martial Arts*, pp. 21-27, 322 n.64.

8. See *ibid.*, p. 46. Bibliographical surveys are given by Carl Albert Thimm, *A Complete Bibliography of Fencing and Duelling as Practised by All European Nations from the Middle Ages to the Present Day* (London: John Lane, 1896); Henry William Pardoel, *The Complete Bibliography of the Art and Sport of Fencing* (Kingston, Ontario: Queen's University, 1996); Charles Richard Cammell, "Frühe Bücher über Fechtkunst," *Philobiblon* 9.9-10 (1936), pp.

353-75; Hellmut Helwig, "Die deutschen Fechtbücher: Eine bibliographische Übersicht," *Börsenblatt für den deutschen Buchhandel: Frankfurter Ausgabe* 2 (1966), pp. 1407-1416; *Chronik alter Kampfkünste. Zeichnungen und Texte aus Schriften alter Meister, entstanden 1443-1674*, 5th ed. (Berlin: Weinmann, 1997); Carlo Bascetta (ed.), *Sport e giuochi: Trattati e scritti dal XV al XVIII secolo* (Milan: Il Polifilo, 1978).

9. According to Anglo, only around one hundred manuals on wrestling and fighting with swords or lances, on foot or on horseback, printed in Europe up to 1620 survive today (thirty-nine Italian, twenty-three German, sixteen Spanish, ten French, and eight English). Anglo, *The Martial Arts*, p. 322 n.63.

10. On the rise of firearms, see Geoffrey Parker, *The Military Revolution: Military Innovation and the Rise of the West 1500-1800* (Cambridge: Cambridge University Press, 1988), pp. 16-20; A. R. Hall, *Ballistics in the Seventeenth Century: A Study in the Relations of Science and War with Reference Principally to England* (Cambridge: Cambridge University Press, 1952); Max Jähns, *Geschichte der Kriegswissenschaften vornehmlich in Deutschland* (1890; Hildesheim: G. Olms, 1997), vol. 2, pp. 972-1013.

11. The key outcome of the new weapons technology was long-range combat, which led to a complete reorientation of military tactics. On military history in general, see the standard works such as Jähns, *Geschichte der Kriegswissenschaften*; Hans Delbrück, *History of the Art of War, IV: The Modern Era*, trans. Walter J. Renfroe (1920; Westport, CT: Greenwood Press 1985); Karl-Volker Neugebauer (ed.), *Grundzüge der deutschen Militärgeschichte* (Freiburg: Rombach, 1994); Hall, *Ballistics*; Parker, *The Military Revolution*; Eberhard Kessel, *Militärgeschichte und Kriegstheorie in neuerer Zeit: Ausgewählte Aufsätze*, ed. Johannes Kunisch (Berlin: Duncker & Humblot, 1987).

12. Hall, *Ballistics*, p. 29.

13. These comments are based on Gerhard Oestreich, "Der römische Stoizismus und die Oranische Heeresreform," in *Geist und Gestalt des frühmodernen Staates: Ausgewählte Aufsätze* (1953; Berlin: Duncker & Humblot, 1969); Werner Hahlweg, *Die Heeresreform der Oranier und die Antike* (1941; Osnabrück: Biblio, 1987); Parker, *The Military Revolution*, pp. 18-23; Jähns, *Geschichte der Kriegswissenschaften*, vol. 2, pp. 869-76. On the reception and implementation of the reforms, see Oestreich, "Zur Heeresverfassung der deutschen Territorien von 1500 bis 1800," in *Geist und Gestalt des frühmodernen Staates*; Hahlweg, *Die Heeresreform der Oranier*, pp. 140-90; Jan Piet Puype and A. A. Wiekart, *From Prince Maurice to the Peace of Westphalia: Tactics and Triumphs of the Dutch Army* (Delft: Legermuseum, 1998); also Michael Sikora, "Die Mechanisierung des Krieges," in Rebekka von Mallinckrodt (ed.), *Bewegtes Leben: Körpertechniken in der Frühen Neuzeit* (Wiesbaden: Harrassowitz, 2008).

14. See Oestreich, "Der römische Stoizismus," p. 20.

15. On isolated earlier efforts in this direction, see Hahlweg, *Die Heeresreform der Oranier*, pp. 26-28.

16. Michel Foucault, *Discipline and Punish: The Birth of the Prison*, trans. Alan Sheridan

NOTES TO PAGES 191-195

sport see, for example, Horst Ueberhorst (ed.), *Geschichte der Leibesübungen*, 6 vols. (Berlin: Bartels und Wernitz, 1980–1989); Henning Eichberg, "Geometrie als barocke Verhaltensnorm: Fortifikation und Exerzitien," *Zeitschrift für historische Forschung* 4.1 (1977), pp. 17–50; Eichberg, *Leistung, Spannung, Geschwindigkeit: Sport und Tanz im gesellschaftlichen Wandel des 18./19. Jahrhunderts* (Stuttgart: Klett-Cotta, 1978); Peter Kühnst, *Sport: Eine Kulturgeschichte im Spiegel der Kunst* (Dresden: Verlag der Kunst, 1996). From the perspective of anthropology and cultural history, see Georges Vigarello, "The Upward Training of the Body from the Age of Chivalry to Courtly Civility," in Michel Feher, with Ramona Naddaff and Nadia Tazi (eds.), *Fragments for a History of the Human Body: Part Two* (New York: Zone Books, 1989), pp. 148–99; Vigarello, *Le corps redressé: Histoire d'un pouvoir pédagogique* (Paris: Delarge, 1978); Grégory Quin, Nicolas Bancel, and Vincent Barras (eds.), "Médecines du mouvement, XIXe–XXe siècles," special issue, *Gesnerus* 70.1 (2013); Rudolf zur Lippe, *Naturbeherrschung am Menschen*, 2 vols. (Frankfurt am Main: Suhrkamp, 1974); Lippe, *Die Geometrisierung des Menschen* (Oldenburg: Bis, 1983); Lippe, *Vom Leib zum Körper: Naturbeherrschung am Menschen in der Renaissance* (Reinbek bei Hamburg: Rowohlt, 1988); Günther Lottes, "Die Zähmung des Menschen durch Drill und Dressur," in Dülmen, *Erfindung des Menschen*. August Nitschke, especially, has studied human movement behavior in order to examine political, societal, and cultural transformation on the basis of historical systems of body and movement. See, for example, Nitschke, *Fremde Wirklichkeiten. II: Dynamik der Natur und Bewegungen der Menschen* (Goldbach: Keip, 1995); Nitschke, *Körper in Bewegung: Gesten, Tänze und Räume im Wandel der Geschichte* (Stuttgart: Kreuz, 1989).

48. Wallhausen, *Kriegskunst zu Fuß*, p. 36.

49. Johann Georg Paschen, *Kurtze iedoch gründliche Beschreibung des Voltiger* (Halle: Melchior Oelschlegeln, 1664), quoted in Eichberg, *Leistung, Spannung, Geschwindigkeit*, p. 118. Vaulting originated in exercises to practice mounting and dismounting a horse with or without weapons. It was taught by fencing masters, sometimes practiced using a table, and in the course of time developed into an art of elegant leaps and figures on the horse.

50. See Anglo, *The Martial Arts*, pp. 7–18. On dancing masters, see Walter Salmen, *Der Tanzmeister: Geschichte und Profile eines Berufs vom 14. bis zum 19. Jahrhundert* (Hildesheim: Olms, 1997); on German dancing manuals, see Marie-Thérèse Mourney, "Galante Tanzkunst und Körperideal," in Mallinckrodt, *Bewegtes Leben*.

51. On the history of fencing, see Anselm Schubert, "Aufgeklärtes Fechten: Anton Friedrich Kahns 'Anfangsgründe der Fechtkunst' (1739) und die ältere deutsche Fechtschule," in Mallinckrodt, *Bewegtes Leben*; Richard Cohen, *By the Sword: A History of Gladiators, Musketeers, Samurai, Swashbucklers, and Olympic Champions* (New York: Random House, 2002); Egerton Castle, *Schools and Masters of Fence: From the Middle Ages to the Eighteenth Century* (1885; Mineola, NY: Dover, 2003).

52. Diderot and d'Alembert, *Encyclopédie*, s.v. "Escrime: Explication des Planches."

53. Diderot and d'Alembert, *Encyclopédie*, s.v. "Escrime." Translation from The

Encyclopedia of Diderot & d'Alembert Collaborative Translation Project, entry trans. Johnathan Hoffman (Ann Arbor: Michigan Publishing, University of Michigan Library, 2008), http://hdl.handle.net/2027/spo.did2222.0000.896.

54. Diderot and d'Alembert, *Encyclopédie*, s.v. "Escrime: Explication des Planches."

55. Diderot and d'Alembert, *Encyclopédie*, s.v. "Escrime." Translation from The Encyclopedia of Diderot & d'Alembert Collaborative Translation Project, entry trans. Johnathan Hoffman (Ann Arbor: Michigan Publishing, University of Michigan Library, 2008), http://hdl.handle.net/2027/spo.did2222.0000.896.

56. On the history of dance see, for example, Selma Jeanne Cohen (ed.), *International Encyclopedia of Dance* (New York: Oxford University Press, 1998); Curt Sachs, *World History of the Dance*, trans. Bessie Schönberg (1933; New York: Norton, 1963); Karl Heinz Taubert, *Höfische Tänze: Ihre Geschichte und Choreographie* (Mainz: Schott, 1968); Sarah R. Cohen, *Art, Dance and the Body in French Culture of the Ancien Régime* (Cambridge: Cambridge University Press, 2000); Dorion Weickmann, *Der dressierte Leib: Kulturgeschichte des Balletts (1580–1870)* (Frankfurt: Campus, 2002); also Nitschke, *Körper in Bewegung*; Lippe, *Vom Leib zum Körper*; Eichberg, *Leistung, Spannung, Geschwindigkeit*.

57. Pierre Rameau, *The Dancing-Master*, trans. John Essex (1725; London: J. Brotherton, 1728), p. 238.

58. Eichberg, *Leistung, Spannung, Geschwindigkeit*, p. 299.

59. See the literature listed in note 47.

60. On Chodowiecki, see Willi Geismeier, *Daniel Chodowiecki* (Leipzig: Seemann, 1993); Ernst Hinrichs and Klaus Zernack (eds.), *Daniel Chodowiecki (1726–1801): Kupferstecher, Illustrator, Kaufmann* (Tübingen: Niemeyer, 1997). On Chodowiecki and Philantropism, see Hanno Schmitt, "Der Beitrag Chodowieckis zum Philantropismus," in Hinrichs and Zernack, *Daniel Chodowiecki*.

61. Johann Bernhard Basedow, *Elementarwerk: Mit den Kupfertafeln Chodowieckis u.a.*, ed. Theodor Fritsch, (1774; Hildesheim: Olms 1972), vol. 3, p. 39. On the history of the art of jumping, see Sandra Schmidt, "Zur Historischen Anthropologie des Sprungs: Die 'inventio' der Kubistik durch Arcangelo Tuccaro (ca. 1530–vor 1616)," in Mallinckrodt, *Bewegtes Leben*.

62. On Vieth, see Gerhard Lukas, *Gerhard Ulrich Anton Vieth: Sein Leben und Werk* (Berlin: Sportverlag, 1964); Walter Kluehe, "Ein technisch-methodischer Vergleich der Werke GutsMuths: 'Gymnastik für die Jugend'; Vieth: 'Versuch einer Enzyklopädie der Leibesübungen II. Teil' und Jahn-Eiselen 'Deutsche Turnkunst,'" Master's thesis, University of Berlin, 1932.

63. The plate was reproduced again in 1796, in Krünitz's *Oekonomische Enzyklopädie* of 1773–1858.

64. On GutsMuths, see Kluehe, "Ein technisch-methodischer Vergleich."

65. The continuity of physical postures and movements from the early modern period

to the early nineteenth century is also revealed in the ideals propounded by etiquette books. See Kirsten O. Frieling, "Haltung bewahren: Der Körper im Spiegel frühneuzeitlicher Schriften über Umgangsformen," in Mallinckrodt, *Bewegtes Leben.*

66. Johann Christoph Friedrich GutsMuths, *Gymnastik für die Jugend* (1793; Schnepfenthal: Buchhandlung der Erziehungsanstalt, 1804), p. 220.

67. *Ibid.*, p. 224.

68. Eichberg, *Leistung, Spannung, Geschwindigkeit*, p. 71.

69. In this respect, too, the reforms drew on classical writers on warfare, especially Aelian. See Hahlweg, *Die Heeresreform der Oranier*, pp. 32-41, 51-100.

70. See Gerhard Oestreich, "Graf Johanns VII. Verteidigungsbuch für Nassau-Dillenburg 1595," in *Geist und Gestalt des frühmodernen Staates*, also regarding the influence of the soldierly tradition of French Calvinism on the forms of military drill.

71. In the Thirty Years' War, an offensive variation of this technique was developed: instead of the soldier falling back, he reloaded where he stood while the rank advanced past him. See Lottes, "Die Zähmung des Menschen," p. 226. On the countermarch, see also Hahlweg, *Die Heeresreform der Oranier*, pp. 70-82.

72. On the basic principles of warfare, see Neugebauer, *Grundzüge der deutschen Militärgeschichte.*

73. Wallhausen, *Ritterkunst*, p. 112.

74. Quoted in Neugebauer, *Grundzüge der deutschen Militärgeschichte*, vol. 2, p. 67.

75. Eichberg, *Leistung, Spannung, Geschwindigkeit*, p. 121.

76. Marching in step became prevalent in the early seventeenth century. *Ibid.*, p. 126.

77. See Eichberg, "Geometrie als barocke Verhaltensnorm," pp. 35-36.

78. War as a variant of the game is discussed by Philipp von Hilgers, *War Games: A History of War on Paper*, trans. Ross Benjamin (Cambridge, MA: MIT Press, 2012); Hilgers, "Vom Einbruch des Spiels in die Epoche der Vernunft," in Pablo Schneider and Horst Bredekamp (eds.), *Visuelle Argumentationen: Die Mysterien der Repräsentation und die Berechenbarkeit der Welt* (Munich: Fink, 2006).

79. Diderot and d'Alembert, *Encyclopédie*, s.v. "Art militaire: Explication des Planches."

80. *Ibid.*

81. Johann Moritz David Herold, *Entwickelungsgeschichte der Schmetterlinge, anatomisch und physiologisch bearbeitet von Dr. Herold* (Kassell: n.p., 1815), explanation of Tab. 2.

82. See Robert Darnton, *The Business of Enlightenment: A Publishing History of the Encyclopédie 1775-1800* (Cambridge, MA: Belknap Press of Harvard University Press, 1979); Cynthia Koepp, "The Alphabetical Order: Work in Diderot's Encyclopédie," in Steven Laurence Kaplan and Cynthia J. Koepp (eds.), *Work in France: Representations, Organization, and Practice* (Ithaca, NY: Cornell University Press, 1986); John R. Pannabecker, "Representing Mechanical Arts in Diderot's Encyclopédie," *Technology and Culture* 39 (1998), pp. 33-73.

83. The classic account is Darnton, *The Business of Enlightenment.*

84. In recent years, the *Encyclopédie*'s plates have attracted increasing attention. Works examining the draftsmen, engravers, source material, studios, and evolution of the plates supply the material foundations for all further interpretations: Jacques Proust, *Marges d'une utopie: Pour une lecture critique des planches de l'Encyclopédie* (Cognac: Temps qu'il fait, 1985); Richard N. Schwab with Walter E. Rex, *Inventory of Diderot's Encyclopédie, 7: Inventory of the Plates* (Oxford: Voltaire Foundation at The Taylor Institution, 1984); Madeleine Pinault, "Les planches de l'Encyclopédie: Catalogue des dessins gravés dans l'Encyclopédie," Master's thesis, École du Louvre, 1972; Pinault, "A propos des planches de l'Encyclopédie," *Studies on Voltaire and the Eighteenth Century* 254 (1988), pp. 351–62; Frank A. Kafker and Madeleine Pinault-Soerensen, "Notices sur les collaborateurs du Recueil de Planches de l'Encyclopédie," *Recherches sur Diderot et l'Encyclopédie* 18–19 (1995), pp. 200–30; Jean-Pierre Séguin, "Courte histoire des planches de l'Encyclopédie," in Roland Barthes, Robert Mauzi, and Jean-Pierre Séguin (eds.), *L'univers de l'Encyclopédie* (Paris: Les Libraires associés, 1964). Among the best-known interpretations are Roland Barthes, *Les planches de l'Encyclopédie de Diderot et d'Alembert* (Pontoise: Les Amis de Jeanne et Otto Freundlich, 1989); Stephen Werner, *Blueprint: A Study of Diderot and the Encyclopédie Plates* (Birmingham, AL: Summa, 1993); Jacques Proust (ed.), *L'Encyclopédie, Diderot et d'Alembert: Planches et commentaires* (Paris: Hachette, 1985).

85. Madeleine Pinault, "Sur les planches de l'Encyclopédie de d'Alembert et Diderot," in Annie Becq (ed.), *L'Encyclopédisme: Actes du Colloque de Caen 12–16 janvier 1987* (Paris: Aux Amateurs de Livres, 1991), p. 362.

86. Gombrich, "Pictorial Instructions," p. 236. To my knowledge, this interpretation was never pursued.

87. The portrayal of the *arts et métiers* is the most extensively studied aspect of the plates. See Pannabecker, "Representing Mechanical Arts." For a survey of the literature, see William H. Sewell, "Visions of Labor: Illustrations of the Mechanical Arts before, in, and after Diderot's *Encyclopédie*," in Kaplan and Koepp, *Work in France*; Jacques Proust, "L'article 'Bas' de Diderot," in Michèle Duchet and Michèle Galley (eds.), *Langue et Langages de Leibniz à l'Encyclopédie* (Paris: Union générale d'éditions, 1977); Proust, "L'image du peuple du travail dans les planches de l'Encyclopédie," in *Images du peuple au dix-huitième siècle: Colloque Aix-en-Provence 25–26 octobre 1969* (Paris: A. Colin, 1973); Georges Benrekassa, "Didactique encyclopédique et savoir philosophique: L'ensemble épingle-épinglier dans l'Encyclopédie," in Becq, *L'Encyclopédisme*; and more recently, David Pullins, "Techniques of the Body: Viewing the Arts and Métiers of France from the Workshop of Nicolas I and Nicolas II de Larmessin," *Oxford Art Journal* 37.2 (2014), pp. 135–55; Celina Fox, *The Arts of Industry in the Age of Enlightenment* (New Haven, CT: Yale University Press, 2009); Charles Kostelnick, "Visualizing Technology and Practical Knowledge in the Encyclopédie's Plates: Rhetoric, Drawing Conventions, and Enlightenment Values," *History and Technology: An International Journal* 28.4 (2013), pp. 443–54.

88. Sewell, "Visions of Labor," p. 276; see also Antoine Picon, "La mesure du travail

humain," *Les Cahiers de Sciences* et Vie 47 (1998), pp. 41-42.

89. The plates of the Encyclopédie show a utopia of work, not the reality of eighteenth-century labor. See, for example, Sewell, "Visions of Labor"; Proust, "L'image du peuple du travail"; Proust, *Marges d'une utopie*; Darnton, *The Business of Enlightenment*, pp. 174-90.

90. On the *Encyclopédie*'s artists and engravers, see Georges Dulac, "Louis-Jacques Goussier, encyclopédiste et 'original sans principes,'" in Jacques Proust (ed.), *Recherches nouvelles sur quelques écrivains des Lumières* (Geneva: Droz, 1972); Kafker and Pinault-Soerensen, "Notices."

91. Diderot and d'Alembert, *Encyclopédie*, s.v. "Escrime: Explication des Planches."

92. Conservatoire national des Arts et Métiers, Musée national des techniques, Portefeuille industriel no. 174. Exp. Louvre 1984 nos. 119-122. In Madeleine Pinault's view, the representation of hands in the *Encyclopédie* is part of a "theatricalization of the gesture." Pinault, "Les mains dans l'Encyclopédie," *Corps écrit* 35 (1990), p. 32.

93. Pinault believes that these engravings, too, are based on Goussier's original drawings. *Ibid.*, p. 30.

94. Diderot and d'Alembert, *Encylopédie*, s.v. "Pêches: Explication des Planches." The *moule*, a cylinder made of willow wood, is a netting tool used alongside the shuttle, as seen in planche 21, fig. 4.

CHAPTER EIGHT: EPIGENETIC ICONOGRAPHY

1. Interest in visual representation in science has grown rapidly in the last twenty years. Martin Rudwick's work on geology was among the first to point out the importance of images: Rudwick, "The Emergence of a Visual Language for Geological Science 1760-1840," *History of Science* 14 (1976), pp. 149-95. In the abundant literature, see, for example, Michael Lynch and Steve Woolgar (eds.), *Representation in Scientific Practice* (Cambridge, MA: MIT Press, 1990); Catelijne Coopmans et al. (eds.), *Representation in Scientific Practice Revisited* (Cambridge, MA: MIT Press, 2014); Annamaria Carusi et al. (eds.), *Visualization in the Age of Computerization* (New York: Routledge, 2014); Brian S. Baigrie, *Picturing Knowledge: Historical and Philosophical Problems Concerning the Use of Art in Science* (Toronto: University of Toronto Press, 1996); Michael Ruse and Peter J. Taylor (eds.), "Pictorial Representation in Biology," special issue, *Biology and Philosophy* 6.2 (1991); Caroline A. Jones, Peter Galison, and Amy Slaton (eds.), *Picturing Science, Producing Art* (New York: Routledge, 1998); Gabriele Dürbeck, Bettina Gockel, and Susanne B. Keller (eds.), *Wahrnehmung der Natur: Natur der Wahrnehmung. Studien zur Geschichte visueller Kultur um 1800* (Amsterdam: Verlag der Kunst, 2001). As part of the expansion of art history into visual studies, art historians have recently begun to turn to pictures outside the art-historical canon. The work of Martin Kemp, Barbara Stafford, and James Elkins is of prime importance here: Kemp, *Visualizations: The Nature Book of Art and Science* (Berkeley: University of California Press, 2000); Kemp, "Taking It on Trust: Form and Meaning in Naturalistic Representation,"

Archives of Natural History 17 (1990), pp. 127-88; Kemp and Marina Wallace (eds.), *Spectacular Bodies: The Art and Science of the Human Body from Leonardo to Now* (Berkeley: University of California Press, 2000); Stafford, *Body Criticism: Imaging the Unseen in Enlightenment Art and Medicine* (Cambridge, MA: MIT Press, 1991); Stafford, *Artful Science: Enlightenment, Entertainment, and the Eclipse of Visual Education* (Cambridge, MA: MIT Press, 1994); Elkins, "Art History and Images That Are Not Art," *The Art Bulletin* 77.4 (1995), pp. 553-71; Elkins, *The Domain of Images* (Ithaca, NY: Cornell University Press, 1999). In the German setting, this kind of approach is anchored in the iconological tradition of Aby Warburg. See Erwin Panofsky and Fritz Saxl, *Dürers "Melencolia I": Eine quellen- und typengeschichtliche Untersuchung* (Leipzig: B. G. Teubner, 1923); Raymond Klibansky, Erwin Panofsky, and Fritz Saxl, *Saturn und Melancholie: Studien zur Geschichte der Naturphilosophie und Medizin, der Religion und der Kunst* (Frankfurt am Main: Suhrkamp, 1990); also Andreas Beyer (ed.), *Die Lesbarkeit der Kunst: Zur Geistes-Gegenwart der Ikonologie* (Berlin: Wagenbach, 1992), and Horst Bredekamp's work, e.g., Bredekamp, *The Lure of Antiquity and the Cult of the Machine: The Kunstkammer and the Evolution of Science, Art, and Technology*, trans. Allison Brown (Princeton, NJ: M. Wiener, 1995). On the formative epoch of the Renaissance, see Samuel Y. Edgerton, "The Renaissance Development of the Scientific Illustration," in David Hoeniger and John W. Shirley (eds.), *Science and the Arts in the Renaissance* (Washington, DC: Folger Shakespeare Library, 1985); Thomas DaCosta Kaufmann, *The Mastery of Nature: Aspects of Art, Science, and Humanism in the Renaissance* (Princeton, NJ: Princeton University Press, 1993); on the links between book and graphic prints: William Mills Ivins, *Prints and Visual Communication* (Cambridge, MA: MIT Press, 1969).

2. The phenomenon of the series has been much studied in the history of art, architecture, and science, yet the concept remains ill defined and is often used differently in different disciplines. Art history uses the term "serial art" to describe twentieth-century American Pop and minimalist art, but has located the origins of serial imagery in the late nineteenth century, with a central role accorded to Claude Monet's *Haystacks* paintings. At the same time, the emergence of the series is associated with industrial mass production, new reprographic technologies, and experimental series in the natural sciences. See Uwe M. Schneede (ed.), *Monets Vermächtnis: Serie—Ordnung und Obsession* (Hamburg: Hamburger Kunsthalle, 2001); Katharina Sykora, *Das Phänomen des Seriellen in der Kunst: Aspekte einer künstlerischen Methode von Monet bis zur amerikanischen Pop Art* (Würzburg: Königshausen und Neumann, 1983). For a survey of art-historical thinking on the phenomenon of time in the visual arts, see Götz Pochat, *Bild-Zeit: Zeitgestalt und Erzählstruktur in der bildenden Kunst von den Anfängen bis zur frühen Neuzeit* (Vienna: Böhlau, 1996). The history of science has primarily attended to the series in the form of chronophotography. On the work of Muybridge and Marey, see, for example, Marta Braun, *Picturing Time: The Work of Etienne-Jules Marey (1830-1904)* (Chicago: University of Chicago Press, 1992); Michel Frizot (ed.), *Neue Geschichte der Fotografie* (Cologne: Könemann, 1998). This is also

the setting of the nineteenth-century predilection for optical-illusionist techniques evoking movement, such as phenakistoscopes, the Mutoscope, zoetrope, and so on. See Henning Schmidgen, "Lebensräder, Spektatorien, Zuckungstelegraphen: Zur Archäologie des physiologischen Blicks," in Helmar Schramm (ed.), *Bühnen des Wissens: Interferenzen zwischen Wissenschaft und Kunst* (Berlin: Dahlem University Press, 2003). The process was continued in film, yet, remarkably, the history of cinema has largely neglected the episteme of seriality. The standard historiography of film regards the emergence of moving pictures almost exclusively as a history of projection techniques and the refinement of optical acceleration. See, for example, Carl Forch, *Der Kinematograph und das sich bewegende Bild: Geschichte und technische Entwicklung der Kinematographie bis zur Gegenwart* (Leipzig: A. Hartleben, 1913); Paul Liesegang, *Wissenschaftliche Kinematographie: Einschließlich der Reihenphotographie* (Leipzig: Liesegang, 1920); Friedrich Zglinicki, *Der Weg des Films: Die Geschichte der Kinematographie und ihrer Vorläufer* (Berlin: Rembrandt, 1956); Hermann Hecht and Ann Hecht (eds.), *Pre-Cinema History: An Encyclopaedia and Annotated Bibliography of the Moving Image before 1896* (London: Bowker Saur, 1993); Mary Ann Doane, *The Emergence of Cinematic Time: Modernity, Contingency, and the Archive* (Cambridge, MA: Harvard University Press, 2002). Instead of this history of technology, it would be interesting to approach the history of the cinema as the history of a visual form, namely, the series. This is the objective of Jörg Jochen Berns, *Film vor dem Film: Bewegende und bewegliche Bilder als Mittel der Imaginationssteuerung in Mittelalter und Früher Neuzeit* (Marburg: Jonas, 2000). Another domain of serial representation and modern visual form is the comic: David Kunzle, *The Early Comic Strip: Narrative Strips and Picture Stories in the European Broadsheet from 1450 to 1825* (Berkeley: University of California Press, 1973); Scott McCloud, *Understanding Comics: The Invisible Art* (New York: Harper Perennial, 1993).

3. Although Leonardo da Vinci tried out a variety of graphic resources to represent the movements of the human body, the representation of animal movements remained a history of animal mechanics. See Ugo Baldini, "Animal Motion before Borelli, 1600–1680," in Domenico Bertoloni Meli (ed.), *Marcello Malpighi: Anatomist and Physician* (Florence: L. S. Olschki, 1997).

4. The use of picture series in embryology has not yet been systematically investigated. On the late nineteenth century, see Nick Hopwood, "Producing Development: The Anatomy of Human Embryos and the Norms of Wilhelm His," *Bulletin for the History of Medicine* 74 (2000), pp. 29–79; Hopwood, "'Embryonen auf dem Altar der Wissenschaft zu opfern': Entwicklungsreihen im späten neunzehnten Jahrhundert," in Barbara Duden, Jürgen Schlumbohm, and Patrice Veit (eds.), *Geschichte des Ungeborenen: Zur Erfahrungs- und Wissenschaftsgeschichte der Schwangerschaft, 17.–20. Jahrhundert* (Göttingen: Vandenhoeck & Ruprecht, 2002); Hopwood, "Visual Standards and Disciplinary Change: Normal Plates, Tables and Stages in Embryology," *History of Science* 43 (2005), pp. 239–303. On the iconography and aesthetics of embryology more generally, see F. Scott Gilbert and Marion

Faber, "Looking at Embryos: The Visual and Conceptual Aesthetic of Emerging Form," in Alfred I. Tauber (ed.), *The Elusive Synthesis: Aesthetics and Science* (Dordrecht: Kluwer, 1996); Tatjana Buklijas and Nick Hopwood, *Making Visible Embryos*, www.hps.cam.ac.uk/ visibleembryos; Hopwood, *Haeckel's Embryos: Images, Evolution, and Fraud* (Chicago: University of Chicago Press, 2015). On human embryos, see Lynn M. Morgan, *Icons of Life: A Cultural History of Human Embryos* (Berkeley: University of California Press, 2009). On the function and aesthetics of three-dimensional models of embryos, see Hopwood, *Embryos in Wax: Models from the Ziegler Studio. With a Reprint of "Embryological Wax Models" by Friedrich Ziegler* (Cambridge: Whipple Museum of the History of Science, University of Cambridge, 2002); Hopwood, "Plastic Publishing in Embryology," in Soraya de Chadarevian and Nick Hopwood (eds.), *Models: The Third Dimension of Science* (Stanford, CA: Stanford University Press, 2004).

5. The artist is unknown. Accompanying the first three plates is a description that according to Howard B. Adelmann may have been written by the book's editor, possibly on the basis of Fabricius's notes; Fabricius's text refers only to the first two plates. Adelmann, "Introduction," in Hieronymus Fabricius of Aquapendente, *The Embryological Treatises of Hieronymus Fabricius of Aquapendente*, ed. and trans. Howard B. Adelmann, 2 vols. (1942; Ithaca, NY: Cornell University Press, 1967), vol. 1, p. 95. On the date of the treatise's composition, see *ibid.*, pp. 74–76; for an analysis of the illustrations, see pp. 95–99. See also Adelmann, *Marcello Malpighi and the Evolution of Embryology*, 5 vols. (Ithaca, NY: Cornell University Press, 1966), vol. 2, p. 757.

6. On Malpighi, see the standard work: Adelmann, *Marcello Malpighi*, especially vol. 2, which also looks in detail at Malpighi's notion of development. In addition, Adelmann surveys microscopical knowledge in chick embryology other than Malpighi's. See also Domenico Bertoloni Meli, *Mechanism, Experiment, Disease: Marcello Malpighi and Seventeenth-Century Anatomy* (Baltimore, MD: Johns Hopkins University Press, 2011); Meli, *Marcello Malpighi*; Walter Bernardi, *Le metafisiche dell'embrione: Scienze della vita e filosofia da Malpighi a Spallanzani (1672–1793)* (Florence: L. S. Olschki, 1986).

7. They contrast with the later illustrations by Antoine Maître-Jan of 1722: Maître-Jan, *Observations sur la formation du poulet: Où les divers changemens qui arrivent à l'oeuf à meure qu'il est couvé, sont exactement expliqués et représentés en figures* (Paris: n.p., 1722).

8. Here Haller advocates a preformationist approach, but he had previously taken other positions. As Hermann Boerhaave's pupil in Leyden, he argued for preformation in the male sperm; in the 1740s, with the discovery of the regeneration of polyps, he moved to an epigenetic view of development, only to favor preformation in the female egg in the late 1750s. See Shirley A. Roe, *Matter, Life and Generation: Eighteenth-Century Embryology and the Haller-Wolff Debate* (Cambridge: Cambridge University Press, 1981); Richard Toellner, *Albrecht von Haller: Über die Einheit im Denken des letzten Universalgelehrten* (Wiesbaden: Steiner, 1971); Maria Teresa Monti, "Introduction," in Albrecht von Haller, *Commentarius*

de formatione cordis in ovo incubato, ed. Maria Teresa Monti (1767; Basel: Schwabe, 2000). See also Haller, *Sur la formation du cœur dans le poulet sur l'œil, sur la structure du jaune &c.*, 2 vols. (Lausanne: Marc-Mich. Bousquet & Comp., 1758), vol. 2, pp. 64–117, 174.

9. On the empirical observations, see Monti, "Introduction," p. clxiii; Elizabeth B. Gasking, *Investigations into Generation 1651–1828* (Baltimore, MD: Johns Hopkins University Press, 1967), pp. 102–104. On the development of the heart and blood vessels in the area vasculosa, see Adelmann, *Marcello Malpighi*, vol. 3, pp. 1104–1153; Roe, *Matter, Life and Generation*, pp. 45–88, 149.

10. The controversy has hitherto been discussed solely from the perspective of the history of ideas and the history of biology: Roe, *Matter, Life and Generation*; Joseph Needham, *A History of Embryology* (New York: Abelard-Schuman, 1959), pp. 193–204; Gasking, *Investigations into Generation*, pp. 97–116; Monti, "Introduction"; Karen Detlefsen, "Explanation and Demonstration in the Haller-Wolff Debate," in Justin E. H. Smith (ed.), *The Problem of Animal Generation in Early Modern Philosophy* (Cambridge: Cambridge University Press, 2006).

11. It is only quite recently that Haller's extensive notes on his embryological studies have appeared in Monti's excellent scholarly edition (Haller, *Commentarius*), presenting a new and as yet almost completely unexploited fund of material for evaluating his position on developmental history.

12. For more detail on Haller's journals, see Monti, "Introduction," pp. cxxxiv–cxxxv.

13. There is just one exception, a single illustration that Haller included in the first volume. To discover what the illustration shows, the development of the eye, the reader must turn to the "Explication des figures" in the second volume. Haller, *Sur la formation*, vol. 1, p. 171; vol. 2, pp. 367–68. The plate itself does little to explain why it is included at this particular point. The caption refers to the Nuremberg engraver Johann Michael Seligmann (1720–1762), but not to the artist. It cannot have been Haller himself, since he wrote to Bonnet that he had commissioned the drawing; Haller, *The Correspondence between Albrecht von Haller and Charles Bonnet*, ed. Otto Sonntag (Bern: Hans Huber, 1983), p. 104. Seligmann's career yields no further clues. His only other work for Haller was a copper engraving in the botanical study *Historia stirpium indigenarum Helvetiae inchoatae*, which Haller found unsatisfactory. See Baldur Gloor, *Die künstlerischen Mitarbeiter an den naturwissenschaftlichen und medizinischen Werken Albrecht von Hallers* (Bern: Haupt, 1958), pp. 11 and 44. Furthermore, as Haller himself remarked, the formation of the eye was only secondary to his research on the heart; Haller, *Sur la formation*, vol. 1, p. 13. Writing to Bonnet, Haller described it as one of the "other discoveries that crossed my path," Haller, *Correspondence*, p. 104. Vanity may have been one reason why this peripheral observation became the subject of the only visual representation in all of Haller's physiological writings: Haller was the discoverer of a tiny membrane in the eye, which he named the "zona ciliaris." Monti, "Introduction," pp. cxxxi– cxxxii.

14. Haller, *Sur la formation*, vol. 1, p. 63.

15. *Ibid.*, p. 78.

16. *Ibid.*, p. 26.

17. *Ibid.*, p. 31.

18. *Ibid.*, p. 28; further examples on pp. 32, 34, 39, 42, 43, 82, 85.

19. *Ibid.*, pp. 26, 84, 39, 118, 310.

20. Monti, "Introduction," pp. cxxxiv-cxxxvii.

21. See Gloor, *Die künstlerischen Mitarbeiter.*

22. Samuel Thomas Soemmerring, *Icones embryonum humanorum* (Frankfurt am Main: Varrentrapp & Wenner, 1799), facsimile reprint with a German translation by Ferdinand Peter Moog in Soemmerring, *Schriften zur Embryologie und Teratologie*, ed. Ulrike Enke (Basel: Schwabe, 2000).

23. Ulrike Enke, "Vorwort," in Soemmerring, *Schriften zur Embryologie und Teratologie*, p. 104; Nick Hopwood, "A History of Normal Plates, Tables and Stages in Vertebrate Embryology," *International Journal of Developmental Biology* 51 (2007), pp. 1-26, here p. 3.

24. See Ulrike Enke's extensively contextualized edition of Soemmerring's writings, Soemmerring, *Schriften zur Embryologie und Teratologie*; also Enke, "Von der Schönheit der Embryonen: Samuel Thomas Soemmerrings Werk 'Icones embryonum humanorum,'" in Duden, Schlumbohm, and Veit, *Geschichte des Ungeborenen*; Barbara Duden, "The Fetus on the 'Farther Shore': Toward a History of the Unborn," in Lynn M. Morgan and Meredith W. Michaels (eds.), *Fetal Subjects, Feminist Positions* (Philadelphia: University of Pennsylvania Press, 1999). Before Soemmerring, representations of the human embryo had shown not fetuses, but infants. On the history of the unborn and of pregnancy, see Duden, Schlumbohm, and Veit, *Geschichte des Ungeborenen.*

25. August Johann Rösel von Rosenhof, *Historia naturalis ranarum nostratium in qua omnes earum proprietates, praesertim quae generationem ipsarum pertinent, fusius enarrantur* (Nürnberg: Johann Joseph Fleischmann, 1758), p. 3. The historical literature has interpreted Rösel's images as a developmental series: Marc J. Ratcliff, "Wonders, Logic, and Microscopy in the Eighteenth Century: A History of the Rotifer," *Science in Context* 13 (2000), pp. 93-119.

26. On Lavater's "physiognomic cabinet," see Gerda Mraz (ed.), *Das Kunstkabinett des Johann Caspar Lavater* (Vienna: Böhlau, 1999); Pieter Camper, *Über den natürlichen Unterschied der Gesichtszüge in Menschen verschiedener Gegenden und verschiedenen Alters*, trans. S. Th. Sömmerring (Berlin: In der Vossischen Buchhandlung, 1792).

27. This was also the purpose of the technique of stereometric projection. See the discussion below.

28. The opposite interpretation is offered by Ulrike Enke, "Vom Präparat zur Bilderfolge: Die Visualisierung von Regelhaftigkeit im Werk Samuel Thomas Soemmerrings," in Rüdiger Schultka and Josef N. Neumann (eds.), *Anatomie und anatomische Sammlungen im 18. Jahrhundert* (Münster: LIT, 2007).

29. My comments on Soemmerring's collection follow the groundbreaking research by

Enke, "Vorwort," and on the structure of the collection, *ibid.*, p. 92.

30. On Kaltschmied's collection, see *ibid.*, pp. 58–68 and 94.

31. On the catalogue and the individual specimens, see *ibid.*, pp. 92–97, especially the overview on p. 96.

32. Soemmerring, *Icones*, p. 173.

33. *Ibid.*

34. See, for example, the explanation of the second illustration: "in its size it accords with the embryos drawn by: Trioen, pl. v, fig. 3, in which however no one part is clearly demarcated. Albinus in the *Annot. Acad.*, pl. v, fig. 5. Blumenbach in *Specimen Physiol. compar.*, Fig. 1, of which he claims that, respecting age, he certainly does not go beyond the fifth week; Denman in the sixth plate, 1783." *Ibid.*, p. 181.

35. Koeck collaborated with these engravers for other works, as well. The first plate in *Icones* was engraved by the brothers Ignaz Sebastian Klauber (1753–1817) and Josef Xaver Wolfgang Klauber (1740–1813), with whom Koeck also worked for the naturalist Johann Gotthelf Fischer von Waldheim (1771–1853). Enke, "Vorwort," p. 85 n. 445. The second plate was made by the Leipzig engraver Gottlieb Wilhelm Hüllmann (1765–after 1828), while the title and end vignettes are by the Frankfurt engraver Friedrich Ludwig Neubauer (1767–1828).

36. On Koeck, see Armin Geus, "Christian Koeck (1758–1818), der Illustrator Samuel Thomas Soemmerrings," in Gunter Mann and Franz Dumont (eds.), *Samuel Thomas Soemmerring und die Gelehrten der Goethezeit* (Stuttgart: G. Fischer, 1985). On the models, see Soemmerring, *Abbildungen des menschlichen Hoerorganes* (Frankfurt am Main: Varrentrapp und Wenner, 1806), preface.

37. Soemmerring, *Icones*, p. 171.

38. *Ibid.*, p. 173.

39. *Ibid.*

40. *Ibid.*

41. *Ibid.*, p. 179.

42. *Ibid.*, p. 185.

43. *Ibid.*, p. 175.

44. *Ibid.*, p. 177.

45. *Ibid.*, p. 173.

46. *Ibid.*

47. *Ibid.*, p. 175.

48. See Michael Hagner, "Enlightened Monsters," in William Clark, Jan Golinski, and Simon Schaffer (eds.), *The Sciences in Enlightened Europe* (Chicago: University of Chicago Press, 1999), p. 207; more generally, see also Hopwood, "Plastic Publishing in Embryology."

49. See Enke's comments on the reception of the work, "Vorwort," pp. 104–10.

50. Soemmerring, *Icones*, p. 177.

51. Enke has also pointed out the priority of anatomical questions in Soemmerring's embryological studies. "Vorwort," pp. 81 and 90.

52. Soemmerring, *Icones*, p. 173.

53. *Ibid.*, p. 172–73.

54. Soemmerring, *Abbildungen und Beschreibungen einiger Misgeburten* (Mainz: Universitätsbuchhandlung, 1791), facsimile in Soemmerring, *Schriften zur Embryologie und Teratologie*, p. 116.

55. Soemmerring, *Icones*, p. 173.

56. Caspar Friedrich Wolff, *Theoria generationis: Ueber die Entwicklung der Pflanzen und Thiere. I., II. und III. Theil*, ed. and trans. Paul Samassa (1759; Leipzig: Engelmann, 1896), Part II, p. 3.

57. See Boris Evgen'evic Raikov, "Caspar Friedrich Wolff," *Zoologische Jahrbücher / Abteilung für Systematik, Ökologie und Geographie der Tiere* 91.4 (1964), pp. 564–65 and 577–78; William Morton Wheeler, "Caspar Friedrich Wolff" (1898), in Jane Maienschein (ed.), *Defining Biology: Lectures from the 1890s* (Cambridge, MA: Harvard University Press, 1986), p. 205. For a different view, see Ilse Jahn, "Caspar Friedrich Wolff (1743-1794)," in Ilse Jahn and Michael Schmitt (eds.), *Darwin & Co.: Eine Geschichte der Biologie in Portraits* (Munich: Beck, 2001), p. 112; Olaf Breidbach, "Einleitung: Zur Mechanik der Ontogenese," in Caspar Friedrich Wolff, *Theoria generationis: Ueber die Entwicklung der Pflanzen und Thiere. I., II. und III. Theil*, ed. and trans. Paul Samassa (1759; Thun: Deutsch, 1999), p. xv; Gasking, *Investigations into Generation*, p. 100.

58. A total of five drawings by Wolff have been discovered separately from his letters in Haller's library, today held in Milan. See Luigi Belloni, "Embryological Drawings Concerning His *Theorie von der Generation* Sent By Caspar Friedrich Wolff to Albrecht von Haller," *Journal of the History of Medicine* 26 (1971), pp. 205-208. The relevant letter is translated into German in Julius Schuster, "Der Streit um die Erkenntnis des organischen Werdens im Lichte der Briefe C. F. Wolffs an A. von Haller," *Sudhoffs Archiv* 34 (1941), pp. 205-207. In the letter, Wolff refers to other observations that he included in the appendix of the German publication *Theorie von der Generation*. Wolff, *Theorie von der Generation: In zwei Abhandlungen erklärt und bewiesen. Theoria generationis* (1759; Hildesheim: Olms, 1966), pp. 260–61.

59. Wolff, *Theoria generationis*, part 2, p. 9; see also pp. 88–89.

60. *Ibid.*, part 2, p. 10.

61. Wolff, *Theorie von der Generation*, p. 83.

62. Haller, *Sur la formation*, vol. 1, pp. 12-13.

63. Wolff, *Theorie von der Generation*, p. 81.

64. Wolff, *Theoria generationis*, part 2, p. 10.

65. Rösel von Rosenhof, *Historia naturalis ranarum*, p. 3.

66. Sebastian von Tredern, *Dissertatio inauguralis medica sistens ovi avium historiae et*

incubationis prodromum (Jena: Etzdorf, 1808). We know little of Tredern other than the information given by Lorenz Oken and Karl Ernst von Baer: von Baer, "Bitte um eine Nachricht über die Litteraturgeschichte unseres Vaterlandes, besonders an diejenigen Herren gerichtet, welche in den Jahren 1806–1808 in Jena oder Göttingen studirt haben," *Das Inland: Eine Wochenschrift für Liv-, Esth- und Curländische Geschichte, Geographie, Statistik und Litteratur* 15 (1836), pp. 253–56; von Baer, "Wegen des Grafen von Tredern zweite Aufforderung," *Das Inland: Eine Wochenschrift für Liv-, Esth- und Curländische Geschichte, Geographie, Statistik und Litteratur* 23 (1836), pp. 391–92; von Baer, "Biographische Nachrichten über den Embryologen Grafen Ludwig Sebastian Tredern," *Bulletin de l'Académie Impériale des Sciences de St. Pétersbourg* 19 (1874), pp. 67–76. At the beginning of the twentieth century, Ludwig Stieda published Tredern's Latin dissertation with a German translation and all the information available on the researcher's life: Stieda, *Der Embryologe Sebastian Graf von Tredern und seine Abhandlung über das Hühnerei* (Wiesbaden: J. F. Bergmann, 1901). Only recently has new evidence on Tredern been recovered. See Jean-Claude Beetschen, "Louis Sébastien Tredern de Lézérec (1780–18?): A Forgotten Pioneer of Chick Embryology," *International Journal of Developmental Biology* 39 (1995), pp. 299–308; Jean-Claude Beetschen and Pierre Baudrier, "New Insights into Life and Death of the Self-Styled Estonian Embryologist, Louis Sébastien Marie de Tredern de Lézérec (1780–1818)," *Trames* 14.2 (2010), pp. 107–19.

67. See Stieda, *Der Embryologe*; Beetschen, "Louis Sébastien Tredern de Lézérec."

68. Von Baer, "Bitte um eine Nachricht," pp. 255–56.

69. Christian Heinrich Pander, *Beiträge zur Entwickelungsgeschichte des Hühnchens im Eye* (Würzburg: n.p., 1817), p. 40.

70. Von Baer, *Über Entwickelungsgeschichte der Thiere: Beobachtung und Reflexion* (1828; Brussels: Culture et civilisation, 1967), vol. 1, p. xv.

71. On Herold's life, see Uwe Runge, *Johann Moritz David Herold (1790–1862)* (Frankfurt am Main: Lang, 1983); on the University of Marburg, see Gisela Altpeter, "Die 'Gesellschaft zur Beförderung der gesamten Naturwissenschaften zu Marburg': Ihre Entstehung, Entwicklung und Bedeutung," PhD thesis, University of Marburg, 1992; on Herold's oeuvre only, see Bernard Balan, *L'ordre et le temps: L'anatomie comparée et l'histoire des vivants au XIXe siècle* (Paris: J. Vrin, 1979), pp. 305–308.

72. William Harvey, "Anatomical Exercises on the Generation of Animals," in *The Works of William Harvey*, trans. Robert Willis (1651; London: Sydenham Society, 1847); Jan Swammerdam, *The Book of Nature, or, the History of Insects: Reduced to Distinct Classes, Confirmed by Particular Instances, Displayed in the Anatomical Analysis of Many Species, and Illustrated with Copperplates*, trans. Thomas Flloyd (1737; London: C. G. Seyffert, 1758); Marcello Malpighi, "Dissertatio epistolica de bombyce," in *Marcello Malpighi: Opera omnia figuris elegantissimis in aes incisis illustrata* (1686; Hildesheim: Olms, 1975); Maria Sybilla Merian, *Metamorphosis insectorum Surinamensium* (Amsterdam: G. Valck, 1705); Francesco Redi, *Esperienze intorno alla generazione degl' insetti* (Florence: All'insegna della Stella, 1668).

73. Neither a modern history of entomology nor a history of its iconography is available. There is no modern historiographical study of the notion of metamorphosis in natural history. On Merian, see Kurt Wettengl (ed.), *Maria Sibylla Merian, 1647-1717: Künstlerin und Naturforscherin* (Ostfildern-Ruit: Hatje, 1997); Elizabeth Rücker and William Stearn (eds.), *Maria Sibylla Merian in Surinam* (London: Pion, 1982). On Swammerdam, see A. Schierbeek, *Jan Swammerdam (1637-1680): His Life and Works* (1947; Amsterdam: Swets & Zeitlinger, 1967); Edward G. Ruestow, *The Microscope in the Dutch Republic: The Shaping of Discovery* (Cambridge: Cambridge University Press, 1996); Meli, *Mechanism, Experiment, Disease.* On the role of visual representation in the history of metamorphosis research, see Janina Wellmann, "Die Metamorphose der Bilder: Die Verwandlung der Insekten und ihre Darstellung vom Ende des 17. bis zum Anfang des 19. Jahrhunderts," *NTM: Zeitschrift für Geschichte der Wissenschaft, Technik und Medizin* 16 (2008), pp. 183-211. On entomology in general and for earlier periods, see Janice Neri, *The Insect and the Image: Visualizing Nature in Early Modern Europe, 1500-1700* (Minneapolis: University of Minnesota Press, 2011); Domenico Bertoloni Meli, "The Representation of Insects in the Seventeenth Century: A Comparative Approach," *Annals of Science* 67.3 (2010), pp. 405-29. Even today, the standard work on the history of entomology, in fact mainly an anthology of sources, is Friedrich Simon Bodenheimer, *Materialien zur Geschichte der Entomologie bis Linné* (Berlin: W. Junk, 1928). On attempts to order insects, see Brian W. Ogilvie, "Order of Insects: Insect Species and Metamorphosis between Renaissance and Enlightenment," in Ohad Nachtomy and Justin E. H. Smith (eds.), *The Life Sciences in Early Modern Philosophy* (Oxford: Oxford University Press, 2014); a list of relevant works can be found in O. E. Essig, *A History of Entomology* (New York: Hafner, 1965).

74. Johann Moritz David Herold, *Entwickelungsgeschichte der Schmetterlinge, anatomisch und physiologisch bearbeitet von Dr. Herold, mit dreyunddreyssig illuminirten und schwarzen Kupfertafeln* (Cassel: n.p., 1815), pp. iii-iv.

75. *Ibid.*, p. v.

76. See Wilhelm Schwemmer, *Nürnberger Kunst im 18. Jahrhundert* (Nürnberg: Stadtbibliothek, 1974), p. 53.

77. Herold, *Entwickelungsgeschichte*, p. 12.

78. *Ibid.*, p. 4.

79. *Ibid.*

80. Although several works on embryology or midwifery appeared in German after Wolff, it was not until the beginning of the nineteenth century that systematic experimental investigation of the development of the hen's egg resumed. Among the writings published after Wolff were Ferdinand Georg Danz, *Grundriß der Zergliederungskunde des ungebohrnen Kindes in den verschiedenen Zeiten der Schwangerschaft: Mit Anmerkungen begleitet von Herrn Hofrath Sömmering in Mainz* (Frankfurt: Krieger, 1792); Johann Heinrich Ferdinand von Autenrieth, *Supplementa ad historiam embryonis humani quibus accedunt observata quaedam circa palatum fissum, verosimillimamque illi medendi methodum* (Tübingen: Jacob.

Frid. Heerbrandt, 1797); Heinrich August Wrisberg, *Descriptio anatomica embryonis observationibus illustrata* (Göttingen: Schultz, 1764); Philipp Adolph Boehmer, *Anatomen ovi humani foecundati sed deformis trimestri abortu elisi figuris illustratum* (Halle: Hendel, 1763); Joseph Mohrenheim, *Abhandlung über die Entbindungskunst von Ioseph Freyherrn von Mohrenheim iter B. mit 46 Kupfertafeln* (St. Petersburg: Kayserliche Akadamie der Wissenschaften, 1792); see also Soemmerring's list in Soemmerring, *Icones,* p. 171.

81. On Döllinger's study group and working methods, see von Baer, *Autobiography of Dr. Karl Ernst von Baer,* 2nd ed., ed. Jane M. Oppenheimer, trans. H. Schneider (1886; Canton, MA: Science History Publications, 1986), pp. 132-33; Philipp Franz von Walther, *Rede zum Andenken an Ignaz Döllinger Dr. in der zur Feier des Allerhöchsten Namens- und Geburtstages Sr. Majestät des Königs am 25. August 1841 gehaltenen öffentlichen Sitzung der königl. Bayerischen Akademie der Wissenschaften* (Munich: Wolff, 1841), pp. 38-39.

82. See von Baer, *Autobiography,* p. 119.

83. For this well-known story, see *ibid.,* pp. 137-42.

84. *Ibid.,* pp. 139, 140-41.

85. Little is known of Pander's biography. He was born in 1794, the oldest son of a Riga merchant. After high school in Riga, in 1812 Pander began studying medicine at the University of Dorpat (today Tartú), where he met Karl Ernst von Baer. In 1814, he went to Berlin and Göttingen, then in 1816 to Würzburg. From an early stage, Pander was interested less in the physician's profession than in the scientific foundations of medicine, and his financial resources enabled him to concentrate on this aspect. Pander's embryological work in Würzburg was only the start of his interest in the natural sciences. After extensive research on the comparative anatomy of fossil remains in Europe, in 1819 he returned to Russia and founded Russian paleontology. See Paul Siegfried and Walter Gross, "Christian Heinrich Pander (1794-1865) und seine Bedeutung für die Paläontologie," *Münstersche Forschungen zur Geologie und Paläontologie* 19 (1971), pp. 101-83. On Pander himself, see von Baer, *Autobiography*; Walther, *Rede zum Andenken*; Ernst Loesch, "Heinrich Christian Pander, sein Leben und seine Werke," *Biologisches Zentralblatt* 40.11-12 (1920), pp. 481-502; Boris Evgen'evic Raikov, *Christian Heinrich Pander: Ein bedeutender Biologe und Evolutionist. An Important Biologist and Evolutionist. 1794-1865,* German translation with commentary and English summaries by W. E. Hertzenberg and P. H. von Bitter (Frankfurt am Main: Kramer, 1984).

86. See Pander, *Dissertatio inauguralis sistens historiam metamorphoseos, quam ovum incubatum prioribus quinque diebus subit* (Würzburg: Nitribitt, 1817), p. 3.

87. Ignaz Döllinger, *Was ist Absonderung und wie geschieht sie?: Eine akademische Abhandlung von Dr. Ignaz Döllinger* (Würzburg: Nitribitt, 1819), pp. 18-19.

88. Döllinger wrote a short essay on three embryological drawings by Malpighi and promised a discussion of Malpighi's other pictures, as well, though this never materialized. Döllinger, "Marcello Malpighi iconum ad historiam ovi incubati spectantium censurae specimen I," *Ordnung der Vorlesungen an der königlichen Universität Würzburg für das*

Winter-Semester 1820/21 (1820), pp. 1–14.

89. Von Baer, *Autobiography,* p. 139.

90. See Pander's thanks to Döllinger and d'Alton in Pander, *Dissertatio inauguralis,* p. 5.

91. *Ibid.,* pp. 3–4.

92. Raikov, *Christian Heinrich Pander,* p. 18.

93. Discussing the copperplates on osteology, Goethe described d'Alton as an "old friend." "Morphologische Hefte" (1817), in *Goethe. Die Schriften zur Naturwissenschaft: Vollständige mit Erläuterungen versehene Ausgabe im Auftrage der Deutschen Akademie der Naturforscher Leopoldina,* vol. 1.9, ed. Dorothea Kuhn (Weimar: Metzler, 1994), p. 251. D'Alton commented on the osteological illustrations in Goethe's morphological notebooks: Eduard d'Alton, "Zur Vergleichenden Osteologie von Goethe, mit Zusaetzen und Bemerkungen," *Verhandlungen der Kaiserlichen Leopoldinisch-Carolinischen Akademie der Naturforscher* 4.1 (1824), pp. 323–32.

94. D'Alton's *Naturgeschichte des Pferdes* is very rare, but the Museum of Prints and Drawings in Berlin holds some of its engravings (S Gal 10).

95. Pander and d'Alton, *Die vergleichende Osteologie. Abth. 1, Lieferungen 1–11* (Bonn: Eduard Weber, 1821–1828).

96. Eduard d'Alton's son, Johann Samuel Eduard d'Alton (1803–1854), studied medicine and in 1834 was appointed to succeed Meckel at the University of Halle, where he was also director of the Anatomy Department's human anatomy collection. He, too, drew and engraved his own copperplates. Lars-Burkhardt Sturm, "Die humananatomische Sammlung des Institutes für Anatomie und Zellbiologie zu Halle/Saale—ihre Geschichte und ihr Präparationsprofil unter den Direktoren Eduard d'Alton (1803–1854), Alfred Wilhelm Volkmann (1801–1877) und Hermann Welcker (1822–1897)," PhD thesis, Martin-Luther-Universität Halle-Wittenberg, 1997.

97. August Wilhelm Schlegel, *Verzeichniß einer von Eduard d'Alton hinterlaßenen Gemälde-Sammlung* (Bonn: C. Georgi, 1840), p. v, including the inventory of the collection.

98. Leopold von Schroeder, "Jugendbriefe K. E. v. Baers an Woldemar v. Ditmar: Vortrag, gehalten am 19. Februar zur Feier des 101. Geburtstages K. E. v. Baers," *Baltische Monatsschrift* 35.40 (1893), pp. 270–71.

99. Von Baer to Altenstein, June 22, 1830, in Hans-Theodor Koch, "Karl Ernst von Baers (1792–1876) Korrespondenz mit den preußischen Behörden," *Wissenschaftliche Beiträge der Universität Halle* 39 (1981), p. 180.

100. Koch, "Karl Ernst von Baers Korrespondenz," p. 177; see also Ilse Jahn, "Die Problematik zeichnerischer Wiedergabe mikroskopischer Beobachtungen am Beispiel Karl Ernst von Baers," in Armin Geus et al. (eds.), *Repräsentationsformen in den biologischen Wissenschaften* (Berlin: Verlag für Wissenschaft und Bildung, 1999).

101. Koch, "Karl Ernst von Baers Korrespondenz," pp. 181–82.

102. See von Baer's letters of July 20, 1830, and June 23, 1831, *ibid.,* pp. 182–83.

103. *Ibid.*, p. 183.

104. See von Baer's letters, *ibid.*, pp. 183, 187, 188; also Jahn, "Die Problematik zeichnerischer Wiedergabe," p. 94.

CHAPTER NINE: FOLDING INTO BEING

1. Despite its importance, Pander's research has been overshadowed by the works of Karl Ernst von Baer, and his embryological oeuvre has attracted almost no attention. See Boris Evgen'evic Raikov, *Christian Heinrich Pander: Ein bedeutender Biologe und Evolutionist. An Important Biologist and Evolutionist. 1794-1865*, German translation with commentary and English summaries by W. E. Hertzenberg and P. H. von Bitter (Frankfurt am Main: Kramer, 1984); Ernst Loesch, "Heinrich Christian Pander, sein Leben und seine Werke," *Biologisches Zentralblatt* 40.11-12 (1920), pp. 481-502; Bernard Balan, *L'ordre et le temps: L'anatomie comparée et l'histoire des vivants au XIXe siècle* (Paris: J. Vrin, 1979), pp. 237-54; Leonid Iakovlevich Blyakher, *History of Embryology in Russia from the Middle of the Eighteenth to the Middle of the Nineteenth Century* (Washington, DC: Smithsonian Institution, 1982), pp. 237-74. A relatively recent French translation of Pander's work exists: Christian Heinrich Pander, *Les textes embryologiques de Christian Heinrich Pander (1794-1865): Édition critique, commentée et annotée*, ed. and trans. Stéphane Schmitt (Turnhout: Brepols, 2003); see also Schmitt, "From Eggs to Fossils: Epigenesis and Transformation of Species in Pander's Biology," *International Journal of Developmental Biology* 49 (2005), pp. 1-8.

2. Christian Heinrich Pander, *Beiträge zur Entwickelungsgeschichte des Hühnchens im Eye* (Würzburg: n.p., 1817), p. 5.

3. *Ibid.*

4. *Ibid.*, pp. 6-7.

5. Pander used the term "blastoderm" in his Latin dissertation, *Dissertation inauguralis sistens historiam metamorphoseos, quam ovum incubatum prioribus quinque diebus subit* (Würzburg: Nitribitt, 1817), p. 21. In his works in German, *Keimhaut* or *Keimblatt* (literally, germ layer or membrane) is used.

6. Pander, *Beiträge zur Entwickelungsgeschichte*, p. 13.

7. *Ibid.*, pp. 13-14.

8. *Ibid.*, p. 6.

9. Schmitt calls the concept of the fold a substantial conceptual advance, but does not go into any further detail: Pander, *Les textes embryologiques*, p. 161. He also suggests that germ layers meant something different to Pander and von Baer: Schmitt, "Pander, d'Alton and the Representation of Epigenesis," in Sabine Brauckmann et al. (eds.), *Graphing Genes, Cells, and Embryos: Cultures of Seeing 3D and Beyond* (Berlin: Max Planck Institute for the History of Science, 2009).

10. Gilles Deleuze, *The Fold: Leibniz and the Baroque*, trans. Tom Conley (Minneapolis: University of Minnesota Press, 1993).

11. Pander, *Beiträge zur Entwickelungsgeschichte*, p. 7.

12. See *ibid.*, pp. 13–14.

13. In the formation of the germ layers during embryogenesis (gastrulation), the blastula (a single-layered, hollow sphere of cells) is reorganized into the two-layer "cup germ" or gastrula, with the entoderm facing inward and the ectoderm facing outward. The entoderm has an opening to the outside, the blastopore. In an avian egg, the ovum consists almost completely of yolk, so that instead of a spherical blastula, a flat germinal disc (blastodisc) forms on the surface of the yolk. At one end of the germinal disc, instead of the blastopore, a long streak forms, the primitive streak. This gives rise to a fissure through which cell material enters the blastocyst cavity.

14. Pander, *Beiträge zur Entwickelungsgeschichte*, p. 8.

15. *Ibid.*, p. 9.

16. *Ibid.* What Pander observed (but only von Baer recognized) was the emergence of the chorda dorsalis or notochord. The notochord is a temporary phenomenon; in most vertebrates, it completely disappears when the vertebral column develops.

17. *Ibid.*, pp. 9–10.

18. *Ibid.*, p. 10.

19. *Ibid.*

20. See the subsequent description, *ibid.*, p. 12.

21. *Ibid.*, pp. 12–13. The first sac is the fovea cardiaca—"the two lateral parts of this sac proceed from the point where they fused together for the formation of the gullet, diverging as folds in the area pellucida [*durchsichtigen Hof*], toward the tail, with undefined boundaries." *Ibid.*

22. *Ibid.*, p. 18. For further details, Pander refers the reader to the work of Haller and the drawings of Malpighi. See Albrecht von Haller, *Sur la formation du cœur dans le poulet sur l'œil, sur la structure du jaune &c.* (Lausanne: Marc-Mich. Bousquet & Comp., 1758); Howard B. Adelmann, *Marcello Malpighi and the Evolution of Embryology*, 5 vols. (Ithaca, NY: Cornell University Press, 1966).

23. Pander, *Beiträge zur Entwickelungsgeschichte*, p. 21, also pp. 23–24.

24. *Ibid.*, p. 24.

25. *Ibid.*

26. *Ibid.*, pp. 24–25.

27. *Ibid.*

28. *Ibid.*, p. 25.

29. Karl Ernst von Baer, *Autobiography of Dr. Karl Ernst von Baer*, ed. Jane M. Oppenheimer, trans. H. Schneider (1886; Canton, MA: Science History Publications, 1986), p. 206.

30. Franz von Paula Gruithuisen, "Beyträge zur Entwicklungsgeschichte des Hühnchens im Eye, von Dr. Pander," *Medicinisch-chirurgische Zeitung* 2 (1818), pp. 309 and 312.

31. See Pander, *Dissertatio inauguralis*, pp. 5–6; also Pander, "Entwicklung des Küchels," *Isis oder Encyclopädische Zeitung* 3 (1818), col. 515; Pander, *Beiträge zur Entwickelungsgeschichte*, p. iii.

32. Pander, *Beiträge zur Entwickelungsgeschichte*, p. iii.

33. Pander's verbal descriptions had not succeeded in giving temporal structure to the processes within the egg. His temporal designations are rare, and certainly not systematic. For the most part, he makes only relational statements: "as soon as," "from time to time," "a short time later," "soon afterward," or "while" are typical. *Ibid.*, pp. 8–9.

34. Describing the collaboration between Pander and Döllinger, von Baer writes: "Pander indeed enjoyed the advantage of being able to utilize Döllinger's previous experience, as well as his proficient methods. But if one wanted to get down to a truly developmental level, one had to begin the entire investigation anew and pursue it persistently. This, to the best of my knowledge, was indeed what had been done primarily by Pander, just as it was he who also first understood Wolff's publication, bore the costs of the project all by himself, and took care of the incubator. Döllinger, as he also did in other research, merely reserved his natural right to be kept fully *au courant* and to check the results to convince himself." Von Baer, *Autobiography*, p. 142. Philipp Franz von Walther, in contrast, claims that Döllinger "stood at the head of the enterprise." Von Walther, *Rede zum Andenken an Ignaz Döllinger Dr. in der zur Feier des Allerhöchsten Namens- und Geburtstages Sr. Majestät des Königs am 25. August 1841 gehaltenen öffentlichen Sitzung der königl. Bayerischen Akademie der Wissenschaften* (Munich: Wolff, 1841), pp. 85–86. Even though the study appeared under Pander's name and we know little about the division of labor, it is probably best regarded as the joint project of Pander, Döllinger, and d'Alton.

35. Von Baer, *Autobiography*, p. 141 (translation emended).

36. *Ibid.*

37. Pander, *Dissertatio inauguralis*, pp. 8–9. On the incubator, see also von Walther, *Rede zum Andenken*, pp. 86–87. In the mid-eighteenth century, René Antoine Ferchault de Réaumur revived the idea of artificial incubation and developed an incubating oven. Réaumur, *Art de faire éclore et d'élever en toute saison des oiseaux domestiques de toutes espèces, soit par le moyen de la chaleur du fumier, soit par le moyen du feu ordinaire* (Paris: Imprimerie Royale, 1749).

38. Pander, "Entwicklung des Küchels," pp. 512–13.

39. Gruithuisen, "Beyträge zur Entwicklungsgeschichte," p. 306.

40. Pander, *Beiträge zur Entwickelungsgeschichte*, p. 29.

41. *Ibid.*, pp. 29–30, emphasis added.

42. Caspar Friedrich Wolff, *Über die Bildung des Darmkanals im bebrüteten Hühnchen*, trans. Johann Friedrich Meckel (1768; Halle: Rengersche Buchhandlung, 1812), p. 152.

43. In this respect, the developmental series differs profoundly from the pictorial series of astronomy, which tracks the chronological progress of observation and can fairly

be described as an observational series. Galileo's argument in favor of the existence of sun spots and the topographical features of the moon's surface rested not on individual observations carefully isolated from the mass and synthesized back into a series only in a second step, but on the sequential observation of the known, periodic orbit of the stars. See Mario Biagioli, *Galileo's Instruments of Credit: Telescopes, Images, Secrecy* (Chicago: University of Chicago Press, 2007), pp. 138 and 142. In contrast, the developmental series *constructs* the temporal succession of development as a pictorial sequence by means of moments chosen arbitrarily or according to the requirements of the visual relationships. On Galileo, see also Erwin Panofsky, "Galileo as Critic of the Arts: Aesthetic Attitude and Scientific Thought," *Isis* 47 (1956), pp. 3–15. On seriality in astronomy, see also Simon Schaffer, "Herschel in Bedlam: Natural History and Stellar Astronomy," *British Journal for the History of Science* 13.45 (1980), pp. 211–39. Observational sequences can also be found in eighteenth-century microbiology. For a different view, see Marc J. Ratcliff, "Temporality, Sequential Iconography and Linearity in Figures: The Impact of the Discovery of Division in Infusoria," *History and Philosophy of the Life Sciences* 21 (1999), pp. 255–92.

44. Pander, *Beiträge zur Entwickelungsgeschichte*, p. 30.

45. *Ibid.*, p. 29.

46. Ignaz Döllinger, "Kreislaufe des Blutes," *Denkschriften der königlichen Academie der Wissenschaften zu München für die Jahre 1818, 1819 und 1821. Classe der Mathematik und Naturwissenschaften* 7 (1821), p. 190.

47. Lorenz Oken, "Dissertatio inauguralis sistens Historiam Metamorphoseos, quam ovum incubatum prioribus quinque diebus subit," *Isis oder Encyclopädische Zeitung* 11/12.192/193 (1817), cols. 1539–1540.

48. Pander, *Beiträge zur Entwickelungsgeschichte*, pp. 31–32.

49. The deviation from the series form is due to the odd number of figures. The fifth figure is placed in the center of the plate for pragmatic reasons, as becomes obvious in Pander's arrangement of the seventh plate. There, he adds a new layer of information to the serial arrangement by inserting actual-size depictions of the embryo between the microscopical views.

50. Pander, *Beiträge zur Entwickelungsgeschichte*, p. 35.

51. In this plate, too, the serial arrangement is disrupted. The explanations say that fig. VI was allocated to this plate only for reasons of space and actually belonged to the final plate, the one with the sections. Pander, *Beiträge zur Entwickelungsgeschichte*, p. 38.

52. *Ibid.*, p. 37.

53. *Ibid.*, p. 36.

54. Pander, "Entwicklung des Küchels."

55. Pander, *Beiträge zur Entwickelungsgeschichte*, p. 40.

56. Pander, "Entwicklung des Küchels," cols. 515, 521–22.

57. Oken, "Dissertatio inauguralis," col. 1529. Gruithuisen, too, praises the "excellent"

NOTES TO PAGES 300-303

copperplates, which are "unparalleled in both refinement and outward elegance" and make the whole thing a "magnificent work." Gruithuisen, "Beyträge zur Entwicklungsgeschichte," pp. 305 and 313.

58. Oken, "Dissertatio inauguralis," col. 1533. On Oken's ownership of the engravings from *Beiträge zur Entwickelungsgeschichte des Hühnchens im Eye*, see also *ibid*., col. 1529.

59. *Ibid*., col. 1534.

60. Sabine Brauckmann follows my emphasis on the visual and on the pictorial resources used in Tredern, Pander, and von Baer: Brauckmann, "Axes, Planes, and Tubes, or the Geometry of Embryogenesis," *Studies in History and Philosophy of Biological and Biomedical Sciences* 42.4 (2011), pp. 381–90.

CHAPTER TEN: KARL ERNST VON BAER AND THE CHOREOGRAPHY OF DEVELOPMENT

1. See Karl Ernst von Baer, *Über Entwickelungsgeschichte der Thiere: Beobachtung und Reflexion*, 2 vols. (1828 and 1837; Brussels: Culture et civilisation, 1967), vol. 1, scholion 5. Von Baer also worked on geography, ecology, and anthropology. On his life, see his autobiography, published in German in 1864 and then in a second edition in 1886, translated into English as *Autobiography of Dr. Karl Ernst von Baer*, ed. Jane M. Oppenheimer, trans. H. Schneider (1886; Canton, MA: Science History Publications, 1986); also Boris Evgen'evic Raikov, *Karl Ernst von Baer (1792–1876): Sein Leben und sein Werk* (Leipzig: Barth, 1968); Ludwig Stieda, *Karl Ernst von Baer: Eine biographische Skizze* (Braunschweig: F. Vieweg, 1878); Wilhelm Haacke, *Karl Ernst von Baer* (Leipzig: Thomas, 1905). On his work, see Frederick B. Churchill, "The Rise of Classical Descriptive Embryology," in Scott F. Gilbert (ed.), *A Conceptual History of Modern Embryology* (New York: Plenum, 1991), pp. 5–12; Jane Marion Oppenheimer, "K. E. von Baer's Beginning Insights into Causal-Analytical Relationships during Development," in Oppenheimer, *Essays in the History of Embryology and Biology* (Cambridge, MA: MIT Press, 1967); Elizabeth B. Gasking, *Investigations into Generation, 1651–1828* (Baltimore, MD: Johns Hopkins University Press, 1967), pp. 148–59; Arthur William Meyer, *Human Generation: Conclusions of Burdach, Döllinger, and von Baer* (Stanford, CA: Stanford University Press, 1956), pp. 49–138; Edward Stuart Russell, *Form and Function: A Contribution to the History of Animal Morphology* (1916; Chicago: University of Chicago Press, 1982), pp. 113–32; Timothy Lenoir, *The Strategy of Life: Teleology and Mechanics in Nineteenth-Century German Biology* (Chicago: University of Chicago Press, 1989), pp. 72–95, and on von Baer and Darwin, pp. 246–75; Florence Vienne, "Seeking the Constant in What Is Transient: Karl Ernst von Baer's Vision of Organic Formation," *History and Philosophy of the Life Sciences* 37.1 (2015), pp. 34–49.

2. Von Baer, *Über Entwickelungsgeschichte*, vol. 1, p. 145. See also vol. 2, p. 96.

3. *Ibid*., vol. 1, p. 19.

4. *Ibid*., vol. 2, p. 46.

5. *Ibid.*, vol. 1, p. 9.

6. Von Baer explains that the animal layer is the same as Pander's serous layer and the vegetative layer the same as Pander's vascular and mucous layer. *Ibid.*, vol. 2 p. 46.

7. *Ibid.*, vol. 1, p. 154.

8. *Ibid.*, vol. 1, p. 164.

9. *Ibid.*, vol. 2, p. 74.

10. *Ibid.*, vol. 1, pp. 154–55.

11. *Ibid.*, vol. 2, p. 79.

12. *Ibid.*, vol. 1, p. 155.

13. *Ibid.*, vol. 2, p. 79.

14. *Ibid.*, vol. 1, p. 155.

15. *Ibid.*, vol. 2, p. 80.

16. *Ibid.*

17. For this description, see *ibid*; on morphological separation, *ibid.*, pp. 78–91.

18. *Ibid.*, vol. 1, p. 154.

19. *Ibid.*, vol. 2, p. 92.

20. See *ibid.*, vol. 2, p. 79.

21. *Ibid.*, vol. 2, p. 94.

22. *Ibid.*, vol. 2, p. 157.

23. *Ibid.*, vol. 2, p. 156.

24. *Ibid.*, vol. 2, p. 94.

25. *Ibid.*, vol. 2, p. xix.

26. Von Baer, *Autobiography*, pp. 209–10.

27. This becomes particularly obvious in the development of vertebrates, the topic of von Baer's fourth scholion.

28. Von Baer, *Über Entwickelungsgeschichte*, vol. 1, p. 162.

29. *Ibid.*

30. *Ibid.*, vol. 1, p. 161.

31. *Ibid.*

32. *Ibid.*, vol. 1, p. 162.

33. *Ibid.*, vol. 1, p. 46.

34. Christian Heinrich Pander, *Beiträge zur Entwickelungsgeschichte des Hühnchens im Eye* (Würzburg: n.p., 1817), p. 40.

35. Pander, "Entwicklung des Küchels," *Isis oder Encyclopädische Zeitung* 3 (1818), col. 524.

36. See von Baer, *Über Entwickelungsgeschichte*, vol. 2, p. 43.

37. *Ibid.*, vol. 1, p. 174.

38. *Ibid.*, vol. 1, p. 102.

39. *Ibid.*, vol. 1, pp. 55–56.

40. *Ibid.*, vol. 1, p. 161.

41. Von Baer also referred to this part as "sketches from my scientific confession of faith on the developmental history of animals." *Ibid.*, vol. 1, p. xvi.

42. *Ibid.*, vol. 1, p. xv.

43. *Ibid.*

44. *Ibid.*, vol. 1, pp. xx and vi.

45. *Ibid.*, vol. 1, p. 5.

46. *Ibid.*

47. *Ibid.*

48. *Ibid.*, vol. 1, p. 7.

49. *Ibid.*, vol. 1, p. 266.

50. For this reason, Nick Hopwood refers to von Baer's illustrations as "mental pictures." Hopwood, "Visual Standards and Disciplinary Change: Normal Plates, Tables and Stages in Embryology," *History of Science* 43.3 (2005), p. 246. On some of von Baer's unpublished drawings, see Erki Tammiksaar and Sabine Brauckmann, "Karl Ernst von Baer's 'Über Entwickelungsgeschichte der Thiere II' and its Unpublished Drawings," *History and Philosophy of the Life Sciences* 26.3-4 (2004), pp. 291-308.

51. Von Baer, *Über Entwickelungsgeschichte*, vol. 1, p. xv.

52. *Ibid.*

53. *Ibid.*, vol. 1, p. xix.

54. *Ibid.*, vol. 1, p. xv.

55. *Ibid.*, vol. 1, p. 266.

56. See, for example, *ibid.*, vol. 1, p. 168.

57. *Ibid.*, vol. 1, p. xv.

58. *Ibid.*, vol. 1, pp. xv–xvi.

59. *Ibid.*, vol. 1, p. 266.

60. *Ibid.*, vol. 1, pp. xv–xvi.

61. *Ibid.*, vol. 1, pp. 147–48.

62. *Ibid.*, vol. 1, p. 265.

63. *Ibid.*, vol. 1, p. 266.

64. *Ibid.*, vol. 1, p. 265.

65. *Ibid.*

66. *Ibid.*, vol. 1, p. 266.

67. *Ibid.*

68. *Ibid.*, vol. 1, p. 265.

69. *Ibid.*

70. Von Baer, "Welche Auffassung der lebenden Natur ist die richtige? Und wie ist diese Auffassung auf die Entomologie anzuwenden?: Zur Eröffnung der Russischen entomologischen Gesellschaft im October 1860 gesprochen," in von Baer, *Reden gehalten in wissenschaftlichen Versammlungen und kleinere Aufsätze vermischten Inhalts. Erster Theil: Reden*

(St. Petersburg: Schmitzdorff, 1864).

71. *Ibid.*, p. 274.

72. *Ibid.*, pp. 274–75.

73. *Ibid.*, p. 276.

74. *Ibid.*, p. 280.

75. *Ibid.*, pp. 280–81.

76. *Ibid.*, pp. 282–83.

CONCLUSION

1. See Reinhart Koselleck, "Vergangene Zukunft der frühen Neuzeit," in Hans Barion et al. (eds.), *Epirrhosis: Festgabe für Carl Schmitt* (Berlin: Duncker & Humblot, 1967); Koselleck, "Historia magistra vitae," in Hermann Braun and Manfred Riedel (eds.), *Natur und Geschichte: Karl Löwith zum 70. Geburtstag* (Stuttgart: Kohlhammer, 1967); Koselleck, *Futures Past: On the Semantics of Historical Time*, trans. Keith R. Tribe (New York: Columbia University Press, 1985), pp. 255–75; Koselleck, "Der neuzeitliche Revolutionsbegriff als geschichtliche Kategorie," *Studium generale* 22 (1969), pp. 825–38; also the relevant articles in Otto Brunner, Werner Conze, and Reinhart Koselleck (eds.), *Geschichtliche Grundbegriffe: Historisches Lexikon zur politisch-sozialen Sprache in Deutschland*, vols. 1 and 2 (Stuttgart: Klett-Cotta, 1972–1975): Wolfgang Wieland, "Entwicklung, Evolution," Koselleck, "Fortschritt" and "Geschichte, Historie," and Horst Stuke, "Aufklärung."

2. See the survey, covering various disciplines, by Stephen Toulmin and June Goodfield, *The Discovery of Time* (Chicago: University of Chicago Press, 1965). For medicine and the natural sciences, Dietrich von Engelhardt identified two categories that together characterized the new quality of knowledge: knowledge of the historicity of nature and knowledge of the historicity of knowledge about nature. Von Engelhardt, *Historisches Bewußtsein in der Naturwissenschaft: Von der Aufklärung bis zum Positivismus* (Freiburg: Alber, 1979); see also Wolf Lepenies, "Die Dynamisierung des Naturbegriffs an der Wende zur Neuzeit," in Jörg Zimmermann (ed.), *Das Naturbild des Menschen* (Munich: Fink, 1982).

3. Arthur O. Lovejoy, *The Great Chain of Being: A Study of the History of an Idea* (Cambridge, MA: Harvard University Press, 1936), p. 259.

4. Michel Foucault, *The Order of Things: An Archaeology of the Human Sciences*, trans. anon (London: Routledge, 1974), p. 152.

5. Wolf Lepenies, *Das Ende der Naturgeschichte: Wandel kultureller Selbstverständlichkeiten in den Wissenschaften des 18. und 19. Jahrhunderts* (Munich: Hanser, 1976); see also Lepenies, "Das Ende der Naturgeschichte und der Beginn der Moderne: Verzeitlichung und Enthistorisierung in der Wissenschaftsgeschichte des 18. und 19. Jahrhunderts," in Reinhart Koselleck (ed.), *Studien zum Beginn der modernen Welt* (Stuttgart: Klett-Cotta, 1977).

6. To list just a selection: François Jacob, *The Logic of Life: A History of Heredity*, trans. Betty E. Spillmann (Princeton, NJ: Princeton University Press, 1993); Jacques Roger, "The

Living World," in Roy Porter and G. S. Rousseau (eds.), *The Ferment of Knowledge: Studies in the Historiography of Eighteenth-Century Science* (Cambridge: Cambridge University Press, 1980); Phillip R. Sloan, "Buffon, German Biology, and the Historical Interpretation of Biological Species," *British Journal for the History of Science* 12 (1979); Peter Hanns Reill, "Bildung, Urtyp and Polarity: Goethe and Eighteenth-Century Physiology," *Goethe Yearbook* 3 (1986), pp. 139–48; Roy Porter, "The History of Time," in John Grant (ed.), *The Book of Time* (Newton Abbot, UK: Westbridge Books, 1980), pp. 26–33; Brian John, "Measuring Time Past," in Grant, *The Book of Time*; Hans-Jörg Rheinberger, "Buffon: Zeit, Veränderung und Geschichte," *History and Philosophy of the Life Sciences* 12 (1990), pp. 202–23; Peter Matussek (ed.), *Goethe und die Verzeitlichung der Natur: Kulturgeschichte der Natur in Einzeldarstellungen* (Munich: Beck, 1998); Robert J. Richards, *The Romantic Conception of Life: Science and Philosophy in the Age of Goethe* (Chicago: University of Chicago Press, 2002).

7. See Janina Wellmann, "Die Metamorphose der Bilder: Die Verwandlung der Insekten und ihre Darstellung vom Ende des 17. bis zum Anfang des 19. Jahrhunderts," *NTM: Zeitschrift für Geschichte der Wissenschaft, Technik und Medizin* 16 (2008), pp. 183–211.

8. Arno Seifert, "'Verzeitlichung': Zur Kritik einer neueren Frühneuzeitkategorie," *Zeitschrift für historische Forschung* 10 (1983), p. 448.

9. *Ibid.*, p. 476.

10. This is the title of a special issue on the diversity of eighteenth-century concepts of time: Stefanie Stockhorst (ed.), "Zeitkonzepte: Zur Pluralisierung des Zeitdiskurses im langen 18. Jahrhundert," special issue, *Das 18. Jahrhundert* 30.2 (2006).

11. In this sense, meter is the simplest form of rhythm. Historically, as we have seen, the concepts of meter and rhythm are often difficult to distinguish.

Illustration Sources and Permissions

Figure 4.1: Wolff, *De formatione intestinorum*, 1768. From Caspar Friedrich Wolff, "De formatione intestinorum praecipue, tum et de amnio spurio: Aliisque partibus embryonis gallinacei, nondum visis. Observationes, in ovo incubatis institutae," *Novi Commentarii Academiae Scientiarim Imperialis Petropolitanae* 12 (1768), plate VII, detail. Courtesy bpk (Bildagentur für Kunst, Kultur und Geschichte) / Staatsbibliothek zu Berlin: Ab 5402.

Figure 4.2: Wolff, *De formatione intestinorum*, 1768, plate VII, detail. Courtesy bpk / Staatsbibliothek zu Berlin: Ab 5402.

Figure 6.1: Ignaz Döllinger, "Kreislaufe des Blutes," *Denkschriften der königlichen Academie der Wissenschaften zu München für die Jahre 1818, 1819 und 1821: Classe der Mathematik und Naturwissenschaften* 7 (1821), plate X. Courtesy bpk / Staatsbibliothek zu Berlin: Ab 714.

Figure 7.1: Manuscript page of Dürer's *Fechtbuch*, early sixteenth century. Reprinted in Friedrich Dörnhöffer, "Albrecht Dürers Fechtbuch," *Jahrbuch der Kunsthistorischen Sammlungen des Allerhöchsten Kaiserhauses* 27, part 2 (1907–1909), plate 5, http://digi.ub.uni-heidelberg.de/diglit/jbksak1907_1909/0394.

Figure 7.2: Hieronymus Mercurialis, *De arte gymnastica* (1569; Venice: Juntas, 1587). Courtesy Herzog August Bibliothek Wolfenbüttel: 6.8 Hist. (1).

Figure 7.3: Jacques de Gheyn, *Wapenhandelinghe van Roers Musquetten ende Spiessen: achtervolghende de ordre van Syn Excellentie Maurits Prince van Orangie Grave van Nassau . . . .* (Amsterdam: n.p., 1608). Courtesy Herzog August Bibliothek Wolfenbüttel: 8° Bell. 2°.

Figure 7.4: From a manuscript by Count John of Nassau-Siegen. Reprinted in Jacob de Gheyn, *The Exercise of Armes: All 117 Engravings from the Classic 17th-century Military Manual*, edited by Bas Kist (1608; Mineola, NY: Dover, 1999), p. XI.

Figure 7.5: De Gheyn, *Wapenhandelinghe van Roers Musquetten ende Spiessen*, 1608, plates 1.5, 1.6, 1.7, 1.8. Courtesy Herzog August Bibliothek Wolfenbüttel: 8° Bell. 2°.

Figure 7.6: Johann Jacob von Wallhausen, *Kriegskunst zu Fuß. Darinnen gelehret und gewiesen werden: I. Die Handgrieff der Musquet und des Spiesses; II. Das Exercitium, oder wie man es nennet, das Trillen; III. Schöne neue Batailie, oder Schlachtordnungen; IV. Der Ungerischen biszhero geführten Regimenten Kriegs Disciplin zu Fusz* (Frankfurt: de Bry, 1620), plates B, C, D. Courtesy Herzog August Bibliothek Wolfenbüttel: Jb 4° 60 (1).

Figure 7.7: Hans Conrad Lavater, *Kriegs-Büchlein, das ist grundtliche Anleitung zum Kriegswesen* (Zurich: Bodmer, 1651), plates III, IV, V. Courtesy Herzog August Bibliothek Wolfenbüttel: 83.1 Quod. (3).

Figure 7.8: Johann Georg Paschen, *Kurtze iedoch deutliche Beschreibung / handlend Vom Fechten auf den Stoß und Hieb/mit sonderbahren Fleiss auffgesetzt und mit mehrern nothwendigen Kupffern ausgebildet* (Halle: Melchior Oelschlegeln, 1664), figs. 8–23. Courtesy Herzog August Bibliothek Wolfenbüttel: Hn 4° 39 (3).

Figure 7.9: Wallhausen, *Kriegskunst zu Fuß*, 1620, plate B, detail. Courtesy Herzog August Bibliothek Wolfenbüttel: Jb 4° 60 (1).

Figure 7.10: Johann Jacob von Wallhausen, *Ritterkunst* (Frankfurt am Main: n.p., 1616), plate 8. Courtesy Bayerische Staatsbibliothek München: Res/4 Gymn. 42; Hannss Friedrich von Fleming, *Der vollkommene teutsche Soldat* (Leipzig: Martini, 1726), plate "Comando der Tamboure." Courtesy Universitäts- und Landesbibliothek Sachsen-Anhalt in Halle (Saale): AB 176538.

Figure 7.11: Wallhausen, *Ritterkunst*, 1616, plate 7. Courtesy Bayerische Staatsbibliothek München: Res/4 Gymn. 42.

Figure 7.12: Wallhausen, *Ritterkunst*, 1616, plate 1. Courtesy Bayerische Staatsbibliothek München; Res/4 Gymn. 42.

Figure 7.13: Ludwig Andrea, Graf von Khevenhüller, *Observations-Puncten*, 3rd ed. (Vienna: Joh. Paul Krauß, 1749), pp. 34 and 80. Courtesy bpk / Staatsbibliothek zu Berlin: Hw 16715 <a>.

Figure 7.14: Fleming, *Der vollkommene teutsche Soldat*, 1726, plate R. Courtesy Herzog August Bibliothek Wolfenbüttel: Jb 2° 5.

Figure 7.15: *Regulament und Ordnung nach welchem sich gesammtes Kaiserlich-Königliches Fuß-Volck, in denen in diesem ersten Theil enthaltenen Hand-Grieffen und allen andern Kriegsexercitien sowohl als in denen in dem zweyten Theil vorgeschriebenen Kriegs-Gebräuchen zu Feld, Besatzungen und überall gleichförmig zu achten haben/1* (Vienna: van Gehlen, 1749), pp. 92–95, 95–101. Courtesy Staatsbibliothek Bamberg: Bip.R.mil.q.18–1.

Figure 7.16: *Instructions militaires simples et faciles* (Paris: J. B. Delamollière, 1792). Courtesy bpk / Staatsbibliothek zu Berlin: 50 MA 6992.

Figure 7.17: A. von Rhein, *Das Bajonettfechten* (Wesel: Bagel, 1840), plate 1; Hugo Rothstein: *Anleitung zum Bajonettfechten*, 2nd ed. Berlin: Schroeder, 1853). Courtesy Humboldt-Universität zu Berlin, Zweigbibliothek Campus Nord: Ga 221 (2).

Figure 7.18: Johann Georg Paschen, *Kurtze iedoch gründliche Beschreibung des Voltiger / so wohl auf den Pferde als über den Tisch. Darinnen gehandelt wird von allen Sprüngen / als in Sattel zu springen / wieder herauß / Rever, Droicts, halben Pomaden, halben Pomaden mit der Rever, ganzen Pomaden, Verwechseln / anderthalb Pomaden, Beinspringen / etc. wie solches heutiges*

Tages in Gebrauch. Mit mehren Lectionen und sehr vielen Kupffern abgebildet von Johann Georg Paschen / S. M. Pagen Hoffmeistern (Halle: Melchior Oelschlegeln, 1664), p. 5, figs. 1–9. Courtesy Herzog August Bibliothek Wolfenbüttel: Hn 4° 39 (3).

Figure 7.19: Domenico Angelo, *École des Armes, avec l'explication générale des principales attitudes et positions concernant l'escrime* (London: R. and J. Dodsley, 1763), plate 15. Courtesy bpk / Staatsbibliothek zu Berlin: Os 8204.

Figure 7.20: Denis Diderot and Jean Le Rond d'Alembert, *Encyclopédie ou dictionnaire raisonné des sciences, des arts et des métiers* / Par une société de gens de lettres. Mis en ordre & publié par M. Diderot . . . par M. d'Alembert . . . & la Société Royale de Londres (Paris: Briasson, 1751–1780), planches vol. 4, plates VI–XIII. Courtesy Herzog August Bibliothek Wolfenbüttel: KA 80-2610.

Figure 7.21: Pierre Rameau, *Le maître à danser: Qui enseigne la maniere de faire tous les differens pas de danse dans toute la régularité de l'art, & de conduire les bras à chaque pas. Enrichi de figures en taille-douce, servant de démonstration pour tous les differens mouvemens qu'il convient faire dans cet exercice, Nouvelle édition* (Paris: Jean Villette Fils, 1734), pp. 208 and 212. Courtesy SUB Göttingen: 8 Art Ill 608 Rara.

Figure 7.22: Johann Bernhard Basedow, *Des Elementarbuchs für die Jugend und für ihre Lehrer und Freunde in gesitteten Ständen Stück. Kupfertafeln Bd. 2* (Altona: Der Verfasser, 1774), plate LXIV. Courtesy Herzog August Bibliothek Wolfenbüttel: Wi 15: Tafelbd. 2.

Figure 7.23: Gerhard Ulrich Anton Vieth, *Versuch einer Encyclopädie der Leibesübungen: Zweyter Theil* (Berlin: Carl Ludwig Hartmann, 1795), plate 11. Courtesy bpk / Staatsbibliothek zu Berlin: Nf 6704.

Figure 7.24: J. C. F. GutsMuths, *Gymnastik für die Jugend, enthaltend eine praktische Anweisung zu Leibesübungen. Zweyte durchaus umgearbeitete und stark vermehrte Ausgabe mit 12 vom Verfasser gezeichneten Tafeln* (Schnepfenthal: Buchhandlung der Erziehungsanstalt, 1804), p. 229. Courtesy bpk / Staatsbibliothek zu Berlin: BXXI, 163.

Figure 7.25: Vieth, *Versuch einer Enzyklopädie für Leibesübungen*, 1795, plate III. Courtesy bpk / Staatsbibliothek zu Berlin: Nf 6704.

Figure 7.26: Carl Heinitz, *Unterricht in der Schwimmkunst, nach der in der k.k. Militär-Schwimmanstalt in Wien eingeführten Lehrmethode, dargestellt vorzüglich zum Behufe des k.k. Militärs* (Vienna: Strauß, 1816), plate III. Courtesy Martha Muchow Bibliothek, Universität Hamburg: EPB Ka 35.

Figure 7.27: Raoul Anger Feuillet, *Choréographie, ou l'art de decrier la danse par caractères, figures et signes démonstratifs* (1700; Hildesheim: Olms, 1979), unpaginated.

Figure 7.28: Diderot and d'Alembert, *Encyclopédie*, 1751–1780, planches vol. 7, plates XVII–XVIII. Courtesy Herzog August Bibliothek Wolfenbüttel: KA 80-2610.

Figure 7.29: Count John of Nassau-Siegen's notes. Reproduced in Werner Hahlweg, *Die Heeresreform der Oranier: Das Kriegsbuch des Grafen Johann von Nassau-Siegen* (Wiesbaden: Historische Kommission für Nassau, 1973), p. 238.

Figure 7.30: Claude Bottée, *Études militaires, Qui comprend le plan général de tout l'ouvrage et l'exercice de l'infanterie avec des figures* (Paris: Claude Jombert, 1731), p. 237. Courtesy Staatsbibliothek Bamberg: 22/Bip.R.mil.o.4.

Figure 7.31: *Règlement concernant l'exercice et les manœuvres de l'infanterie: Du 1er. Août 1791* (Paris: Bureau du Journal militaire, 1791) , plate XVI. Courtesy Staatsarchiv des Kantons Bern: C 662.

Figure 7.32: Wallhausen, *Ritterkunst*, 1616, plate 13. Courtesy Bayerische Staatsbibliothek München: Res/4 Gymn. 42.

Figure 7.33: Khevenhüller, *Observations-Puncten*, 1749, p. 18. Courtesy bpk / Staatsbibliothek zu Berlin: Hw 16715 <a>.

Figure 7.34: Bottée, *Études militaires*, 1731, pp. 282 and 438. Courtesy Staatsbibliothek Bamberg: 22/Bip.R.mil.o.4.

Figure 7.35: *The General Review Manoeuvres; or, the Whole Evolutions of a Battalion of Foot; As Now Practised by Order of His Majesty. Illustrated with Copperplates. By an Officer of the Army. To Which Is Annexed, the Manual Exercise with Explanations* (London: J. Millan, 1779), pp. 13-14. Courtesy Institute of Historical Research London: W. 6112/Bri/Off.

Figure 7.36: John Russell, *A Series of Military Experiments, of Attack and Defence, Made in Hyde Park in 1802, under the Sanction of His Royal Highness the Cmdr in Chief, with Infantry, Cavalry, and Artillery; and in the Island of Jersey in 1805, by Permission of Lieutenant-General Andrew Gordon. With Notes, Remarks, and Illustrations* (London: T. Egerton, 1806), plate 1. Courtesy Zentral- und Hochschulbibliothek Luzern: E1.571.k.8.

Figure 7.37: *Règlement concernant l'exercice et les manœuvres de l'infanterie*, 1791, plate XXVIII. Courtesy Staatsarchiv des Kantons Bern: C 662.

Figure 7.38: Diderot and d'Alembert, *Encyclopédie*, 1751-1780, planches vol. 1, plates I-II. Courtesy Herzog August Bibliothek Wolfenbüttel: KA 80-2610.

Figure 7.39: Diderot and d'Alembert, *Encyclopédie*, 1751-1780, planches vol. 1, plate XV. Courtesy Herzog August Bibliothek Wolfenbüttel: KA 80-2610.

Figure 7.40: Campbell Dalrymple, *A Military Essay: Containing Reflections on the Raising, Arming, Cloathing, and Discipline of the British Infantry and Cavalry with Proposals for the Improvement of the Same* (London: D. Wilson, 1761), pp. 178 and 180. Courtesy SUB Göttingen: ARS MIL 270/5:2.

Figure 7.41: Johann Moritz David Herold, *Entwickelungsgeschichte der Schmetterlinge, anatomisch und physiologisch bearbeitet von Dr. Herold, mit dreyunddreyssig illuminirten und schwarzen Kupfertafeln* (Kassell: n.p., 1815), plate 11. Courtesy SUB Göttingen: 4 ZOOL VI, 5226:TAF.

Figure 7.42: Diderot and d'Alembert, *Encyclopédie*, 1751-1780, planches vol. 1, plate II. Courtesy Herzog August Bibliothek Wolfenbüttel: KA 80-2610.

Figure 7.43: Diderot and d'Alembert, *Encyclopédie*, 1751-1780, planches vol. 8, plates XXI-XXIII. Courtesy Herzog August Bibliothek Wolfenbüttel: KA 80-2610.

Figure 7.44: Diderot and d'Alembert, *Encyclopédie*, 1751–1780, planches vol. 8, plates XXIV and XXV. Courtesy Herzog August Bibliothek Wolfenbüttel: KA 80-2610.

Figure 8.1: Hieronymi Fabritii ab Aquapendente, *Opera physica anatomica: De formato foetu, venarum ostiolis, formatione ovi, et pulli, locutione et eius instrumentis, brutorum loquela; Cum indicibus capitum et rerum notatu dignarum* (Padua: Roberto Meglietti, 1625), plate 3. Courtesy bpk / Staatsbibliothek zu Berlin: Kt 3536.

Figure 8.2: Marcello Malpighi, *De formatione pulli in ovo* (London: Joannis Martyn, 1673), plate II. Courtesy bpk / Staatsbibliothek zu Berlin: Lk 8201.

Figure 8.3: Haller's notes. Courtesy Burgerbibliothek Bern: N Albrecht von Haller 6, f. 32v.

Figure 8.4: Haller's notes, reproduced in Albrecht von Haller, *Commentarius de formatione cordis in ovo incubato*, edited by Maria Teresa Monti (1767; Basel: Schwabe, 2000), p. 86.

Figure 8.5: Haller's notes, reproduced in Haller, *Commentarius de formatione cordis in ovo incubato*, 2000, p. 158.

Figure 8.6: August Johann Rösel von Rosenhof, *Historia naturalis ranarum nostratium in qua omnes earum proprietates, praesertim quae generationem ipsarum pertinent, Fusius enarrantur . . .* (Nürnberg: Johann Joseph Fleischmann, 1758), plate x. Courtesy Klassik Stiftung Weimar, Herzogin Anna Amalia Bibliothek: Th G 1: 7 (a).

Figure 8.7: Samuel Thomas Soemmerring, *Icones embryonum humanorum* (Frankfurt am Main: Varrentrapp & Wenner, 1799), plate 1. Courtesy bpk / Staatsbibliothek zu Berlin: L 7250.

Figure 8.8: Soemmerring, *Icones embryonum humanorum*, 1799, plate II. Courtesy bpk / Staatsbibliothek zu Berlin: L 7250.

Figure 8.9: Caspar Friedrich Wolff, *Theoria generationis* (Halle: Litteris Hendelianis, 1759), plate II, detail. Courtesy bpk / Staatsbibliothek zu Berlin: L 420.

Figure 8.10: Wolff, *Theoria generationis*, 1759, plate II, detail. Courtesy bpk / Staatsbibliothek zu Berlin: L 420.

Figure 8.11: Drawing by Wolff in a letter to Haller, 1764. Courtesy Biblioteca Nazionale Braidense: AE.XI. 15/6.

Figure 8.12: Louis Sebastian Tredern, *Dissertatio inauguralis medica sistens ovi avium historiae et incubationis prodromum* (Jena: Etzdorf, 1808). Courtesy Herzog August Bibliothek Wolfenbüttel: Nx 4°5 (5).

Figure 8.13: Herold, *Entwickelungsgeschichte der Schmetterlinge*, 1815, plates VI, VIII, X, XII, XIV, XVI, XVIII, XX, XXII, XXIV, XXVI, XXVIII, XXXII. Courtesy SUB Göttingen: 4 ZOOL VI, 5226:TAF

Figure 9.1: Christian Heinrich Pander, *Beiträge zur Entwickelungsgeschichte des Hühnchens im Eye* (Würzburg: n.p., 1817), plate I. Courtesy of the Library, Max Planck Institute for the History of Science, Berlin.

Figure 9.2: Pander, *Beiträge zur Entwickelungsgeschichte des Hühnchens im Eye*, 1817, plate

II and schematic outline. Courtesy of the Library, Max Planck Institute for the History of Science, Berlin.

Figure 9.3: Pander, *Beiträge zur Entwickelungsgeschichte des Hühnchens im Eye*, 1817, plate III and schematic outline. Courtesy of the Library, Max Planck Institute for the History of Science, Berlin.

Figure 9.4: Pander, *Beiträge zur Entwickelungsgeschichte des Hühnchens im Eye*, 1817, plate VII and schematic outline. Courtesy of the Library, Max Planck Institute for the History of Science, Berlin.

Figure 9.5: Pander, *Beiträge zur Entwickelungsgeschichte des Hühnchens im Eye*, 1817, plate V. Courtesy of the Library, Max Planck Institute for the History of Science, Berlin.

Figure 9.6: Pander, *Beiträge zur Entwickelungsgeschichte des Hühnchens im Eye*, 1817, "Tafel der Durchschnitte." Courtesy of the Library, Max Planck Institute for the History of Science, Berlin.

Figure 9.7: Christian Heinrich Pander, "Entwicklung des Küchels," *Isis oder Encyclopädische Zeitung* 3 (1818), cols. 512–24, plate 8. Courtesy of the Library, Max Planck Institute for the History of Science, Berlin.

Figure 10.1: Karl Ernst von Baer, *Über Entwickelungsgeschichte der Thiere: Beobachtung und Reflexion. Erster Theil. Mit drei colorirten Kupfertafeln* (Königsberg: bei den Brüdern Bornträger, 1828), vol. 1, plate 1. Courtesy bpk / Staatsbibliothek zu Berlin: Ll 8252-1.

Figure 10.2: Von Baer, *Über Entwickelungsgeschichte der Thiere*, 1828–1837, vol. 1, plate II. Courtesy bpk / Staatsbibliothek zu Berlin: Ll 8252-1.

Index

Zone Books series design by Bruce Mau
Typesetting by Meighan Gale
Image placement and production by Julie Fry
Printed and bound by Maple Press